Disaster Resilience and Green Growth

Over the years, the relationship between environment and disasters has received significant attention. This is largely due to the emerging recognition that environmental changes - climate change, land-use and natural resource degradation make communities more vulnerable to disaster impacts. There is a need to break this nexus through environment based and sustainability inclusive interventions. Science – technology and economic measures for disaster risk management, hence, need to adapt more integrated approaches for infrastructure and social resilience. Environmental and anthropogenic factors are key contributors to hazard, risk, and vulnerability and, therefore, should be an important part of determining risk-management solutions.

Green growth approaches have been developed by emphasizing sustainability inclusion and utilizing the benefits of science-technology interventions along policy-practice linkages with circular economy and resource efficiency. Such approaches recognize the perils of traditional material-oriented economy growth models that tend to exploit natural resources, contribute to climate change, and exacerbate disaster vulnerabilities, Green growth integrated approaches are rapidly becoming as preferred investment avenue for mitigating climate change and disaster risks and for enhancing resilience. This includes ecosystem-based and nature-based solutions with potential to contribute to the resilience of infrastructure, urban, rural and peri-urban systems, livelihoods, water, and health. They can lead to food security and can further promote people-centric approaches.

Some of the synergistic outcomes of green growth approaches include disaster risk reduction, climate change mitigation and adaptation, resilient livelihoods, cities, businesses and industry. The disaster risk reduction and resilience outcome of green growth approaches deserve special attention, both for the academic and policy communities. Scholars and professionals across the domains of DRR, CCA, and green growth are in need of publications that fulfill their knowledge needs concerning the disaster resilience outcomes of green growth approaches. Keeping the above background in view, the book series offers comprehensive coverage combining the domains of environment, natural resources, engineering, management and policy studies for addressing disaster risk and resilience in the green growth context in an integrated and holistic manner. The book series covers a range of themes that highlight the synergistic outcomes of green growth approaches.

The book series aims to bring out the latest research, approaches, and perspectives for disaster risk reduction along with highlighting the outcomes of green growth approaches and including Science-technology-research-policy-practice interface, from both developed and developing parts of the world under one umbrella. The series aims to involve renowned experts and academicians as volume-editors and authors from all the regions of the world. It is curated and developed by authoritative institutions and experts to serve global readership on this theme.

Anil Kumar Gupta • Manish Kumar Goyal •
S. P. Singh
Editors

Ecosystem Restoration: Towards Sustainability and Resilient Development

Springer

Editors
Anil Kumar Gupta
National Institute of Disaster Management (Ministry of Home Affairs, Government of India)
New Delhi, India

Manish Kumar Goyal
Department of Civil Engineering
Indian Institute of Technology Indore
Indore, Madhya Pradesh, India

S. P. Singh
HNB Garhwal University
Srinagar, Uttarakhand, India

ISSN 2662-4885 ISSN 2662-4893 (electronic)
Disaster Resilience and Green Growth
ISBN 978-981-99-3686-1 ISBN 978-981-99-3687-8 (eBook)
https://doi.org/10.1007/978-981-99-3687-8

This Springer imprint is published by the registered company Springer Nature Singapore Pte Ltd.
The registered company address is: 152 Beach Road, #21-01/04 Gateway East, Singapore 189721, Singapore

Foreword

Environment, disasters and development are inextricably linked. Environmental changes primarily climate change, landscape changes and natural resource degradation are the key drivers of risk in the domain of climate crisis and disasters. Climate change has impacted, disrupted and altered basic human necessities like food production and security, water supply, biodiversity and forests, human health and settlements, etc. Its manifestations have also resulted in an increase in the frequency and intensity of extreme events and disasters. For example, drought events in 2016 affected 330 million people, accounting for an economic loss of $100 billion in India.

Extreme weather events such as droughts, heatwaves, heavy rain and violent storms are now part of our daily news. The ecosystems take much of the brunt of such challenges. Only with healthy ecosystems can we enhance people's livelihoods, counteract climate change and stop the collapse of biodiversity.

Ecosystem services and values are being acknowledged now globally. Almost all sectors including health, development, finance, tourism, education, etc. are dependent on the four types of ecosystem services for the smooth functioning of their businesses and reaping continual economic benefits. Ecosystem-based Disaster Risk Reduction holds much relevance in the current context of the global climatic scenario for understanding, implementing and averting disaster risk. The UN Decade on Ecosystem Restoration is a rallying call for the protection and recovery of ecosystems all around the world, for the benefit of people and nature. It aims to halt the degradation of ecosystem and restore them to achieve global goals.

I congratulate the editorial team led by Professor Anil Gupta and all the contributors of this book for taking out such a relevant publication to showcase the case studies from South Asia and other developing nations. This will be an added advantage to the existing knowledge of the ecosystem's support services to climate change. I hope this book will be an interesting read for academicians, policy planners, researchers, practitioners and subject experts.

Chair, IUCN Commission on Ecosystem Management, Gland, Switzerland

Dr. Angela Andrade

Preface

The world is facing severe challenges of instability and deceleration—mainly due to increasing disasters, extremes and stresses that result from anthropogenic pressures, ecosystem service recession, inequalities and uncertainties. Environmental changes—particularly climate change, land-use and landscape changes and natural resource degradation—are known to be among the key drivers of connected hazards and vulnerabilities leading to such complexities and losses. The recent G-20 dialogues, under India's Presidency tenure and otherwise too in recent times, considered land degradation and ecosystem dimensions of disasters—particularly of forest fire, health risks, mined-out lands, etc.—a key facet of ecosystem degeneration calling significant strides of restoration, the way paved by the UN's Decade on Eco-restoration. The UNCCD in its COP-14 stressed on global water action agenda as key means of driving returning land back to nature—a thought process for ensuring future sustainability.

Millions of people are suffering from the catastrophic effects of extreme disasters, climate emergencies, water and food insecurity, and the COVID-19 pandemic taught big lessons of synergy between development and natural drivers. Ecosystems are an essential partner to meet these challenges. Hence, managing and protecting ecosystem resources in a sustainable manner is very crucial. ***Ecosystem Restoration: Towards Sustainability and Resilient Development*** is a book providing comprehensive information on the fundamentals and latest developments in the field of ecosystem restoration, resilience and sustainability. The book is comprised of 20 chapters covering almost all the aspects of ecosystem restoration and sustainability including agricultural and ecosystem resilience, the role of biodiversity, climate change and water resource systems more particularly from tools, techniques and modelling perspectives, besides strategic ways to deal with extreme events, disaster risk and management through sustainable policy-making facets of disaster management.

Some important topics like water management technology, water quality and pollution also have been discussed. This book collects wide-ranging contributions such as reviews, output of research studies, case studies, reports on technological developments, presenting latest findings and raising awareness about ecosystem restoration and climate change. It brings forth the theme of ecosystem restoration as ecosystems support all life forms on Earth. Written for scientists and researchers, academics and students, ***Ecosystem Restoration: Towards Sustainability and Resilient Development*** is a comprehensive reference that covers the basic knowledge of the topic combined with practical applications. The editors hope that the book is useful to readers of diverse backgrounds as it brings a lateral collection of perspectives from engineering, modelling and analytical perspectives of ecosystem resources combined with policy and strategy aspects for understanding 'Ecosystem Restoration'.

New Delhi, India — Anil Kumar Gupta
Indore, Madhya Pradesh, India — Manish Kumar Goyal
Srinagar, Uttarakhand, India — S. P. Singh

Acknowledgements

The editors acknowledge the contribution of authors on various key themes of the book on ecosystem restoration as a key strategy for addressing present-day problems of sustainability and resilience. The editors also extend sincere thanks to Springer Nature, the commissioning editor and the publishing coordinator team.

Acknowledgements are due to Dr. Angela Andrade, Chair of the Commission on Ecosystem Management, at the International Union for Conservation of Nature (IUCN-CEM), for writing the foreword of the book. Thanks are also due to Mr. Rajendra Ratnoo, Executive Director, National Institute of Disaster Management, New Delhi, and Prof. Suhas Joshi, Director, Indian Institute of Technology, Indore, for encouragement and cooperation. Team also extends thanks to UNEP PEDRR team and UNCCD G-20 group for all support and cooperation in undertaking the cause of ecoDRR and NBS promotion at all the levels in resilience and adaptation.

Support and inputs from NIDM's ECDRM-CECR team and in particular Ms. Pritha Acharya and Ms. Fatima Amin are acknowledged.

Contents

Editors and Contributors

About the Editors

Anil Kumar Gupta is a noted resilience and sustainability strategist, crisis management professional, and thought leader. Presently he is Head of Division, Director of Projects & COE, in charge of International Cooperation and Advisory Services, and Full Professor of policy planning and strategies at India's National Institute of Disaster Management at New Delhi. He is known for pioneering national and sectoral planning, national human resource planning, ebA/ecoDRR & NBS, adaptation strategies, and consortium projects, in DRR and sustainability areas. He implemented projects of action research, policy planning, advocacy, knowledge management, etc. with many international and national partners.

Manish Kumar Goyal is Professor of Civil Engineering and Dean of Infrastructure Development at Indian Institute of Technology, Indore, India. His research interests include climate change, environment, water resources, hydrology, biodiversity, artificial intelligence and GIS-remote sensing applications.

S. P. Singh is among the top-notch Himalayan ecologists. His research systems range from the blade of a leaf to inherently fragile and disaster-prone landscapes. He along with his associates have studied almost all dimensions of the Himalayan forest ecosystems, including biomass and productivity, nutrient cycling, carbon sequestration, phenology, tree water relations, forest fire adaptation and ecosystem services.

Contributors

Pritha Acharya ECDRM Division, National Institute of Disaster Management (NIDM), Delhi, India

Ishtiyaq Ahmad Department of Civil Engineering, National Institute of Technology, Raipur, India

Masood Zafar Ansari Department of Civil Engineering, National Institute of Technology, Raipur, India

Vartika Awasthi Forest Research Institute (FRI), Dehradun, Uttarakhand, India

Sweta Baidya ECDRM Division, National Institute of Disaster Management (NIDM), New Delhi, India
Mitigation Division, National Disaster Management Authority, Ministry of Home Affairs, New Delhi, India

Anjali Barwal ECDRM Division, National Institute of Disaster Management (Ministry of Home Affairs), New Delhi, India

Shubham Bhardwaj Department of Civil Engineering, National Institute of Technology, Hamirpur, India

Kavita Bramhanwade National Environmental Engineering Research Institute (CSIR-NEERI), Nagpur, Maharashtra, India

Pritha Chakraborty ESG Book, New Delhi, India

Prabhat Chandra Central Water and Power Research Station, Pune, India

R. B. Deogade Central Water and Power Research Station, Pune, India

Shalini Dhyani National Environmental Engineering Research Institute (CSIR-NEERI), Nagpur, Maharashtra, India

Venkatesh Dutta Department of Environmental Science, Babasaheb Bhimrao Ambedkar University, Lucknow, Uttar Pradesh, India

Manish Kumar Goyal Department of Civil Engineering, Indian Institute of Technology Indore, Indore, Madhya Pradesh, India

Anil Kumar Gupta National Institute of Disaster Management (Ministry of Home Affairs, Government of India), New Delhi, India

Mayur Shirish Jain Department of Civil Engineering, Indian Institute of Technology Indore, Indore, Madhya Pradesh, India

Vijay Jain Department of Civil Engineering, Indian Institute of Technology Indore, Indore, Madhya Pradesh, India

Sashindra Kumar Kakoty Mechanical Engineering, IIT Guwahati, Guwahati, India

Chander Kant Department of Civil Engineering, National Institute of Technology, Hamirpur, India

Dhananjay Kumar Department of Environmental Science, Babasaheb Bhimrao Ambedkar University, Lucknow, Uttar Pradesh, India

Narendra Kumar Department of Environmental Science, Babasaheb Bhimrao Ambedkar University, Lucknow, Uttar Pradesh, India

Sanjeev Kumar Department of Environmental Science, Central University of Jharkhand, Ranchi, India

Prerit Machiwar Department of Civil Engineering, National Institute of Technology, Hamirpur, India

Aditi Madan Institute for Human Development, New Delhi, India

Bhabesh Mahanta School of Agro and Rural Technology, IIT Guwahati, Guwahati, India

Nivedita Mani Research and Communications, Gorakhpur Environmental Action Group (GEAG), Gorakhpur, Uttar Pradesh, India

Ray Singh Meena Department of Civil Engineering, National Institute of Technology, Hamirpur, India

Supriya Nath Central Water and Power Research Station, Pune, India

S. Nithiyanantham Department of Physics (Ultrasonic/NDT and Bio-Physics Divisions), Govt Arts and Science College (Bharathidasan University, Thiruchirapalli), Thiruvarur, Tamil Nadu, India

Shivakumar J. Nyamathi Water Resources Engineering, Department of Civil Engineering, UVCE, Bangalore University, Bangalore, India

Padma Parija L and T Reality, Powai, Mumbai, Maharastra, India

Pooja Patle Department of Civil Engineering, National Institute of Technology, Raipur, India

Sagarika Patowary Department of Civil Engineering, Goalpara Polytechnic, Goalpara, India

Paras Pujari National Environmental Engineering Research Institute (CSIR-NEERI), Nagpur, Maharashtra, India

Shivukumar Rakkasagi Department of Civil Engineering, Indian Institute of Technology Indore, Indore, Madhya Pradesh, India

Kuldeep Singh Rautela Department of Civil Engineering, Indian Institute of Technology Indore, Indore, Madhya Pradesh, India

Sinchana M. Sakalesh Water Resources Engineering, Department of Civil Engineering, UVCE, Bangalore University, Bangalore, India

Arup Kumar Sarma Department of Civil Engineering, Indian Institute of Technology Guwahati, Guwahati, India

Urvashi Sharma Department of Environmental Science, Babasaheb Bhimrao Ambedkar University, Lucknow, Uttar Pradesh, India

Jayshree Shukla National Environmental Engineering Research Institute (CSIR-NEERI), Nagpur, Maharashtra, India

Pushpendra Kumar Singh Division of Water Resources, National Institute of Hydrology, Roorkee, India

Shivam Singh Department of Civil Engineering, Indian Institute of Technology Indore, Indore, Madhya Pradesh, India

Sunidhi Singh National Environmental Engineering Research Institute (CSIR-NEERI), Nagpur, Maharashtra, India

Swati Singh United Nations Development Programme (UNDP), Delhi, India

Atisha Sood ECDRM Division, Ministry of Home Affairs, Government of India, New Delhi, India

J. S. Sudarsan School of Energy and Environment, NICMAR University, Pune, Maharashtra, India

Deepak Kumar Tiwari Department of Civil Engineering, GLA University, Mathura, Uttar Pradesh, India

H. L. Tiwari Department of Civil Engineering, Maulana Azad National Institute of Technology, Bhopal, Madhya Pradesh, India

L. Ar. Poonam Upadhyay AIT - School of Architecture and Planning, Greater Noida, Uttar Pradesh, India

Parikshit Verma National Environmental Engineering Research Institute (CSIR-NEERI), Nagpur, Maharashtra, India

N. Vivekanandan Central Water and Power Research Station, Pune, Maharashtra, India

Jitesh N. Vyas Central Water and Power Research Station, Pune, India

Shiraz Wajih Gorakhpur Environmental Action Group (GEAG), Gorakhpur, Uttar Pradesh, India

Chapter 1
Ecological Restoration: An Overview of Science and Policy Regime

Vijay Jain, Kuldeep Singh Rautela, and Manish Kumar Goyal

Abstract Ecological restoration projects require science-based assessment in association with policy development for making a strategic framework and prioritizing the intervening actions. It will ensure the long-term future health and stability of the ecosystems. Ecological restoration science aims to provide ecosystems in pre-disturbance conditions. It will give an approach to improve the regional biodiversity and structural and functional integrity with surroundings and diminish the regional climatic and economic value degradation. The futuristic tasks associated with ecological restoration across different regions worldwide can be classified into social, ecological, economic, and political categories. Here, we have discussed several aspects of ecological restoration plans for urban, wetlands, and Himalayan ecosystems and their associated ecological and socioeconomic impacts. Therefore, through proper implementation of restoration principles, guidelines, and maximum involvement of stakeholders, we can achieve maximum ecological and sustainable aims during the 2021–2030 decade of ecosystem restoration.

Keywords Restoration science · Policy · Cities · Wetlands and Himalayan ecosystem

1.1 Introduction

The World Environment Day on 5 June 2021 marked the beginning of the United Nation decade for ecosystem restoration from 2021 to 2030 to discontinue and invert the deterioration of the natural world. The initiative of this decade begins with the

V. Jain (✉) · K. S. Rautela
Department of Civil Engineering, Indian Institute of Technology Indore, Indore, Madhya Pradesh, India
e-mail: phd2101104008@iiti.ac.in

M. K. Goyal
Department of Civil Engineering, Indian Institute of Technology Indore, Indore, Madhya Pradesh, India

A. K. Gupta et al. (eds.), *Ecosystem Restoration: Towards Sustainability and Resilient Development*, Disaster Resilience and Green Growth,
https://doi.org/10.1007/978-981-99-3687-8_1

observance of the climatic emergency conditions due to losses of natural ecosystems and higher levels of pollution, which threatens the existence of millions of species on our planet. This initiative leads to science and policy discussions among all the United Nations member country for focusing on future ecological restoration programs (UNEP 2021a). The key objective of this program is to restore at least 0.35 billion hectares of degraded ecosystems across the globe by 2030. Its projected cost will be 1 trillion US dollars, equivalent to 0.1% of worldwide financial gains between 2020 and 2030. The above restoration plan creates 9 trillion US dollars in natural ecosystem services and will decrease society's poverty (United Nations 2022). Ecological restoration projects require science-based assessment in association with policy development for making a strategic framework and prioritizing the intervening actions. It will ensure the long-term future health and stability of the ecosystems. Scientific research groups have to assess the prioritization of the regions for restoration in association with national and local policymakers for making sustainable landscapes to sustain biodiversity and ecosystem services. Several studies estimated that restoring 15% of transformed land in critical priority areas would avert at least 60% of future decline in biodiversity and the severity of climate change. For the achievement of this aim, wetlands and forests are significant landscapes. However, if the objective is to achieve restoration with minimum economic inputs, then arid ecosystems and grasslands are the regions of interest (Crossman 2017; Csákvári et al. 2022).

Furthermore, cost-benefit and cost reduction methods will be used to optimize the monetary and resource allocation for restoration techniques during spatial development. For example, in Europe, steppe habitats have community significance for the drier part of the aridity gradient. These habitats were included in the Eurasian forest-steppe areas, with great species and functional diversity. The grassland's presence in the region offers additional societal and financial significance, which has been utilized as extensive pastures and meadows since earlier times. Despite this, residual semi-natural habitats face risk due to anthropogenic activities like enormous land degradation and habitat fragmentation. Therefore, it is a massive task for the scientific community, policymakers, practitioners, and activists to keep and restore the lasting semi-natural steppe regions. It should be simplified by executing restoration primacies and assessing the optimum restoration techniques locally and countrywide. The historical and futuristic data science assessment for all the ecological and economic factors is required to make an optimized restoration action plan. Ecological restoration plans are of two types, i.e., passive and active. The passive restoration plan strategy focused on allowing natural processes with minimal external interference. It applies to locations with minimum abiotic stress and low to reasonable disturbances in landscapes due to humans.

In contrast, the active restoration strategy's key objective is to enhance the landscape's abiotic and biotic factors. It is worth mentioning that the interference cost in an active restoration plan is higher than passive restoration. The prioritization of ecosystem restoration will be based on the consideration of an optimized ecological and economical passive or active restoration plan through the regeneration capacity of regional habitats, neighboring vegetation covers, and long-standing fields. Restoration of the ecological system comprised a wide range of approaches. It will vary with geographical, topographical, hydrological and type of ecosystem. It is mentioned as a go-to technique for resolving upcoming environmental changes (Csákvári et al. 2022).

There are three types of restoration science practices followed presently worldwide, which are as follows:

- **Rapid environmental, ecological, and cultural changes:** It will focus on providing climate change mitigation and adaptation solutions, minimizing the regional flora and fauna invasive species, and increasing the ecosystem benefits and valuation compared to traditional restoration practices.
- **Innovation in combination with ecological and engineering design:** It will provide a newly designed ecosystem through green infrastructures in urban regions and the making of agro-ecosystems and natural landscapes in peri-urban and rural regions.
- **International and national agreements for restoration investments:** This will focus on providing transboundary restoration projects and achieving common sustainable goals through cooperation at different levels for the betterment of humankind and nature.

Therefore, ecological restoration science aims to provide ecosystems in pre-disturbance conditions. It will give an approach to improve the regional biodiversity and structural and functional integrity with surroundings and diminish the regional climatic and economic value degradation. The United Nations governance flow chart for ecosystem restoration is shown in Fig. 1.1.

1.2 Need for the Ecological Restoration

The understanding of ecological restoration in our day-to-day life requires an understanding of ecological terms and their significance. Therefore, we have to understand first the word ecology, derived from the Greek word *Oekologie.* It can be separated into *Oikos (home)* and *logos (to study).* Therefore, ecology is defined as the study of life at home with an emphasis on the interrelations between organisms and their environment. The German biologist "Ernst Haeckel" defined ecology in 1866 as "the complete knowledge of all the relations of organisms and its surrounding outer world in the presence of certain organic and inorganic conditions of existence." Ecology studies interactions at organisms, populations, communities, ecosystems, biomes, and the biosphere levels. Thus, ecological restoration for any region and species will improve its, directly and indirectly, associated surrounding conditions and complex interrelationships with different biotic and abiotic components.

The subset of ecology is an ecosystem, and usually, the terms "ecological restoration" and "ecosystem restoration" are used identically despite certain distinctions. Ecologically regions comprised of several ecosystems like forests and rivers,

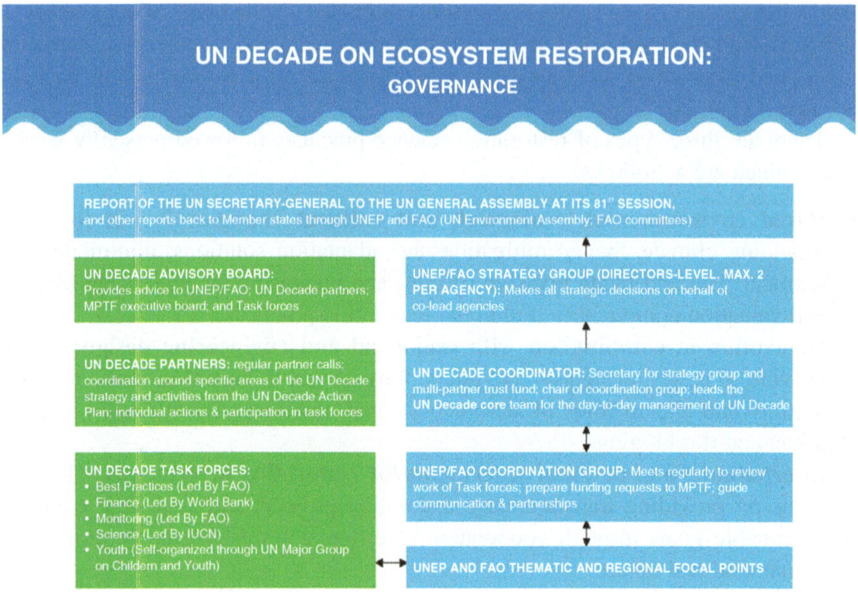

Fig. 1.1 UN decade on ecosystem restoration governance (United Nations 2022). (*Source:* https://www.decadeonrestoration.org/about-un-decade)

which are two distinct ecosystems in a region. Ecologically they can be studied in a combined/isolated manner to understand the interrelations between them. The ecological restoration of any region is usually subdivided into various ecosystem restorations present in the region for proper management and monitoring for more extended periods. The ecosystem is a vital subunit of the ecology that contains various essential interactions between different life forms on Earth. It associates all the living organism interactions within themselves and their surroundings. It has variable scales ranging from small soil grains to the entire Earth planet (United Nations 2021a).

The broader categories of ecosystem types include natural and artificial or man-made ecosystems. Natural ecosystems function naturally with minimum or no anthropogenic interference. The natural ecosystems will be further classified as terrestrial (forests, deserts, grasslands) and aquatic ecosystems (oceans, rivers, lakes). Artificial ecosystems were operated and managed entirely through anthropogenic activities like crop cultivation fields, aquariums, cities, etc. Ecosystem stability in a region determines the structural and rate functions. Structure signifies the composition of living organisms and their distribution and amount, the distribution of non-living materials like nutrients, water, and the range of physical conditions like temperature and light time, while rate refers to the energy flow in the community, rate of nutrient cycles, and biological and ecological regulations for biotic and abiotic components combined. Therefore, ecological restoration is necessary to make a sustainable relationship with our ecosystems. It will provide sustainability through climate change mitigation, assuring food availability for rising populations, and minimizing the decline in biodiversity. To achieve the target of keeping the mean temperature of our planet below 2 °C, ecological restoration will be required in combination with renewable energy systems. In this manner, it will benefit the present and future generations by conserving and preserving natural resources (UNEP 2021a; United Nations 2021b).

Ecological restoration is "the process of assisting the recovery of an ecosystem that has been degraded, damaged, or destroyed." It is the international agreement for accelerating the process of ecosystem recovery in terms of structure, functional utility, and its exchanges with surrounding landscapes and seascapes. It will enhance ecological integrity and ecosystem resilience to climate change (SER 2022).

There are various reasons for explaining the requirements of ecological restoration at a large scale globally; some of them are presented below:

- It is estimated that there will be 10 trillion US dollars in economic loss globally by 2050 if the ecosystems continuously decline at the present rate.
- Restoration of 15% of converted land will prevent 60% of biodiversity loss on the Earth.
- Globally, 81% of urban regions would prevent water pollution through efficient water management practices for their population, in combination with forest protection and better agricultural practices (UNEP 2021b).

Ecological restoration can be used to reduce pressure on natural resources with significant interventions like afforestation, removal of invasive flora and fauna species, and re-establishment of locally extinct species. Thus, ecological restoration works will benefit us by providing physical and mental health benefits and long-term food and water security, enhancing biodiversity and economy, reducing future war and migration conflicts for natural resources, and providing climate change mitigation and adaptation solutions (Fischer et al. 2021; Howell et al. 2013).

1.3 Restoration Ecology

Restoration ecology is the scientific discipline for applying the ecological restoration concept in the field. Thus, restoration ecology is distinguished through ecological restoration in terms of research and development applications for on-site ecological restoration projects. The concept of restoration has been in practice through earlier centuries, but in the present time, the development of restoration ecology began in the twentieth century through the Aldo Leopold conservationist movement. It has been grown at different scales with more minor scales like afforestation to large-scale projects like Louisiana wetlands. The success of the ecological restoration projects depends on the initiation from the regional population and implementation through community members voluntarily. As restoration is a complex process in terms of collaborations with a wide range of participation of people from diverse fields, therefore, social science will be an integral part of the restoration at each development level.

The restoration ecology primarily focused on community and ecosystem ecology specific to vegetation development in the region. Recently, it has been utilizing the concept of conservational biology to increase certain animals' presence in the region. In the latter part of the twentieth century, restoration ecology was recognized as a scientific field and studied by several organizations worldwide. The widespread applicability of restoration ecology began with the establishment of the International Society for Ecological Restoration in 1988. Research work in association with journals like *Ecological Restoration* (1981) and *Restoration Ecology* (1993) increases the popularity of the restoration ecology concept among the scientific community and policymakers.

There are certain concepts associated with restoration science that have widespread applicability, which are detailed below:

Disturbance Disturbance events can occur at various scales in severity and with variable temporal resolution. These disturbances cause the alteration in the species, nutrient cycles, soil, and water properties. These disturbances can be classified as natural or anthropogenic. Natural disturbances are extreme weather events, such as wildfires, volcanic eruptions, etc., while anthropogenic disturbances cause alteration through habitat destruction and fragmentation through converting land for agricultural usage or dam reservoir construction across the water bodies. Anthropogenic activities influence climate change and extreme weather events' severity and temporal variability. The restoration project aims to initiate ecosystem recovery after the disturbances and recreate the natural cycles of the pre-disturbance.

Genetics The long-term sustainability of the restoration plan considers the genetic diversity of the ecosystem. Regional flora and fauna, with lots of consideration, are used in vast numbers for a successful establishment to enhance genetic diversity in the re-established ecosystem. It is a decisive factor in ecological restoration as it will decide the evolving and recovering ability from disturbances for extended periods.

Succession It is the process through which the community structure varies in terms of numbers and proportions of the wide range of species found in the ecosystem due to the recovery time post-disturbance event. Passive restoration will allow succession to naturally occur in the ecosystem after eliminating all the disturbing sources. For example, the rise of deciduous forests in the Eastern United States was a successful restoration after finishing all the agricultural practices in the region. In contrast, active restoration increases the restoration process with changes in the natural succession path. For example, like mine tailings, the passive restoration will take a considerable time; thus, active restoration is the appropriate choice.

Community Assembly Theory It is the concept involved in developing sites like pre-disturbed events. It will depend on the initial arrival of the wide range of species in the region and their mutual biological interactions. This is a challenging task in restoration projects as each site does not develop ideally to the pre-disturbed event with desired species composition or ecosystem utility. It will depend on the seed mixes, the planting order, and the time when there is a requirement for redeveloping specific ecological communities with the plans to prevent them from invasion by pests.

Landscape Ecology Landscape ecology involves several concepts for restoration. The restored regions are usually divided into smaller parts, making them sensitive to several disturbance events due to habitat fragmentation. A smaller habitat region will support a lesser number and proportion of species due to the high risk associated with interbreeding and regional extinction. Biogeography island theory estimates that diverse species populations will develop in huge landscapes with better connectivity to the people in the surrounding patches. This theory presumes that the area between habitat patches is uniform and hostile. These patches decided for restoration are less hostile to the targeted restoration species and desired ecosystem utilities. The restoration projects will improve the connectivity among the fragmented habitat regions for restoring paths. These paths can be termed in different forms based on the requirement like wildlife corridors and stepping stones for allowing better movement across the region. Corridors are defined as thin linear strips between isolated habitat regions, while stepping stone is having low connectivity with allowable movement paths across different patches.

Application The restoration application has a large number of steps that consider all or specific points described below:

Site Assessment Complete appraisal of the restoring site's present conditions and determine all necessary activities. It is the initial step for assessing all the disturbance events and techniques required for halting or reversing them.

Assessing the Aims of the Restoration Project This will be based on the requirements of the restored community. The decision of the aims will be based on the literature review for a similar environment, assessing the historical datasets for detailing the pre-disturbance region and its biological interactions. It will identify the best species to sustain present and future climatic changes.

Removing Sources of Disturbance Initiation of the restoration project will be based on the prior removal of all the disturbance-inducing factors in the region, for example, halting all mining and farming practices, restricting cattle entry, and removing the area's toxicity and invasive species.

Restoring Processes/Disturbance Cycles There are specific restoring processes like natural flooding or fires that can restore regional ecological integrity, as these will provide the ability for native species to evolve and tolerate natural disturbances and return to the initial state without actions by restoration people.

Substrate Rehabilitation It will involve improving the local environmental conditions, like soil, water, and air quality, for pre-disturbed events.

Vegetation Restoration These seem to be a simple solution but require an assessment to identify native species in the region, as invasive species plantation in the region may cause a complete failure of the restoration project. It can be done in the form of seeds or seedlings.

Monitoring and Maintenance The continuous monitoring of the restoration site till the achievement of the restoration target is a critical task that can be useful for futuristic restoration strategy and management. The ideal restoration projects achieve self-sustainability for the ecosystem without anthropogenic interference. Still, in some cases, like periodic removal of pests have to be carried out until the planted saplings are not fully developed for making restoration efficient for longer-term (Howell et al. 2013; Palmer et al. 2016).

Restoration efforts are required for each ecosystem, but critical attention has been given to severely impacted ecosystems by anthropogenic actions like forests, wetlands, and key biodiversity areas worldwide (Vaughn et al. 2010).

1.4 Ecological Theory

The restoration ecology is a complicated conservation process and can't be considered a substitute for conserving or preserving the present natural resources and ecosystems. The primary objective of ecological restoration is to establish the native species on the restoration site in the pre-disturbed environment according to the community model. The community or the ecosystem model concept is based on the prior understanding of the ecological theory and the nature of earlier communities in the restoration site. Thus, the concept of ecological theory will be considered the heart of each restoration project in the mind of the restorationists. The ecological theory utilizes for understanding human actions during the restoration process. The restoration team, scientific community, and policymaker members' combined activities and interactions will decide restoration efficiency for more extended periods.

1.5 Restoration Science: Principles, Strategy, and Guidelines

In a general manner, each ecological restoration plan for any site has three principles to be followed, which are described in detail below:

Get Everyone on Board Identify each stakeholder associated with the restoration project. Provide complete information on the restoration plan to each person impacted or affected by this directly or indirectly. Acknowledge all the diverse views from different backgrounds of the population on nature to achieve maximum benefits with the restoration plan. Get maximum involvement from various persons who can contribute through their knowledge and skills to get valuable support to the

Fig. 1.2 Restoration principles (UNEP 2022b). (***Source:*** https://www.unep.org/news-and-stories/story/panel-unveils-10-guiding-principles-campaign-revive-earth)

restoration project, minimize the opposing points, or overcome the upcoming challenges.

Setting Goals and Measuring Progress Through prior assessment of the restoration site and learning on ecosystems, we have to establish a protected and self-sustainable restoration site. It will be based on continuous monitoring of the time-based achievement of mid-term and ultimate goals. Determine all the measurable indicators for monitoring the progress and regulating the project's completion time.

Help Nature Help Itself It is a wholly known fact that we can't be able to restore the degraded or destroyed ecosystem to its original state. However, we can enhance its ecological and functional integrity naturally and sustainably. Therefore, the efficient restoration project plan should consider all the natural processes like introducing all the native species, improving the nutrient cycling regimes in different environmental components, etc.

The broad principles as decided by the UNEP (2022b) are shown in Fig. 1.2 and described below:

Principle 1 Global Contribution Successful ecosystem restoration requires global contribution, which can assist us in adaptation to climate change, the decline in biodiversity loss, and land degradation.

Principle 2 Broad Engagement Individuals are the core of the restoration plan. It is estimated that more than 40% of the Earth's population is impacted by ecological degradation. Thus, we have to engage all the marginalized people in restoration plans, like indigenous people, minorities, women, etc.

Principle 3 Continuum of Activities No single approach can be followed for all the restoration tasks, as it is decided based on the severity of degradation and its self-sustainability for restoration on its own.

Principle 4 Benefits to Nature and People Restoration should be carried out in such a manner that under any scenario it should not lead to further degradation of regional environment and people who depend on it.

Principle 5 Addressing Causes of Degradation Restoration is not a substitute for conservation. Thus, we required a proper assessment of degradation's natural and anthropogenic drivers.

Principle 6 Knowledge Integration Integrate all the forms of knowledge, i.e., scientific sources and traditional and local community information, for making an efficient restoration plan.

Principle 7 Measurable Goals Ecological, economic, and social aims of the restoration plan with a timeline should be decided before the initiation of the restoration plan.

Principle 8 Local and Land/Seascape Contexts The restoration plan's impact varies from place to place. Thus, it should be made in the context of the local site and its surrounding population.

Principle 9 Monitoring and Management Restoration is a complex process; thus, continuous monitoring with proper management is required regularly for long-term sustainability. Ecological theory and adaptive restoration will be valuable approaches.

Principle 10 Policy Integration Regional- and national-level government cooperation is needed for policymaking and supporting the ecosystem's restoration plan for extended periods.

The strategy adopted for the worldwide greater utilization of ecosystem restoration will focus on the following points, which are shown in Fig. 1.3.

Restoration guideline preparation will follow the approach for making restoration plan-associated principles and strategies effective in re-establishing the site, engaging all the stakeholders with maximum participation, and producing maximum benefits with minimum resources, time, costs, and efforts (IUCN 2012; UNEP 2021a).

Fig. 1.3 Strategy points for ecosystem restoration (SER 2022). (*Source:* https://www.ser.org/page/DecadeonRestoration)

1.6 Adaptive Restoration

An adaptive approach to ecological restoration will provide several tasks for reorganizing and redesigning the aims, strategies, and technologies involved at each stage of the restoration project. This approach in restoration provides steps to the restoration team for daily evaluation of the project progress and identifying the corrections scientifically for modifying our expectations in achieving the ecological restoration objectives. In supplement to this, we can determine the impact of any special disturbance event or the involvement of any new innovative techniques for preparation to meet the upcoming challenges. This approach has certain limitations; it is time-consuming and requires extensive documentation of each day and activity during the completion of the restoration project. However, project documentation with continuous collection and assessment of information for planning, implementation, and management is crucial for assisting in long-term sustainable success for the restoration project and adopted practices in it. The research work involved in developing restoring practices will contribute to the ecological theory. Therefore, scientific experiments and on-site trials were the main components of adaptive restoration used to determine the critical information for site designing and selecting restoration implementation and management strategies.

1.7 Case Studies of Ecological Restoration

This section discusses case studies of ecological restoration for wetlands, Himalayan, and urban (artificial) ecosystems in detail below.

1.7.1 Urban Ecological Restoration

Urban regions are classified as artificial ecosystems with the maximum anthropogenic influence in their creation and regulation. The urban regions occupy less than 1% of the Earth's land surface but provide household facilities to more than half of the world's inhabitants. These regions are created after replacing the natural ecosystems and are generally observed as having high degradation due to informal planning, less vegetation, huge waste, and greenhouse gas emissions. These phenomena lead to the degradation of more natural ecosystems and fertile land in surrounding urban regions (UNEP 2021a). In this context, Varanasi, the oldest city in India, has recently observed several changes in urban design to restore urban ecology. The city has a population of 1.3 million and generates about 445 tons of municipal waste per day, which is collected and transported along with different forms of waste. Thus, to improve the city's waste management, Varanasi Nagar Nigam prepared a "holistic waste management strategy" with support from United Nations Environment Programme, ICLEI – Local Governments for Sustainability, South Asia, and International Environmental Technology Centre (IETC). The primary objective of this waste management planning is to make the Varanasi city "*Nirmal Kashi*" by enhancing integrated solid waste management practices that are self-sustainable, benefit society, and enhance the quality of life. It will target waste reduction, reuse, and complete city coverage with street sweeping, waste collection, transportation, and scientific disposal. Thus, this waste management practice will provide a clean city and its surrounding ghats for religious pilgrimages (UNEP 2018). Several developments were initiated to implement a waste management plan in the city. The construction of three transfer stations for the whole city is to provide better collection and segregation of waste, as shown in Fig. 1.4.

This waste management approach will lead to maximum resource utilization and increase waste treatment and disposal efficiency, leading to a circular economy's creation. Similarly, cities have a high population density and built-up regions, which provide very little vegetation in the urban region. Varanasi city is estimated to have less than 5% vegetation, which is relatively less than the 20% requirement for better living standards. The region's urban forest or green belts will minimize the urban heat island effect and air, noise, and soil pollution, increase microclimate regulation, etc. (Das et al. 2020; Hinge et al. 2018; Poonia et al. 2021a). Therefore, several plantation programs are presently initiated with the integration of the Indian smart cities mission and through several NGOs to increase the vegetation in the city (Goswami and Kumar 2020). The city is located on the west side of the Ganga

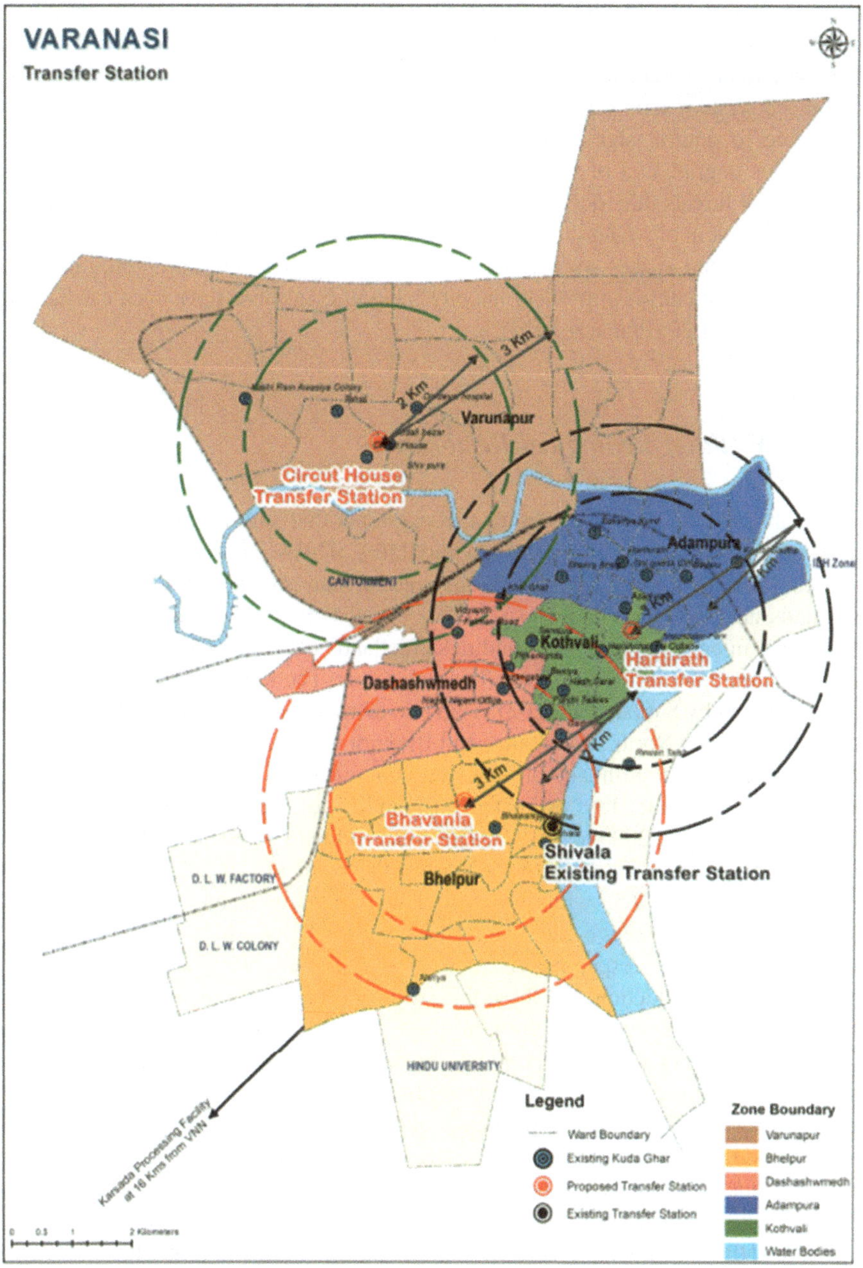

Fig. 1.4 Three proposed transfer stations and their buffer zones (UNEP 2018)

river. Due to the presence of the Ganga river, the city has religious significance, and many rituals occur across its banks over several ghats daily. Regular use and improper management will lead to the degradation of water quality, kunds, ghats, and its ecosystem across the city. "Kunds and talav" are small water bodies developed due to natural significance and will face threats of extinction due to the continuous degradation of water quality. One of the talavs, i.e., "Pushkar Talav," faces severe threats due to the continuous dumping of solid and liquid waste. Thus, the rejuvenation of water bodies like "kunds and talav" and the restoration of community open spaces over ghats under several policy programs through local public involvement for improving the city's living conditions have been carried out (Bansal et al. 2020; Praharaj 2014).

1.7.2 Wetland Ecological Restoration

These are classified as natural aquatic ecosystems that sustain unique biodiversity and play a vital part in climate change variability. This section discusses ecological restoration case studies for floodplain, coastal, and saline wetlands. Floodplain wetlands are highly ecologically sensitive ecosystems that provide structural and functional integrity to their surroundings. Wetlands occupy 6% of the land surface and consume 12% of the Earth's carbon emissions. The floodplain wetland restoration presented by Sarkar et al. (2021) is in two districts of West Bengal, i.e., Mathura and Chandania in North 24 Parganas and Bhomra in Nadia. Ecological degradation, encroachment, and shrinkage are causing ecological value, biodiversity, and fish production to decline. It is estimated through spatiotemporal assessment of these three wetlands from 2000 to 2017 by GIS tools and field assessment. The significant outcomes of the evaluation provide that all the wetlands are alkaline, with a pH ranging from 7 to 9.5, with increased macrophyte invasion. All the wetlands shrinking at the effective rate, i.e., Bhomra at 57.68%, Mathura 37.20%, and Chandania 44.07%, respectively. Bhomra and Mathura wetlands are showing a decline in fish production. All these three wetlands have lost their connection with the parent river channel. Therefore, all these factors combined cause fisheries and diversity decline of wetlands. Probable threats to wetland diversity are water pressure (95%), sedimentation (85%), macrophyte invasion (70%), and the absence of river linkage (65%). Apart from these, climate change significantly impacts the wetlands due to variability in temperature, rainfall patterns, and extreme weather events. Thus, to make an ecological restoration plan for these wetlands, we have to monitor land use changes, aquatic weed invasion, industrial runoff, solid waste, and climatic variability in temperature and rainfall (Sarkar et al. 2021; Sharma and Goyal 2020).

Climate change is estimated to significantly impact 84% of coastal wetlands and 13% of saline wetlands due to a rise in the 1 m sea level. Sea level rise and storm surge will significantly impact the coastal wetlands (Sarkar et al. 2021; Wetlands International 2022). Here we discuss the case study of assessing the restoration practices for Ashtamudi lake Ramsar site coastal wetland in Kerala. Major threats to these wetlands are a decline in water quality, mangrove vegetation, and fisheries production. The risk of rising population, urbanization, over-exploitation of natural resources, and less environmental management is causing significant threats to wetlands (Goyal et al. 2022). Therefore, restoration economic feasibility is determined through the assessment of ecosystem services of wetlands in market values, determining the marginal willingness to pay (WTP) for the wetland attributes by the local population, and comparing WTP with restoration costs and benefits. Based on this assessment, it is determined that WTP for the restoration project is comparable to the average weekly wages of Kerala people in 2019. These cost evaluations can be minimized through financial support from the government in subsidies and NGOs, involvement of corporate industries through corporate social responsibility, etc. (Poonia et al. 2021b; Sinclair et al. 2021).

The Sambhar Salt Lake wetland is the playa wetland, i.e., shallow circular-shaped wetlands. The salinity in this wetland contributed due to the presence of ions like calcium, sodium, potassium, carbonates, bicarbonates, etc. This is declared a Ramsar site due to its importance in reference to migratory birds, as nearly 0.1 million flamingos come to these wetlands in winter annually. It is determined that due to illegal saltpan encroachment, wetlands' water level and quality are declining. This leads to a falling migratory bird population in the winter season. Thus, this site's restoration will require a complete stoppage of all salt mining in the region and providing systems for raising the water level in the wetland during fewer rainfall seasons (Goyal et al. 2018; Goyal and Ojha 2010, 2012; Naik and Sharma 2022).

1.7.3 Himalayan Ecological Restoration

These are classified as a natural terrestrial ecosystem that spread in several Asian nations, contains the highest mountain peaks, and is a source of many freshwater perennial rivers in the region. In this study, we discuss in detail the central Himalayan ecosystem restoration project carried out in the Uttarakhand region of India.

The Himalayan region (HR), with a geographical covering of 5.95 square km, stretches over 2500 km between the Indus and Brahmaputra River systems (Rautela et al. 2022). The HR physio-graphically stretches from the southern foothills (Siwaliks) to the northern Tibetan plateau. The three significant geographical groupings of the Himalayan regions are greater (Himadri), lesser (Himachal), and outer (Siwaliks), which are continuous throughout their length and are divided by critical geological fault lines. Excessive usage of resources, substantial land modification, and climate change affect mountain ecosystems across the Himalayas (Goswami et al. 2018; Yadav et al. 2021). Mountain ecosystems are essential for their natural beauty, as places of serenity and spiritual refuge, and for the ecosystem products and services, which they supply to almost half of the worldwide population. The Himalayan ecosystems are stressed by environmental factors such as climate fluctuation, altitude, and soil. The winter pastures are transferred to higher elevations

Fig. 1.5 Ecological restoration by mixed forest approach used by Pandey et al. (2017)

every year and gradually recede. Various ecological impacts in the Himalayan environment include the conservation of flora and fauna, restoration and recharging of groundwater and surface streams, land restoration and soil enrichment, mitigation of geohazards and forest fires, and so on.

A study was carried out by Pandey et al. (2017) in Kot Malla village, Rudraprayag district, Uttarakhand state of India. They studied ecological restoration and livelihood sustainability through mixed forestry in a man-made forest called JUNGALEE, made by Jagat Singh. Through the surge in demographic evolution of the anthropogenic presence, the mid-Himalayan region observes an increase in the pressure on forest ecosystems. The probable impacts due to this approach are shown in Fig. 1.5.

1.7.4 *Ecological Impact*

The forest offers many resources today that support many ecological services and functions. Because there are more than 75 different types of trees, plants, and herbs in this jungle, flora and animals may thrive there. A wide diversity of trees has accelerated carbon sequestration and filtered filthy air, allowing the entire world to breathe easier. A few years ago, the mountainside was wholly desolate and had a rocky surface. However, Jagat's strenuous efforts have covered the mountain with greenery. The forest's abundant vegetation has drawn local fauna. Wild cocks can be seen hanging out in the forest during the day and running through it, as well as birds chirping from the branches and jackals howling at night. A rich floral unit may naturally develop into a rich hydrological unit, as Jagat Singh Jungalee's hard work has demonstrated. A forest is thought of as a possible supply of water since it aids in processes like evapotranspiration, infiltration, and percolation that are either directly or indirectly connected to rainfall. The village of Kot Malla, which had previously had sparse vegetation and a bare stony surface insert, had almost all its streams dry up, and the groundwater table had been falling yearly. However, Jagat's persistence had recharged the streams surrounding his forest, and today the streams flow for 6–8 months of the year.

To restore soil and land, Jagat created terraces that sloped inward, allowing water to be retained for a while and also providing nutrients to early-planted tiny plants. Allowing the community animals to graze on his newly planted grasses makes their feces available and improves the soil's nutritional content. He also ensured that bushes degraded and that the ground eventually sprouted new life with compost, manure, and trash. This voyage is distinguished by a significant investment in soil improvement and land restoration. It was essential to plant a variety of trees and to reject monocropping to meet the nutrient demands of trees without misbalancing soil composition by excessive pressure on a few nutrients. Several pits with litter and decomposed materials were dug within the jungle to prepare compost and spread it throughout. The mid-Himalayan region is susceptible to geohazards like landslides, rockslides, rock falls, etc., because of their fragile ecosystem. A steep mountain slope with little to no vegetation is particularly vulnerable to landslides and rock falls, which can seriously harm people and property. Due to the variety of plants in Jagat's forest, his jungle has proven essential in preventing landslides. While a single tree variety may partially cause a landslide, many tree varieties have unique root networks intertwined together that hold soil in their way and prevent it from loosening up.

In contrast to mixed forests, mono-plantations typically have loose roots. The massive monoculture of pine trees, one of the leading causes of forest fires, has been significantly reduced by diversified forestry. Pine, an oily pine, burns fast, but the diversified characteristic of mixed forestry controls forest fires and makes them burn slowly. The mixed forest provides enough time to put out fires with human effort and can help to avoid serious risks.

1.7.5 Socioeconomic Impact

Jagat Singh was the one who originally planted the seeds of this forest, prioritizing socioeconomic issues over everything else. He created this forest using his traditional knowledge and scientific technologies that may be used in mixed forestry to address the local problems. For solving several ecological and socioeconomic issues, a global solution has emerged. Without assistance from any governmental or non-governmental organization, a local-level solution to the problems of fuel, feed, and water has been found. The inhabitants have easy access to the very nutritious fodder grasses produced in this area. Now, women in this region do not have to walk tens of kilometers with bundles of fodder on their backs through the dangerous, untamed forest. The Jagat's forest model has been applied throughout the neighborhood, where residents who previously had little interest in growing trees on their agricultural property willingly do so. The cultivation of feed and medicinal plants has begun. Although milk sales are still uncommon in this region, they may be promoted, and a source of income is created because fodder is now readily accessible and will help the local economy. People are economically prosperous and growing increasingly independent.

On the other hand, a study of ecological restoration in the higher-altitude ecosystems of the Himalayan region was conducted by Kuniyal et al. (2021). As these regions are affected by global warming and climate change activities, their rational management has been a matter of excessive concern (Trant et al. 2020). Many eco-services are provided by Himalayan alpine pastures, like carbon sequestration, water availability, biodiversity preservation, and food security. Several factors contribute to degradation, including the emergence of invasive species, poor management and development strategies, soil erosion, overgrazing, and the harvesting of medicinal plants. Therefore, these regions require both the preservation of natural vegetation and, when needed, their restoration. In addition to increasing native vegetation coverage and connectivity, eco-restoration also provides a range of socioeconomic benefits, as well as increasing ecosystem goods and services and promoting socioeconomic development. Still, there are no clear guidelines or orders for restoration initiatives in high-altitude degraded grasslands. To overcome this, Kuniyal et al. (2021) completed an on-site assessment of eco-restoration in high-elevation grassland ecosystems in Dayara alpine pasture (Bugyal) (3501 m), Uttarkashi district, Uttarakhand state of India. The case study tries to assess how the project will affect, among other things, the amount of grazing space available; the number of tourists in the area that can be supported; the general stability of soil erosion; the vegetation profile, mainly plant development; and the availability of soil nutrients. The ecosystem restoration evaluation involved computing the restoration success index while seeing 3 classes, including direct management measure (M), environmental desirability (E), and socioeconomic feasibility (SE), while considering 22 different factors. Both biotic and abiotic stresses were included in "M."

In alpine regions, soil erosion was considered an abiotic pressure due to topographic fragility, while grazing and tourism were seen as biotic pressures. The degradation factors considered for the current investigation were discovered to be consistent with other Himalayan region impacted locations. According to reports, overgrazing has occurred in high-altitude meadows of Himalayan areas (Byers 2005; Cao et al. 2013; Khan et al. 2019; Nautiyal et al. 2004). In the instance of the Dayara Bugyal, it was discovered that local livestock populations were growing at a rate of 33 CU annually and were also thought to be the maximum at fault for overgrazing. After the rehabilitation phase, it was discovered that the migratory livestock unit was under control. Significant processes and deterioration characteristics in alpine ecosystems, such as soil cracking and plant disturbance, are significantly correlated with overgrazing. Ecosystem restoration will promote tourism and is predicted to be the primary source of future development in the Himalayan region. Several factors hinder sustainable tourism growth in Uttarakhand, including a lack of thoughtful plans, an outdated legislative framework, and an influx of visitors. Ecotourism may now be encouraged in the region, which may induce changes in the socioeconomic status of two nearby rural regions because of the low tourism activity in the area.

Homestays are recommended for visitors, tour guides are trained, and birdwatching is encouraged.

Consequently, eco-restoration techniques and their assessment model may have a role in making restoration practices more effective in the future. Other erodible high-altitude locations can also be restored with geo-coir mats and pirul (leaves of pine trees) check dams. Several of the recommendations made by the case study, including grazing pattern and herb establishment prevention, tourism management strategies, and other recommendations, are also applicable to other degraded pastures in the Indian Himalayan and other mountain alpine and subalpine ecosystems.

1.8 Policy Planning and Implementation for Ecological Restoration

The concept of ecological restoration has gained more attention on the public policy agenda. The implementation of ecological restoration typically takes place at the more minor levels. However, several political actors, both nationally and internationally, have declared they intend to engage in it (Corcoran 2010). Restoration has many benefits, one of which is that it helps combat global environmental change. Reforestation plans for carbon sequestration and wetland restoration for flood safety are two examples of climate protection and adaptation policies increasingly relying on restoration. In addition, it is used to protect the delivery of ecosystem services. The European Union, among others, emphasizes restoration as a strategy to achieve the 2020 biodiversity goals, also known as the Aichi goals. As an implementing tool for various resource-specific policies, such as the EU Water Framework Directive, which aims to restore surface water and groundwater to an "ecological good state" (ECD EC2000/60), it is also becoming increasingly crucial in agricultural policy frameworks and for enhancing the food security. Likewise, restoration is also used as a planning technique to compensate for disturbances to the aesthetic or cultural values of the landscape, often including elements of traditional methods and local culture (European Commission 2011, 2022).

Promoting ecological restoration activities, including vegetation, water, wildfires, wildlife, and recreation management, will be the responsibility of policymakers and related parties. Some examples of possible actions are monitoring the health of resources; maintaining, repairing, or improving terrestrial and marine ecosystems; or controlling human use. The following initiatives will be supported: the eradication of invasive species; restoration of grasslands and riparian areas to improve river basin functionality; practicing environmentally and ecologically friendly fire management; and prescribed burning and thinning of forests to reduce fuel load and increase forest heterogeneity and enhance the habitat of wildlife and fish. The focus will be on growth and building partnerships to build organizational capacity and utilize large-scale management contracts at the landscape level and improving and developing consulting work with tribal governments to leverage their conventional expertise in land management. Partnerships with stakeholders, communities, local authorities, volunteers, and residents will be prioritized to promote communication and reduce conflict when developing and implementing ecological restoration activities.

Restoration initiatives can be carried out in cities and rural regions, focusing on various ecological systems or landscapes. Such activities can take place at multiple scales, ranging from small-scale, vastly localized experiments and tentative trials to remediation of industrial or mining locations to what is referred to as "megaprojects," like the Kissimmee River restoration in Central Florida (Whalen et al. 2002), prairie restoration in the United States (Ryan 2005), or current water preservation practices in the Netherlands (Drenthen 2009). Projects may also entail the purposeful restoration of desired species that have been eradicated or rendered extinct locally due to pressures from other development pressures and changes in land usage. For instance, wolf reintroduction programs have generated controversy both in Scandinavia and North America, not least because of local worries about the possibility of livestock loss (Gross 2008). Another project priority is river restoration, which may involve the removal of dams, re-routing and re-bouldering of rivers (as was done, for instance, in Sweden; see Lejon et al. 2009), "daylighting" of culverted rivers, or biological restoration of city riverbanks (e.g., in the United Kingdom; see Eden and Tunstall 2006).

1.8.1 Policy Planning and Implementation for Ecological Restoration in India

The natural resources and ecosystems have been protected and conserved in India by various acts and interventions held by the government. On the other hand, local conservation initiatives have been successful thanks to grassroots community movements such as the Chipko movement in Uttarakhand, Silent Valley Movement in Kerala, etc. However, India's main ecological issues include 96 million hectares of degraded land, an unparalleled decline in biodiversity, and catastrophic weather occurrences. Through greenhouse gas emissions and biodiversity decline, this ecosystem degradation accelerates the degradation of natural ecosystems (Bellard et al. 2012). India can significantly contribute through its policy framework to attaining socioeconomic and environmental sustainability by viewing restoration and conservation as the primary objective and long-term funding (Fleischman et al. 2020). For instance, the MGNREGA program helps jobless migrant workers and other rural people find employment while working on damaged land to establish different forests, wetlands, botanical gardens, and economic models through environmental restoration. Rural regions can promote ecological and rural tourism by creating new ecosystems based on biodiversity on degraded soil. The innovation in biodiversity-based ecosystems can promote ecological and rural tourism on degraded soils in rural regions. This will provide local communities with new small-business options and advance conservation via the care of these individuals. It is anticipated that India's rural or nature tourist industries will expand in the post-COVID future (Roy et al. 2020). Furthermore, using community and wastelands accessible to various regional and nationwide departments, like the forest, revenue, and railway departments, to progress, for example, specialized biomass production facilities can help with carbon sequestration and bioenergy and benefit laborers and other stakeholders financially.

1.9 Future Challenges

The futuristic tasks associated with ecological restoration across different regions worldwide can be classified into social, ecological, economic, and political categories. Here we discuss specific challenges in detail as follows:

1.9.1 Ecological Challenges

In a recent assessment, the Space Applications Centre (SAC) of the Indian Space Research Organisation (ISRO) found that 96.3 Mha of land, or 29.32% of the country's total geographic area, has been degraded or turned into desert. However, deforestation (29.3 Mha), wind erosion (18.23 Mha), and salinity are the leading causes of ecosystem degradation (3.67 Mha). According to Díaz et al. (2006), habitat loss is a significant cause of the decline in biodiversity, food insecurity, climate variability, and unemployment in rural regions. Protected or not, most of India's forest ecosystems are degraded and suffer from biodiversity decline caused by biological encroachment, anthropogenic influence, and excessive logging (Bawa et al. 2020; Pandey 2008). India has contributed 6255 animal and 2235 plant species to the IUCN's revised red list (IUCN 2020). One hundred fifty-nine vulnerable, 197 endangered, and 93 severely endangered floral species are among the 449 that have been classified as threatened, with varying conservation statuses. Five plant species in all have already vanished. A total of 722 species of animals, including 398 vulnerable, 230 endangered, and 94 severely endangered species, face threats (see IUCN 2020). Ecosystem degradation has sped up the rate at which wild species disappear from the environment, in addition to tiny population sizes and restricted ranges. The extinction of different indigenous species in the Himalayas (a biodiversity hotspot) has been seriously threatened by habitat degradation for various development schemes (Pandit 2020).

1.9.2 Social Challenges

India has historically struggled to include rural residents in development efforts because they lack access to necessities like health care, education, and transportation to urban regions (Bhattacharya 2020). The most difficult facing India in its severe economic downturn is delivering livelihood options for mobile workforces who come back to their areas in state-wide lockdown (Bhagat et al. 2020). According to a survey on employment in rural regions of India, more than 25 million families demanded effort in August 2020, which is a 66% increase over the demand levels in August 2019 (Ministry of Law and Justice 2005; MoEFCC 2022). Maximum time, these minor farmers work as laborers in the fields of progressive farmers or adjacent towns since they lack land (Singh et al. 2020). But because of land use policies (1950–1972), these backward, poorer, and minor farmers have been awarded a tiny plot of typically deteriorated or fertile land (Nielsen and Oskarsson 2016). Therefore, they rely entirely on the government's nutritional aid distribution policies or definite work initiatives, especially women and the elderly.

1.9.3 Political Challenges

Politicians believe pledges, planning, and planting will solve all problems (Fleischman et al. 2020). This is being implemented as a political ruse in several states of India, where in a day, he plants millions or 500 million seedlings in a year (Adlakha 2022). These plantation activities are essential to the country's efforts to sequester carbon. However, restoring degraded ecosystems by increasing vegetation cover instead of following conventional thinking will accelerate biodiversity loss and lead to near-term deterioration. It may deteriorate further in the coming time (Holl and Brancalion 2020). The IPCC Panel has cautioned against creating more than a few million square kilometers of a single species-based landscape, even for bioenergy production (Stokstad 2019). This leads to easier land use changes for ecosystem services (Dong et al. 2020). Recent scientific advances in restoration and conservation indicate that floral diversity is the goal and should be seen as the main force behind restoration plans (Palmer et al. 2016). Restoring a destroyed ecosystem or creating a new ecosystem in a degraded area takes time and effort. As defined by restoration experts, conventional ecological concepts and practices should be used to conserve biodiversity and restore degraded ecosystems (IUCN 2018). We cannot combat land degradation and protect biodiversity unless the actions of policymakers are grounded in the restoration science and the functioning of ecosystems (Palmer et al. 2016).

1.9.4 Opportunities

Restoring the ecosystem could be a practical, affordable, and lasting solution to India's ecological problems. In addition, it provides an opportunity for economies to recover from the COVID-19 epidemic and for humanity to be protected from expected future more deadly pandemics (Tollefson 2020). The United Nations Environment Programme calculates that for every $1 invested in ecosystem restoration, $9 is supported in a range of ecosystem services. India has great potential to make restoration and conservation programs one of its top priorities and incorporate these activities into its current socioeconomic policies. In addition, India's degraded land represents approximately 27.5% of the world's degraded land area. The United Nations Decade of Ecosystem Restoration could greatly benefit from India's improved access to foreign funding for biodiversity and ecosystem restoration projects.

1.10 Conclusion

The ecological restoration approach provides the path for restoring the site to its pre-disturbed conditions, which usually depends on several climatic, ecological, and anthropogenic factors. Ideally, each restoration plan can never restore the site's ecosystem in its initial state with the same species assemblage. Thus, restoration plans for low to moderately degraded sites will provide maximum benefits with minimum costs rather than a wholly degraded ecosystem. Ecological restorations act as supplementary paths rather than thinking as a complete substitute for the conservation approach. It has high uncertainty in providing its outcomes which need continuous monitoring and management throughout its implementation with the integration of knowledge, adaptive restoration, ecological theory, etc. However, it has the enormous ability to provide the foundational changes in our surrounding natural environment (Fischer et al. 2021; Vaughn et al. 2010). The United Nations' strategy for ecosystem restoration involves leadership and a shift in the behaviors toward restoration, requiring involvement from social, political, corporate, and spiritual leaders. Their combined action will help to involve a vast population through social responsibility, faith belief, and financial benefits by governments and corporates (UNEP 2022a). Ecological restoration for each site in this decade from 2021 to 2030 may allow us to prevent the severity of climate change and extreme weather events and achieve sustainable goals for future generations (United Nations 2022). The severity and impact on the population due to ecological degradation are variable with the type of ecosystem. For example, the urban ecosystems are observing severe changes due to an increase in population, and its impact is observed instantly by humans, while wetlands are shrinking annually, which impacts the migratory bird's population and fish production for local people. Thus, species consideration for restoration is the critical point for a restoration plan to provide maximum benefits for highly impacted species populations. Therefore, through restoration principles, guidelines, and maximum involvement of stakeholders, we can achieve maximum ecological and sustainable aims during the decade of ecosystem restoration (UNEP 2022c).

References

Adlakha N (2022) Enough with token tree plantation. The Hindu. https://www.thehindu.com/sci-tech/energy-and-environment/how-tree-planting-drivesoften-mask-an-economic-aim-as-a-climate-saving-solution/article32869567.ece

Bansal V, Bansal S, Sen J (2020) Assessing the role of urban design in a rapidly urbanizing historical city and its contribution in restoring its urban ecology: the case of Varanasi, India. In: Urban ecology. Elsevier, Amsterdam, pp 413–434. https://doi.org/10.1016/B978-0-12-820730-7.00021-5

Bawa KS, Nawn N, Chellam R, Krishnaswamy J, Mathur V, Olsson SB, Pandit N, Rajagopal P, Sankaran M, Shaanker RU, Shankar D, Ramakrishnan U, Vanak AT, Quader S (2020) Envisioning a biodiversity science for sustaining human well-being. Proc Natl Acad Sci 117(42):25951–25955. https://doi.org/10.1073/pnas.2018436117

Bellard C, Bertelsmeier C, Leadley P, Thuiller W, Courchamp F (2012) Impacts of climate change on the future of biodiversity. Ecol Lett 15(4):365–377. https://doi.org/10.1111/j.1461-0248.2011.01736.x

Bhagat RB, Reshmi RS, Sahoo H, Roy AK, Govil D (2020) The COVID-19, migration and livelihood in India: challenges and policy issues. Migration Lett 17(5):705–718. https://doi.org/10.33182/ml.v17i5.1048

Bhattacharya H (2020) Environmental and socio-economic sustainability in India: evidence from CO2 emission and economic inequality relationship. J Environ Econ Policy 9(1):57–76. https://doi.org/10.1080/21606544.2019.1604267

Byers A (2005) Contemporary human impacts on alpine ecosystems in the Sagarmatha (Mt. Everest) National Park, Khumbu, Nepal. Ann Assoc Am Geogr 95(1):112–140. https://doi.org/10.1111/j.1467-8306.2005.00452.x

Cao J, Yeh ET, Holden NM, Qin Y, Ren Z (2013) The roles of overgrazing, climate change and policy as drivers of degradation of China's grasslands. Nomadic Peoples 17(2):82–101. https://doi.org/10.3167/np.2013.170207

Corcoran E (2010) Dead planet, living planet: biodiversity and ecosystem restoration for sustainable development: a rapid response assessment. UNEP/Earthprint

Crossman ND (2017) The role of ecological restoration and rehabilitation in production landscapes: an enhanced approach to sustainable development. https://www.unccd.int/sites/default/files/2018–06/16.Ecological%2BRestoration__N_D_Crossmann.pdf

Csákvári E, Molnár Z, Halassy M (2022) Estimates of regeneration potential in the Pannonian sand region help prioritize ecological restoration interventions. Commun Biol 5(1):1136. https://doi.org/10.1038/s42003-022-04047-8

Das J, Jha S, Goyal MK (2020) On the relationship of climatic and monsoon teleconnections with monthly precipitation over meteorologically homogenous regions in India: wavelet and global coherence approaches. Atmos Res 238:104889. https://doi.org/10.1016/j.atmosres.2020.104889

Díaz S, Fargione J, Chapin FS, Tilman D (2006) Biodiversity loss threatens human well-being. PLoS Biol 4(8):e277. https://doi.org/10.1371/journal.pbio.0040277

Dong Z, Zhang Q, Li L, Lu Z, Li C, Ouyang F, Tscharntke T, Yu Y, Men X (2020) Landscape agricultural simplification correlates positively with the spatial distribution of a specialist yet

negatively with a generalist pest. Sci Rep 10(1):344. https://doi.org/10.1038/s41598-019-57077-4
Drenthen M (2009) Ecological restoration and place attachment: emplacing non-places? Environ Values 18(3):285–312
Eden S, Tunstall S (2006) Ecological versus social restoration? How urban river restoration challenges but also fails to challenge the science—policy nexus in the United Kingdom. Environ Plann C Govt Policy 24(5):661–680. https://doi.org/10.1068/c0608j
European Commission (2011) Our life insurance, our natural capital: an EU biodiversity strategy to 2020. European Commission. https://eur-lex.europa.eu/legal-content/EN/TXT/PDF/?uri=CELEX:52011DC0244&from=EN
European Commission (2022) EU biodiversity policy. European Commission. https://ec.europa.eu/environment/nature/biodiversity/
Fischer J, Riechers M, Loos J, Martin-Lopez B, Temperton VM (2021) Making the UN decade on ecosystem restoration a social-ecological endeavour. Trends Ecol Evol 36(1):20–28. https://doi.org/10.1016/j.tree.2020.08.018
Fleischman F, Basant S, Chhatre A, Coleman EA, Fischer HW, Gupta D, Güneralp B, Kashwan P, Khatri D, Muscarella R, Powers JS, Ramprasad V, Rana P, Solorzano CR, Veldman JW (2020) Pitfalls of tree planting show why we need people-centered natural climate solutions. Bioscience 70(11):947–950. https://doi.org/10.1093/biosci/biaa094
Goswami KP, Kumar V (2020) Need of urban forestry for ecological development in Varanasi City. EasyChair
Goswami UP, Hazra B, Goyal MK (2018) Copula-based probabilistic characterization of precipitation extremes over North Sikkim Himalaya. Atmos Res 212:273–284. https://doi.org/10.1016/j.atmosres.2018.05.019
Goyal MK, Ojha CSP (2010) Evaluation of various linear regression methods for downscaling of mean monthly precipitation in arid Pichola watershed. Nat Resour 1(1):11–18. https://doi.org/10.4236/nr.2010.11002
Goyal MK, Ojha CSP (2012) Downscaling of precipitation on a lake basin: evaluation of rule and decision tree induction algorithms. Hydrol Res 43(3):215–230. https://doi.org/10.2166/nh.2012.040
Goyal MK, Panchariya VK, Sharma A, Singh V (2018) Comparative assessment of SWAT model performance in two distinct catchments under various DEM scenarios of varying resolution, sources and resampling methods. Water Resour Manag 32(2):805–825. https://doi.org/10.1007/s11269-017-1840-1
Goyal MK, Gupta AK, Jha S, Rakkasagi S, Jain V (2022) Climate change impact on precipitation extremes over Indian cities: non-stationary analysis. Technol Forecast Soc Chang 180:121685. https://doi.org/10.1016/j.techfore.2022.121685
Gross M (2008) Return of the wolf: ecological restoration and the deliberate inclusion of the unexpected. Environ Polit 17(1):115–120. https://doi.org/10.1080/09644010701643159
Hinge G, Surampalli RY, Goyal MK (2018) Prediction of soil organic carbon stock using digital mapping approach in humid India. Environ Earth Sci 77(5):172. https://doi.org/10.1007/s12665-018-7374-x
Holl KD, Brancalion PHS (2020) Tree planting is not a simple solution. Science 368(6491):580–581. https://doi.org/10.1126/science.aba8232
Howell EA, Harrington JA, Glass SB (2013) Introduction to restoration ecology: instructor's manual, 1st edn. Island Press, Washington, DC. https://islandpress.org/sites/default/files/9781597261890_Howell_instructor_manual.pdf
IUCN (2012) Developing capacity for a protected planet. https://portals.iucn.org/library/efiles/documents/PAG-018.pdf
IUCN (2018) Ecological restoration for protected areas. https://portals.iucn.org/library/efiles/documents/PAG-018.pdf
IUCN (2020) The IUCN red list of threatened species. IUCN. https://www.iucnredlist.org

Khan RWA, Shaheen H, Mehmood A, Naseer Awan S (2019) Grazing intensity impacts on soil carbon stocks of Western Himalayan alpine paddocks. Carbon Manag 10(6):533–540. https://doi.org/10.1080/17583004.2019.1667701

Kuniyal JC, Maiti P, Kumar S, Kumar A, Bisht N, Sekar KC, Arya SC, Rai S, Nand M (2021) Dayara bugyal restoration model in the alpine and subalpine region of the central Himalaya: a step toward minimizing the impacts. Sci Rep 11(1):16547. https://doi.org/10.1038/s41598-021-95472-y

Lejon AGC, Renöfält BM, Nilsson C (2009) Conflicts associated with dam removal in Sweden. Ecol Soc 14(2):4

Ministry of Law and Justice (2005) Mahatma Gandhi National Rural Employment Guarantee Act-2005. https://rural.nic.in/sites/default/files/nrega/Library/Books/1_MGNREGA_Act.pdf

MoEFCC (2022) Development of urban forests. Government of India. https://pib.gov.in/PressReleasePage.aspx?PRID=1629587

Naik R, Sharma LK (2022) Monitoring migratory birds of India's largest shallow saline Ramsar site (Sambhar Lake) using geospatial data for wetland restoration. Wetl Ecol Manag 30(3):477–496. https://doi.org/10.1007/s11273-022-09875-3

Nautiyal MC, Nautiyal BP, Prakash V (2004) Effect of grazing and climatic changes on alpine vegetation of Tungnath, Garhwal Himalaya, India. Environmentalist 24(2):125–134. https://doi.org/10.1007/s10669-004-4803-z

Nielsen KB, Oskarsson P (2016) Industrialising rural India: land, policy and resistance. Taylor & Francis, Milton Park

Palmer MA, Zedler JB, Falk DA (2016) Ecological theory and restoration ecology. In: Foundations of restoration ecology. Island Press, Washington, DC, pp 3–26. https://doi.org/10.5822/978-1-61091-698-1_1

Pandey S (2008) Linking ecodevelopment and biodiversity conservation at the great Himalayan National Park, India: lessons learned. Biodivers Conserv 17(7):1543–1571. https://doi.org/10.1007/s10531-008-9365-9

Pandey B, Saurabh S, Jee Ranjan O, Singh A (2017) Ecological restoration and livelihood sustainability through mixed forestry in Himalaya: a case study of Kot Malla Village, Rudra Prayag Uttarakhand. In: Environmental concerns and sustainable development in Himalaya. Research India Press, New Delhi, pp 176–193

Pandit MK (2020) The Himalaya should be a nature reserve. Nature 583(7814):9. https://doi.org/10.1038/d41586-020-01809-4

Poonia V, Goyal MK, Gupta BB, Gupta AK, Jha S, Das J (2021a) Drought occurrence in different river basins of India and blockchain technology based framework for disaster management. J Clean Prod 312:127737. https://doi.org/10.1016/j.jclepro.2021.127737

Poonia V, Jha S, Goyal MK (2021b) Copula based analysis of meteorological, hydrological and agricultural drought characteristics across Indian river basins. Int J Climatol 41(9):4637–4652. https://doi.org/10.1002/joc.7091

Praharaj S (2014) Rejuvenation of water bodies (Kunds) and restoring active community spaces in the cultural capital of India: Varanasi. In: Neo-international conference on habitable environments

Rautela KS, Kuniyal JC, Alam MA, Bhoj AS, Kanwar N (2022) Assessment of daily streamflow, sediment fluxes, and erosion rate of a pro-glacial stream basin, central Himalaya, Uttarakhand. Water Air Soil Pollut 233(4):136. https://doi.org/10.1007/s11270-022-05567-z

Roy A, Suresh P, Nirmal SS (2020) Impact of the Covid-19 pandemic on travel and tourism industry and future revival strategy with special reference to Indian tourism sector. Hospitality and Tourism Industry, 18

Ryan RL (2005) Exploring the effects of environmental experience on attachment to urban natural areas. Environ Behav 37(1):3–42

Sarkar UK, Das Ghosh B, Puthiyottil M, Das AK, Lianthuamluaia L, Karnatak G, Acharya A, Das BK (2021) Spatio-temporal change analysis of three floodplain wetlands of eastern India in the

context of climatic anomaly for sustainable fisheries management. Sustain Water Resour Manag 7(3):41. https://doi.org/10.1007/s40899-021-00529-5
SER (2022) UN decade on ecosystem restoration. SER. https://www.ser.org/page/DecadeonRestoration
Sharma A, Goyal MK (2020) Assessment of drought trend and variability in India using wavelet transform. Hydrol Sci J 65(9):1539–1554. https://doi.org/10.1080/02626667.2020.1754422
Sinclair M, Vishnu Sagar MK, Knudsen C, Sabu J, Ghermandi A (2021) Economic appraisal of ecosystem services and restoration scenarios in a tropical coastal Ramsar wetland in India. Ecosyst Serv 47:101236. https://doi.org/10.1016/j.ecoser.2020.101236
Singh K, Awasthi A, Singh RP, Tewari SK (2020) Merging the margins for beneficial biofuels. In: Current developments in biotechnology and bioengineering. Elsevier, Amsterdam, pp 163–178. https://doi.org/10.1016/B978-0-444-64309-4.00007-6
Stokstad E (2019) Bioenergy not a climate cure-all, panel warns. Science 365(6453):527–528. https://doi.org/10.1126/science.365.6453.527
Tollefson J (2020) Why deforestation and extinctions make pandemics more likely. Nature 584(7820):175–176. https://doi.org/10.1038/d41586-020-02341-1
Trant A, Higgs E, Starzomski BM (2020) A century of high elevation ecosystem change in the Canadian Rocky Mountains. Sci Rep 10(1):9698. https://doi.org/10.1038/s41598-020-66277-2
UNEP (2018) Holistic waste management strategy for Varanasi City. https://wedocs.unep.org/20.500.11822/31291
UNEP (2021a) Ecosystem restoration playbook. A practical guide to healing the planet. https://www.decadeonrestoration.org/publications/ecosystem-restoration-playbook-practical-guide-healing-planet#:~:text=Released at the start of,ecosystems and foster their recovery
UNEP (2021b) UNEP planetary action. UNEP. https://www.unep.org/annualreport/2021/index.php
UNEP (2022a) A strategy for engagement the role of faith leaders and faith-based organisations in the un decade on ecosystem restoration (2021–2030). UNEP. https://www.decadeonrestoration.org/publications/strategy-engagement-role-faith-leaders-and-faith-based-organisations-un-decade
UNEP (2022b) Panel unveils 10 guiding principles in campaign to revive the earth. UNEP. https://www.unep.org/news-and-stories/story/panel-unveils-10-guiding-principles-campaign-revive-earth
UNEP (2022c) Urban areas. UNEP. https://www.decadeonrestoration.org/types-ecosystem-restoration/urban-areas
United Nations (2021a) Ecosystem restoration for people, nature and climate becoming #GenerationRestoration. http://www.un.org/Depts/Cartographic/
United Nations (2021b) The United Nations decade on ecosystem restoration. https://wedocs.unep.org/bitstream/handle/20.500.11822/31813/ERDStrat.pdf?sequence=1&isAllowed=y
United Nations (2022) About the UN decade. United Nations https://www.decadeonrestoration.org/about-un-decade
Vaughn KJ, Porensky LM, Wilkerson ML, Balachowski J, Peffer E, Riginos C, Young TP (2010) Restoration ecology. Nature education knowledge. https://www.nature.com/scitable/knowledge/library/restoration-ecology-13339059/#:~:text=Ecological restoration aims to recreate, hurricanes%2C floods%2C and fires
Wetlands International (2022) Indian Wetlands. Wetlands International. https://south-asia.wetlands.org/
Whalen PJ, Toth LA, Koebel JW, Strayer PK (2002) Kissimmee River restoration: a case study. Water Sci Technol 45(11):55–62. https://doi.org/10.2166/wst.2002.0379
Yadav JS, Tiwari SK, Misra A, Rai SK, Yadav RK (2021) High-altitude meteorology of Indian Himalayan region: complexities, effects, and resolutions. Environ Monit Assess 193(10):654. https://doi.org/10.1007/s10661-021-09418-y

Chapter 2
Lifestyle for Environment: LiFE Mission and Synergies with Eco-restoration

Aditi Madan, Anil Kumar Gupta, and Manish Kumar Goyal

Abstract The chapter discusses the Lifestyle for Environment (LiFE) mission, a concept introduced by India's Prime Minister at the UNFCCC's COP26, aimed at combating climate change. The mission is centred around the principle of promoting a sustainable lifestyle by leveraging traditional values of conservation and moderation. The chapter explores India's updated Nationally Determined Contribution (NDC), which includes shifting towards greener energy, creating green jobs, and promoting the manufacturing and use of renewable energy. It highlights the importance of sustainable living and ecosystem restoration, noting that current lifestyle choices pose significant threats to the environment. It further outlines the LiFE campaign's approach and goals, including promoting individual eco-friendly behaviour, adopting a global perspective, and using local cultures to advance the cause. The envisaged outcomes of LiFE include creating a global network of "Pro-Planet People" and leveraging the power of social networks to change climate-related cultural norms. Further, it also discusses the synergies of LiFE with Sustainable Development Goals (SDGs) and highlights the importance of disaster risk reduction and climate change adaptation behaviour. It concludes by emphasizing India's potential to lead behaviour change for climate action based on its extensive traditional knowledge and climate-friendly practices.

Keywords Lifestyle for Environment (LiFE) · Sustainable lifestyle · Ecosystem restoration · Climate change adaptation

A. Madan
Institute for Human Development, New Delhi, India

A. K. Gupta (✉)
National Institute of Disaster Management (Ministry of Home Affairs, Government of India), New Delhi, India

M. K. Goyal
Department of Civil Engineering, Indian Institute of Technology Indore, Indore, Madhya Pradesh, India

A. K. Gupta et al. (eds.), *Ecosystem Restoration: Towards Sustainability and Resilient Development*, Disaster Resilience and Green Growth,
https://doi.org/10.1007/978-981-99-3687-8_2

2.1 LiFE Mission

2.1.1 Background (PM's Concept of a Mass Movement for LiFE as a Key to Combating Climate Change)

India provided the United Nations Framework Convention on Climate Change (UNFCCC), with an updated nationally determined contribution (NDC) on October 2, 2015, which has eight objectives, three of which have quantitative targets for 2030. These objectives are to reach 40% of installed non-fossil energy capacity, cut the GDP's emissions intensity by 33–35% from 2005 levels, and add 2.5–3 billion tonnes of CO_2 equivalent of additional forest and tree cover. The NDC aims to increase India's contributions to strengthening the international response to combat the threat of climate change, as agreed upon under the Paris Agreement, and to secure the nation's interests and future development requirements based on the principles and provisions of the UNFCCC.

India pledged to step up its climate action at the COP26 held in Glasgow, UK, by showcasing its five nectar ingredients (panchamrit) to the world. The "panchamrit" stated at COP26 is translated into improved climate targets by modifying to India's current NDC. "To put forward and further disseminate a healthy and sustainable way of life based on traditions and values of conservation and moderation, especially through a mass movement for 'LIFE'—'Lifestyle for Environment' as a cornerstone to addressing climate change," states the amended NDC. The choice of expanded NDCs reflects India's dedication to the decoupling of economic growth from greenhouse gas emissions at the highest level. A "One-Word Movement" was presented to the international community at COP26 in recognition of the significant role that lifestyle plays in climate change. This one word stands for LiFE, which stands for Lifestyle for Environment. Living in harmony with our world and causing no harm to it is the goal of LiFE (PIB Delhi 2022).

India's revised NDC similarly reflects this citizen-centric strategy for addressing climate change. The modified NDC provides a roadmap to switch to greener energy from 2021 to 2030. The updated framework will present a chance to improve India's manufacturing capabilities and increase exports, along with many other government initiatives, such as tax breaks and incentives like the Production Linked Incentive scheme for promoting manufacturing and use of renewable energy. Overall, it will boost the number of green jobs in fields like renewable energy, clean energy industries, the automotive industry, the production of low-emission goods like electric vehicles and incredibly efficient appliances, and cutting-edge technologies like green hydrogen, etc. The new National Development Plan (NDP) of India will be executed over 2021–2030 via programmes and schemes of pertinent ministries/ departments and with the necessary assistance from states and union territories.

2.1.2 Need for a Sustainable Lifestyle and Ecosystem Restoration

The world's population may surpass ten billion by 2050, and with more people, more demand will be generated for things like food, clothing, travel, housing, and other desires. A growing number of individuals are likely to struggle to achieve their necessities, with two to three billion new urban consumers and young people expected to rely heavily on social media for their news. Our existing lifestyle choices are endangering the environment in a world where resources are limited and biodiversity loss and climate change are serious threats.

Understanding how our lifestyle choices affect the environment and figuring out ways for everyone to live better, lighter lives are two key components of sustainable living. A "people lens" approach to sustainability is novel, highly advantageous, and contemporary. The Sustainable Development Goals (4 Education and 12.8 Responsible Consumption) include sustainable living and lifestyles for the first time (UNEP). Sustainable living, in simple words, also means acknowledging day-to-day life choices and reflecting if there can be alternatives that may impact the environment less (PBNS 2022). As people do not alter their behaviour in accordance with what is appropriate, they don't react to data, statistics, or pessimistic predictions of the future. People take action to satisfy their wants and goals. They base their choices on factors including cost, accessibility, effectiveness, and extra standards like trends or well-being. The defining factors do not include sustainability. Even those who desire to live more sustainably frequently don't have access to information or accessible, appealing goods and services. This shows that, in addition to individuals, it is up to businesses and governments (who are also consumers!) to disseminate information, encourage positive behaviour change, and promote the creation of new business models that will make sustainable living the norm (UNEP).

According to SER (Society for Ecological Restoration), ecological restoration is defined as the "process of assisting the recovery of an ecosystem that has been degraded, damaged, or destroyed". To restore an ecosystem, everyone must contribute. It will take a concentrated effort to rebuild the globe since the restoration of ecosystems at a large scale is no easy feat. The UN Decade on Ecosystem Restoration seeks to spark a global movement among citizens, activists, women, young people, indigenous peoples, businesses, financiers, academics, and governments at all levels. The appeal of restoration is that it can occur at any scale, be it a home plot, a municipal park, a river valley, a national forest, or a globally imperilled ecosystem. It also sends a message of action and optimism. This implies that anyone can participate (UNEP 2021).

Need of LiFE Mission:

1. Climate change and environmental degradation are global processes, meaning that decisions made in one region of the world influence ecosystems and inhabitants everywhere else.

2. Global policy changes, financial incentives, and laws have all been put into place during the past 20 years to address environmental degradation and climate change.
3. The necessary acts at the level of people, communities, and institutions haven't gotten much attention despite their immense potential.
4. The environmental and climate crises can be significantly reduced by altering only the behaviour of individuals and communities.

2.1.3 *About the LiFE Campaign: Concepts, Contribution/Approach in Creating Synergies with the Ecosystem, and Envisaged Outcomes of LiFE*

The "use-and-dispose" economy that is currently in place, which is characterized by mindless and harmful consumerism, is to be replaced by a circular economy, which would be characterized by attentive and purposeful utilization. The notion of "Lifestyle for the Environment" was introduced by Prime Minister Narendra Modi at the COP26 in Glasgow in November 2021 in addition to the panchamrit, or five climate-related commitments of the nation (LiFE). Instead of "thoughtful and wasteful consumption", the idea promotes deliberate and mindful use by everyone. PM Modi introduced LiFE on June 5, 2022, the World Environment Day, intending to address the global climate catastrophe by utilizing the strength of both individual and group actions. The movement's goal is to persuade people and communities to incorporate straightforward and focused climate-friendly practices into their daily routines. One could, for instance, carry a reusable cloth bag rather than a plastic one. Global examples: Pro-planet initiatives have been done in the past all around the world. For instance, Denmark encourages cycling by prohibiting parking in the city's core and constructing dedicated bike lanes. Since the early 1950s, Japan has had a unique "walk-to-school" mandate in place. However, LiFE is envisioned as a first-of-its-kind global movement, led by India in collaboration with other nations, that will give the globe a special people-powered platform to continuously focus on bringing individual and group acts to the forefront of the climate action narrative.

LiFE approach focuses on:

1. **Individual behaviours:** Focus on the actions and attitudes of people in communities and individuals to turn life into a mass movement (Jan Andolan).
2. **Adopting eco-friendly behaviours:** The United Nations Environment Programme (UNEP) estimates that if one billion of the world's almost eight billion inhabitants adopted eco-friendly behaviours in their daily lives, global carbon emissions would decrease by around 20%.
3. **Co-create globally:** Use prestigious universities, think tanks, and international organizations to crowdsource scalable, empirical ideas from the brightest minds on earth.

4. **Use local cultures:** To promote the campaign, use the social norms, worldviews, and customs of various cultures around the world.

LiFE aims to:

Encourage at least one billion Indians and other world citizens to pursue both individual and group environmental protection and conservation measures during 2022–2027. By 2028, it is hoped that at least 80% of all villages and urban local bodies in India will be environmentally friendly. This will be accomplished by:

1. Using the strength of group activities to encourage people all over the world to take small, everyday steps towards a more sustainable future.
2. Promoting an environmentally conscious lifestyle that emphasizes "mindful and deliberate utilization" rather than "mindless and wasteful consumption".
3. Establishing and supporting a global network of "Pro-Planet People" (P3) by the year 2027 will bring about a radical change via their common dedication to adopting and promoting eco-friendly lifestyles.
4. Making use of social networks' power to change cultural norms related to climate.

Envisaged outcomes of LiFE:

1. India, the UN, and the other 26 organizations are proud and devoted partners in our effort to give climate action a fresh lease on life.
2. India is a great place to start because LiFE is a global goal. With over 1.3 billion people, the momentum created if we achieve a true Jan Andolan here will be huge.
3. The COP26 summit in Glasgow and the Paris Agreement represent urgent, international efforts to reduce emissions.

Contribution of LiFE in changing people's behaviour:

As a global initiative, Mission LiFE aims to achieve three key paradigm shifts in our society's approach to sustainability throughout time. The COVID epidemic serves as a reminder to us all that, despite our society's technical advancements, we are all still subject to the whims of nature. A threat to one affects all people because we are a community of people living in the same natural environment. In this regard, the LiFE movement can play a crucial role in mainstreaming a peaceful and conscientious way of living in addition to reversing the effects of climate change because of its multidimensional, multicultural, and global approach.

Change in Demand (Phase I): Encouraging people to make small but significant environmental improvements in their daily lives.

Change in Supply (Phase II): As large-scale individual demand shifts, markets and industries should gradually be prodded to adapt supply and procurement practices to the new market demands.

Change in Policy (Phase III): The long-term goal of Mission LiFE is to bring about changes in extensive industrial policy and governmental policy that can support sustainable consumption and production by influencing demand and supply dynamics in India and around the world.

The mission will be developed, curated, and piloted by NITI Aayog before being non-linearly and randomly implemented by the Government of India's Ministry of Environment, Forest and Climate Change (MoEF&CC). All phases of Mission LiFE are concurrent, even though one phase will naturally feed into the next (Ranveer 2022 (mopng.gov.in)).

2.1.4 LiFE and SDGs

One of the 17 SDGs is environmental degradation. It is a threat multiplier that can make some of humanity's most pressing problems—such as illness, poverty, and hunger—worse. The world needs to adopt a new paradigm for development to support billions of livelihoods, encourage growth and investment opportunities, improve living standards, and address the climate issue. All people must adjust their lifestyles to be in harmony with the resources available on the planet, according to the SDGs that are specifically focused on sustainable cities and communities (SDG 11), responsible production and consumption (SDG 12), climate change (SDG 13), life on land (SDG 15), and life under water (SDG 14). Furthermore, the New Climate Economy study shows that by 2030, bold environmental action could generate up to 65 million jobs.

SDG 12 calls for more effective and environmentally friendly management of resources, including increased energy efficiency, sustainable infrastructure, access to essential services, and the creation of green and decent jobs to ensure a higher standard of living for all. Beyond businesses, society has a responsibility to include everyday consumers as active contributors to the achievement of SDG 12.

It is significant to note that Mission LiFE contributes directly OR indirectly to practically all the SDGs given the global commitment to achieve the SDGs by 2030. Additionally, given the size of India's population, any significant behaviour change adopted there will immediately and directly support the global SDGs. For instance, the Swachh Bharat Mission (SDG 6: Clean Water and Sanitation) reduced 60% of the one billion open defecators worldwide by encouraging 550 million Indians to use toilets instead of the open defecation method. (Ranveer 2022 (mopng.gov.in))

2.2 Disaster Risk Reduction and Climate Change Adaptation Behaviours that Can Be Adopted at the Different Levels Under LiFE

2.2.1 Household/Individual Level

Climate-friendly behaviours can be encouraged in a variety of industries, such as water, transportation, food, electricity, waste management, recycling, and reuse, by using behaviour change strategies targeted at individuals, households, and communities.

2.2.1.1 Water Conservation

An average person uses 2.5 L of water every day for things like brushing, bathing, cleaning utensils, and laundry. Water resources can be protected by everyone. Two basic things may be done: collect rainwater and decrease water demand through efficiency and conservation (Subramoniam et al. 2022). These decisions are influenced by socioeconomic position and location, as with all lifestyle decisions (Goyal and Ojha 2010, 2012).

Adopting a no-drip policy is essential because all leaks waste water continuously and require repair. The usage of dry compost toilets or low-flush toilets that utilize less water is a further option. Installing water-efficient showerheads and using washing machines with an energy label or rating are two more choices to consider since they not only use less electricity per load but also consume less water. Water usage can be decreased with easy steps like turning off running faucets while doing dishes, taking a shower, brushing your teeth, collecting any water that isn't being used and using it to water plants. Grey water should be treated and reused before being used to flush toilets. Since native plants typically demand less water, homeowners can cultivate them in their gardens. Owners of cars and bicycles have two options: either they take their vehicles to a carwash that uses recycled water or they wash their vehicles with a sponge and minimal rinsing. Households will be able to increase consumption without incurring additional costs thanks to recycling (Aqua Dataset Query 2011).

2.2.1.2 Electricity Consumption

By using less electricity, individuals can dramatically lower their carbon impact. This could require using better appliances, but much can be accomplished with a greater understanding of the problems at hand and a will to adopt beneficial habits. Up to 282 kW of energy can be saved each day by turning off lights, air conditioners, and heaters while not in use. Most modern household appliances, such as lights, fans, refrigerators, televisions, washers, water heaters, and computers, still consume more energy than is necessary for their normal operation. Most of these appliances are kept in standby mode, making it convenient to turn them on quickly when needed. According to the Centre for Environment Education (2010), turning off appliances at the plug would save a household over 133 kg of CO_2 emissions annually and result in an immediate drop in electricity costs (Mohanty and Scherfler 2012).

Other lifestyle choices in energy are:

- The use of pressure cookers as opposed to conventional cooking vessels has the potential to reduce carbon dioxide (CO_2) emissions by about 125 kilograms (kg) per family.
- Solar-powered kitchenware including stoves, water heaters, and lamps immediately reduce carbon emissions. For instance, a single solar water heater can reduce CO_2 emissions by 687 kg per year (Centre for Environment Education 2010).

India is a nation with about 180 million households. Over 1 million tonnes of CO_2 emissions could be avoided each year if just 1% of these households switched to solar cookers.

- Use energy labels while making selections about your purchases. An energy-efficient refrigerator or air conditioner can reduce CO_2 emissions by over 250 kg annually, saving the average household over $30 in electricity costs. By switching from desktops to notebook computers, each cathode ray tube (CRT) monitor will save over 275 kg of CO_2 annually. The long-term savings from lower operating and maintenance costs offset the high initial costs.
- Optimizing the settings on home appliances can also result in significant reductions in carbon emissions. When washing daily wearables in the washing machine, using the cold cycle over the hot cycle will consume less electricity and save roughly 100 kg of CO_2. In the UK, it has been calculated that lowering the thermostat by 1 °C will result in a 10% reduction in heating costs, saving £50 annually and reducing CO_2 emissions.
- Promote physical activity and outdoor recreation over video games. This can reduce annual CO_2 emissions from children by up to 90 kg.

2.2.1.3 Transport

To achieve emission reductions, driving behaviour changes are essential. According to the UK's Driving Standards Agency, eco-driving instruction results in fuel efficiency improvements of at least 8%, saving drivers more than £2 billion annually in fuel costs. Regular maintenance to ensure fuel efficiency, proper tire inflation to reduce drag, removal of any extra luggage from the car, and combining several short trips into one longer trip are all examples of eco-driving or green driving practices. At 2010 fuel prices, turning off the engine at traffic lights is found to save 63 kg of carbon emissions per car annually or about $30 (Sodhi et al. 2010). Nowadays, a traffic signal change counter may be found in many large cities (a timer that counts down to the next light change). This encourages people to turn off their engines by serving as a helpful reminder of how much longer they will need to wait for the lights to change. Driving smoothly, accelerating smoothly, avoiding idling, paying attention to the air conditioning in the car, and changing gears early are all examples of fuel-efficient driving techniques.

Other lifestyle choices in transportation are:

- Limit air travel; instead, travel small distances by land or rail as these create less carbon. For example, a person taking a trip from London to Paris generates ten times as much carbon dioxide (CO_2) as someone taking the Eurostar (c).
- Practice eco-friendly driving: For the UK, a 7% average savings across all road vehicles through eco-friendly driving would imply a 1.2% annual reduction in all UK CO_2 emissions.
- Take public transportation and lessen your reliance on personal vehicles. For example, a single person can cut their annual CO_2 emissions by 2180 kilograms

(kg) by switching a 32-km round-trip commute to existing public transportation (CUNY and SAIC 2006).

- When making routine purchases, use home delivery. It has been discovered that home delivery is four times more effective than individual shopping excursions, offering a potential worldwide carbon abatement of 13 mega tonnes (World Economic Forum and Accenture 2009).
- Travel short distances on foot or by bicycle. Short routes have significant emissions, especially while the engine is still warm, and the gasoline catalyst is not yet operating at peak efficiency. Therefore, the advantages are especially great if these journeys are replaced by cycling for short distances (Anable 2008).
- Work from home whenever you can to cut down on mileage.
- Teleconferences as opposed to in-person meetings (Mohanty and Scherfler 2012).

2.2.1.4 Food Habits

The water footprint can be greatly reduced by altering dietary practices and consuming less meat and dairy. A significant amount of energy, food, and water is needed for the production of meat, and 70% of the world's water withdrawals are used for agriculture (Aquastat Database Query 2011). According to the Water Footprint Network, a kilogram of beef requires 15,500 L of water, a kilogram of cheese requires 5000 L, a kilogram of chicken meat requires 3900 L, and a kilogram of barley requires 1300 L. Rockström (2003) calculated that a diet consisting of 80% plant-based foods and 20% animal products (the proportion of meat in industrialized countries is 30–35%) requires 1300 m^3 of water annually, while a strictly vegetarian diet only needs around half of this (Rockström et al. 1999).

Numerous personal lifestyle adjustments related to food will have a favourable effect on the environment or GHG emissions. Some of those options call for nutritional changes, while others call for shifting one's regular buying habits. It is well known that the meat industry has a significant carbon footprint due to both its high energy consumption and its effects on land use. According to the Food and Agriculture Organization (FAO), the livestock industry is responsible for 18% of all global GHG emissions, both directly and indirectly. Indians, for example, eat about 1/11th of the average Chinese and 1/25th of the average American's meat consumption. A worldwide shift to a low-meat diet would significantly lower mitigation costs.

The other lifestyle changes that are related to food and diet are:

- Consume food that is farmed nearby to cut down on food miles. One meal per week of locally sourced food can prevent the emission of 2268 kg of carbon dioxide (CO_2). Even if just 1% of Asia's 4 billion people made this shift, more than 100 million tonnes of CO_2 emissions could be avoided.
- Buy organic food instead of regular food because it uses 30% less energy (Rodale Institute 2005).

- Eat less meat and dairy because they contribute 18% of the world's greenhouse gases (GHGs) and significantly deplete water supplies. Grow your food in the garden and do not waste food: combining these actions could reduce our footprint by 11%.
- Promote neighbourhood kitchens to save money on supplies and distribution.
- Encourage the use of food courts in public areas, which have shared maintenance facilities, dining areas, and cutlery and thus help to reduce waste, energy consumption, and transportation needs.

2.2.1.5 Building and Construction

Given how much habitats and buildings contribute to carbon emissions, people who live in buildings should change their lifestyles in a way that reduces CO_2 emissions. Buildings use energy in two different ways, both of which have the potential for mitigation and adaptation measures: (a) during construction, including the embodied energy in the building materials used, and (b) during use and maintenance.

Even though the market determines how structures are made, many lifestyle decisions in this industry are heavily reliant on the legislation, the development of the market, and the capacity building of builders and architects. Individuals seem to find energy cost reduction strategies and retrofitting existing structures with greener technologies to be the simplest carbon reduction activities. The World Business Council for Sustainable Development's 2009 report, which was analysed, concluded that technology alone is unlikely to provide building energy performance. "Conservation behaviour can save a third of the energy that wasteful activity can add to a building's designed energy performance" (WBCSD 2009).

Lifestyle changes in terms of building and construction include:

- Use alternatives to cement and steel in buildings like rammed earth or bamboo as those contain less embodied energy. Switch to green buildings and green technology where possible. Use compact building areas. Reduce energy needs for heating and lighting.
- Lighter-coloured countertops and cabinetry give rooms the appearance of being larger and require less lighting.
- Create distinct zones within the building, each with a different indoor climate and corresponding energy requirement; this will maximize the use of natural resources for heating, cooling, and lighting and can reduce energy use by up to 30% (WBCSD 2009).

2.2.1.6 Waste

Promote environmentally friendly transportation by highlighting cycling and walking and educating consumers about their options for reducing waste. The 3Rs—reduce, reuse, and recycle—must be the focus of these. Emphasis should initially be placed on reducing one's trash whenever possible. Every socioeconomic group has a very different way of life and purchasing preferences. A few such habits are as follows:

Green shopping involves altering one's usual buying routines. The reduction of carbon emissions is largely a result of wise and environmentally friendly shopping practices like buying durable items or equipment and buying locally produced goods. It is possible to provide preference to products that are produced with fewer resources, using non-hazardous ingredients, and without the creation of hazardous waste. Also, buying products with minimum packaging is an important choice a buyer can make.

Waste segregation at the household level is crucial for the later conversion of waste materials into useful resources. Compost can be made from organic waste, including food scraps. Hazardous garbage and electronic waste need to receive extra care. Use the buy-back policy offered by some businesses so that customers can return their used equipment. Individuals and communities can increase their consumption without having to incur additional costs by segregating their waste and recycling (Mohanty and Scherfler 2012).

Other lifestyle changes include:

- Use of reusable, non-hazardous goods that are durable.
- Avoid using bulky packaging. If you reduce your rubbish by 10% per year, you might save 540 kg of carbon dioxide (CO_2) (Earthforce 2011). Over 6 million tonnes of CO_2 might be averted each year in Asia's urban areas if just 1% of people stopped using heavy packaging.
- Take your bags when you shop to reduce your reliance on plastic bags. Each year, the USA consumes 100 billion plastic bags, using 12 million barrels of oil in the process. Only 1% of plastic bags are recycled in total.
- Whenever possible, purchase in bulk and steer clear of excessive packaging.
- Convert to electronic statements and accounts to help cut down on paper waste. Paper goods account for most of the municipal solid trash in the USA, and just printing out invoices annually produces about 2 million tonnes of CO_2.
- Whenever possible, reuse cans, bags, and containers. Separate household waste to make it simple to recycle materials. Recycling paper reduces the need for 4100 kWh of electricity, 2.3 cubic meters of landfill area, and 265,000 gallons of water every tonne.
- Send outdated electronics, such as laptops, back to the business. Electronic waste and equipment make up around 40% of the heavy metals, such as lead, mercury, and cadmium, that are dumped in landfills.

2.2.2 Facilitators/Access: Institutional Arrangements, Enablers, and Actions Under Each Sectoral Department

2.2.2.1 Water Conservation

- Make rainwater collection required.
- Promote the reuse of grey water in homes and businesses.
- Make water-saving fixtures for sinks, toilets, and washing machines a requirement.
- Assist farmers in adopting water-saving techniques like drip irrigation; give small farmers specialized subsidies and creative financing.
- Take action to cut back on water distribution losses.
- To decrease water consumption and waste, think about raising the price (Poonia et al. 2021; Das et al. 2020; Goyal et al. 2018).

2.2.2.2 Electricity

- Promote energy efficiency through green standards and labelling, subsidies, grants, and interest-free loans for products and investments that improve energy efficiency.
- Promote the use of renewable energy sources.
- Energy resource pricing should be adjusted, and products that are wasteful of energy and hazardous to the environment should be taxed.
- Implementing smart grid technology requires the creation of a national framework, financial incentives for investments, and consumer benefits proof.
- Launch public education campaigns about how to preserve electricity (Mohanty and Scherfler 2012).

2.2.2.3 Transport

- Create a national policy for a low-carbon transportation system.
- Promote public transportation by investing in high-speed rail infrastructure, better urban planning, designated lanes, and an improved information system on schedules.
- Make parking spaces in city centres car-free.
- Make bicycles available for rent in the city's core and encourage take-up.
- Labelling regulations for cars include requirements for fuel economy and carbon dioxide (CO_2) emissions, as well as greater levies on large and heavy vehicles.
- Encourage car sharing, raise parking fees, and other measures to discourage individualized motorized transportation.
- Modify the rules governing air travel by capping the annual number of flights and adding fees for extra miles.

- Promote the use of biofuels and electric vehicles by passing legislation requiring all taxi and public transportation firms to do so, offering tax incentives for the purchase of electric vehicles, and supporting infrastructure (loading stations, battery banks, biofuel stations).
- Make driving lessons for eco-friendly vehicles required.
- Start public awareness efforts about the risks associated with using fossil fuels (Mohanty and Scherfler 2012).

2.2.2.4 Construction

- A national policy on the development and use of low-carbon land.
- Enforce/promote green construction requirements for all new public and commercial structures; establish green certification.
- Promote environmentally friendly building materials by developing markets, increasing capacity, and offering training.
- Laws requiring the retrofit of public and commercial buildings and offering fiscal incentives.
- Encourage material recovery from the construction industry by promoting recycling and imposing taxes on construction waste.

2.2.2.5 Urban Planning

- Rather than designing cities around cars, consider how people will be shaped by them.
- Stop sprawling urban areas.
- Include walking and bicycle pathways in all urban planning.
- Create environmentally friendly transportation options to lessen reliance on fossil fuels.
- Create at least one eco-village in every district and encourage its expansion and improvement.
- Aim for a carbon-neutral city and cut energy use by a predetermined amount.
- Commit to utilizing as much renewable energy as possible.
- Promote land-use strategies that will increase urban agriculture and green spaces.

2.2.2.6 Waste

- Introduce laws to enforce eco-design, including tighter environmental regulations for industry and funding for research and information sharing. Adopt zero waste policy at the national level.
- Create eco-industrial parks using the industry.
- Require garbage to be collected and separated at the municipal, industrial, and construction levels.

- Encourage the creation of production methods that incorporate recycled resources.
- Encourage material reclamation for energy or agricultural use, for example, in eco-industrial parks.
- Mandating the treatment of solid waste includes enforcing legal requirements, providing managerial and technical support, promoting biogas production, and fee-for-service collection.
- Start consumer education campaigns to inform people about waste management.
- Shopping enterprises can be forced to charge customers for each plastic bag instead of continuing the practice of giving out free plastic bags at malls.

2.2.2.7 Awareness Campaigns

- Implement a national environmental education curriculum. Inform the populace about problems with the food system, such as food waste, food miles, organic farming, and food security.
- Launch awareness campaigns regarding waste issues, educating the public and businesses about the 3Rs and offering practical solutions.
- Promote water-saving technology and methods for recycling water to increase awareness of challenges related to water use.
- Start awareness campaigns to educate the public about energy-related concerns, including information on energy-saving equipment, carbon dioxide (CO_2) emissions, and climate change. Teach businesses how to reduce their energy use.
- Promote environmentally friendly transportation by encouraging cycling and walking, educating the public on the carbon footprint of the current transportation infrastructure, and informing them of greener options.
- Educate and broaden public understanding of green construction (Mohanty and Scherfler 2012).

2.3 Conclusion

Disposable culture and materialism as well as the "take-make-use-dispose" economy of today is leading to a climate catastrophe. Indian cultures have long engaged in and followed sustainable lives. In addition to adopting eco-friendly habits at the individual, household, and community levels, India has shown leadership in innovative, individually led programmes and efforts to combat climate change. Swachh Bharat Mission, GOBARdhan Scheme, Give It Up, and Catch the Rain are some of the other successful campaigns (Mohanty and Scherfler 2012). India is best positioned to assume a leadership role in promoting behaviour change on climate action due to its extensive traditional knowledge and climate-friendly practices. It is feasible to scale up the country's national or local best practices to influence behaviour change for climate action.

Good Practices of Sustainability in the Indian Context

Indian cultures have long engaged in and followed sustainable lives. Some of the actions are listed below:

1. **Use of energy-efficient methods:** Such as drying clothing in the sun and washing dishes by hand, which eliminates the need for dishwashers and energy-intensive dryers, respectively.
2. **Common Indian family behaviours show that recycling and reuse are important:** Most Indian households keep and reuse bags, bottles, jars, and other goods. Old garments are frequently reused. Old sarees, for instance, can be sewn together to create stunning quilts. Old blankets are used to make rugs, while jute sacks are used to make doormats.
3. **Use of biodegradable utensils:** Throughout the nation, street vendors and public eateries continue to use Sal tree leaves as biodegradable utensils for food preparation and clay pots for serving tea (kulhad).
4. **Use of earthen blocks as a building material:** Earthen blocks have been widely researched and promoted as building materials by the Auroville Earth Institute in India. It has been determined that these technologies are both economical and energy efficient. Finding methods to reduce the use of steel, cement, and reinforced cement concrete while utilizing composite blocks is the primary job (earth, fibres, and stabilizer). The Institute is also looking at stabilized earth as an alternative to cement and "homoeopathic" milk of lime and alum for waterproofing (Mohanty and Scherfler 2012).
5. **Carpooling:** Those in need of a ride can connect with those who have space in their vehicles through an Internet effort in India. With the primary goal of lowering carbon emissions, a company in the USA by the name of GreenXC developed a campaign to get individuals to start carpooling. The objective is to cross the country and only use carpooling to visit different national parks and forests. Their only goal is to be environmentally friendly; therefore they don't plan on driving or renting a car to go where they're going. Individuals and communities can effectively tackle climate change through carpooling.

Other initiatives undertaken for promoting sustainable lifestyle and ecosystem restoration at the community level include:

- In Ahmedabad, India, all structures with a surface area of more than 1500 m^2 are required to collect rainwater.
- Auroville, India, provides lunch and dinner for up to 800 people each day.
- India has a programme to gradually phase out incandescent light bulbs in favour of CFLs. Financing is provided in part by the Clean Development Mechanism.
- Plastic bags are banned in Coorg, India (Mohanty and Scherfler 2012).

References

Anable DJ (2008) The cost-effectiveness of carbon abatement in the transport sector. The Centre for Transport Policy. The Robert Gordon University, and The UK Energy Research Centre. Campaign for Better Transport, London

Aquastat Database Query (2011). http://www.fao.org/nr/water/aquastat/data/query/index.html

Centre for Environment Education (2010) Low carbon lifestyles. Centre for Environment Education, New Delhi

Columbia University, Hunter College (CUNY), Science Applications International Corporation (SAIC) (2006) Mitigating New York City's Heat Island with urban forestry, living roofs, and light surfaces. New York City regional Heat Island initiative. Columbia University Center for Climate Systems Research at the Goddard Institute for Space Studies, New York

Das J, Jha S, Goyal MK (2020) On the relationship of climatic and monsoon teleconnections with monthly precipitation over meteorologically homogenous regions in India: Wavelet & global coherence approaches. Atmos Res 238:104889. https://doi.org/10.1016/j.atmosres.2020.104889

Earthforce (2011) What you can do. www.salesforcefoundation.org/files/Earthforce-WhatYouCanDo.PDF

Goyal MK, Ojha CSP (2010) Evaluation of various linear regression methods for downscaling of mean monthly precipitation in arid Pichola watershed. Nat Resour 1(1):11–18. https://doi.org/10.4236/nr.2010.11002

Goyal MK, Ojha CSP (2012) Downscaling of precipitation on a lake basin: evaluation of rule and decision tree induction algorithms. Hydrol Res 43(3):215–230. https://doi.org/10.2166/nh.2012.040

Goyal MK, Panchariya VK, Sharma A, Singh V (2018) Comparative assessment of SWAT model performance in two distinct catchments under various DEM scenarios of varying resolution, sources and resampling methods. Water Resour Manag 32(2):805–825. https://doi.org/10.1007/s11269-017-1840-1

Mohanty B, Scherfler M (2012) Lifestyle choices and societal behaviour changes as local climate strategy. Asian Development Bank Institute, Tokyo

PBNS (2022) Global launch of Mission LiFe: sustainable future with sustainable lifestyle

PIB Delhi (2022). https://pib.gov.in/PressReleaseIframePage.aspx?PRID=1847812

Poonia V, Goyal MK, Gupta BB, Gupta AK, Jha S, Das J (2021) Drought occurrence in Different River Basins of India and blockchain technology based framework for disaster management. J Clean Prod 312:127737. https://doi.org/10.1016/j.jclepro.2021.127737

Ranveer (mopng.gov.in) (2022) Lifestyle for Environment (LiFE) Movement: an India-led global mass movement to nudge individual and community action to preserve the environment

Rockström J, Gordon L, Folke C, Falkenmark M, Engwall M (1999) Linkages among water vapor flows, food production, and terrestrial ecosystem services. Conserv Ecol 3(2):5. http://www.ecologyandsociety.org/vol3/iss2/art5/

Rodale Institute (2005). http://www.news.cornell.edu/stories/July05/organic.farm.vs.other.ssl.html

Sodhi P, Gopal R, Srinivas SN, Jalan RK, Sarabhai KV (2010) Low carbon lifestyles. Centre for Environmental Education, New Delhi

Subramoniam SR, Ravindranath S, Rakkasagi S, Ram H (2022) Water resource management studies at micro level using geospatial technologies. Springer, Cham, pp 49–74

UNEP (2021) Ecosystem restoration for people, nature and climate

World Business Council for Sustainable Development (WBCSD) (2009) Energy efficiency in buildings: transforming the market. WBCSD, Geneva

World Economic Forum and Accenture (2009) Supply chain decarbonization: the role of logistics and transport in reducing supply chain carbon emissions. Logistics and transport partnership programme. World Economic Forum, Geneva

Chapter 3
Optimal Management of Potential Water and Sediment Yield from Urban Hilly Watershed

Sagarika Patowary and Arup Kumar Sarma

Abstract The sustainable ecological management of soil and water yields from hilly watersheds of an urban area is a real challenge from the perspective of economic and structural feasibility. It is of key importance for future city planning to implement ecological management practices (EMPs) optimally in urban hilly watersheds incorporating future urban growth. To achieve this, here, the Revised OPTimal EMP Model with Linear Programming for Single Ownership (R-OPTEMP-LS) is applied to manage the water and sediment yield calculated concerning a future LULC scenario of a hilly urban watershed of Guwahati city, India. The future LULC of the study area has been derived by using the ASEA (Assessment of Settlements in Eco-sensitive Area) model. The model parameters were input from the "Master Plan for Guwahati Metropolitan Area – 2025". It is found that in 2025, although the sediment yield from the watershed can be managed by executing the optimal combinations of EMPs, the peak runoff will be manageable only if the rainwater harvesting system will be strictly installed along with the EMPs. The model result says that for a 36.73% increase in the urban settlement in the hilly watershed from 2015 to 2025, there is a 33.4% increase in the total cost of EMPs. The high economic investment to control the adverse consequences of urban development in the hilly areas of Guwahati city emphasizes the need for strict conservation of those ecologically sensitive areas.

Keywords Projected urban settlement · ASEA · Soil loss · RUSLE · Peak runoff · Rational method · R-OPTEMP-LS

S. Patowary (✉)
Department of Civil Engineering, Goalpara Polytechnic, Goalpara, India

A. K. Sarma
Department of Civil Engineering, Indian Institute of Technology, Guwahati, India

A. K. Gupta et al. (eds.), *Ecosystem Restoration: Towards Sustainability and Resilient Development*, Disaster Resilience and Green Growth,
https://doi.org/10.1007/978-981-99-3687-8_3

3.1 Introduction

The increased imperviousness not only magnifies the amount of runoff volume but also lowers the time of concentration, which thus produces flash floods (Ng and Marsalek 1989; Viessman et al. 1989; Flinker 2010). Due to the sedimentation in the drainage channels, the soil erosion manifests the urban flash floods manifold (Kondoh and Nishiyama 2000; Ozacar 2013). Therefore, for the ease of hydrological investigation and to achieve a sustainable as well as an environment-friendly solution for hazards like urban flood and soil erosion, watershed management practices are essential (Sarma et al. 2015; Poonia et al. 2021). Urban development must therefore be designed in such a manner that it can minimize adverse effects along with the achievement of the developmental objectives (Mukheibir 2008; Goyal et al. 2018).

Different land covers respond differently to the hydrological hazards with diverse efficiencies of their soil erosion and runoff control (Poff et al. 2006; Goyal and Ojha 2012; Das et al. 2020). This makes the land use-based urban management strategies well-accepted, drawing the continuous attention of researchers from different disciplines (Tong and Chen 2002; Stutter and Lumsdon 2008; Goyal and Ojha 2010). Ecological management practices or EMPs, introduced by Sarma (2011), are considered as appropriate and competent urban watershed management practices if implemented appropriately (Patowary et al. 2019). EMPs are used to define the structural and vegetative means including cultivation or plantation of various shrubs, grass, or trees, contour terracing, retention and detention ponds, rainwater harvesting systems, vegetative buffer zone, stone or boulder pitching, retaining wall, etc. These measures are not only used for controlling soil erosion and runoff but also for controlling water pollution and carbon sequestration (Sarma et al. 2015). However, the expenditure, effectiveness, and suitability of EMPs are generally different for different places based on the condition of the proposed construction area. It claims the use of optimization techniques, which were successfully used in the past to answer the multi-objective environmental or water resources along with the land allocation problems (Chang et al. 1995; Gabriel et al. 2006; Hsieh and Yang 2007; Karterakis et al. 2007; Sadeghi et al. 2009). Sarma et al. (2015) introduced OPTEMP-LS (OPTimal EMP Model with Linear Programming for Single Ownership) for the optimum allocation of EMPs and performed a sensitivity analysis to determine how the project cost is sensitive to the various model parameters such as rainfall intensity, soil erodibility factor, runoff factor, etc. Patowary et al. (2019) revised the OPTEMP-LS by including the hill cut factor to determine the optimal combination of EMPs more accurately, based on GIS-based urban settlement estimation. The hill cut factor assesses the steep hill cut area (caused due to the residential development) that is rarely identifiable in ortho-rectified satellite images (Patowary and Sarma 2018). Again, the dynamic behaviour of the effect of urbanization always highlights the necessity of potential projection of various hydrological hazards to plan the management scheme accordingly. Despite the availability of several studies focusing on the best possible land cover combination for minimizing the undesirable impacts of urban development, this study uses the EMPs in urban hilly watersheds considering the potential urban development there. Therefore, the basic aim of this study is to apply the revised OPTEMP-LS with reference to the future hydrologic data derived by using potential urban settlement. The study has been carried out concerning an urban hilly watershed of Guwahati city, the capital of Assam. The future urban settlement data for the study watershed has been taken from Patowary and Sarma (2020), estimated by using the ASEA model. With the application of these data, the future hydrological parameters like peak runoff and sediment yield have been calculated, and finally, the R-OPTEMP-LS has been run based on these data.

3.2 Materials and Method

3.2.1 Study Area

The area selected in this study is an urban watershed of a hill of Guwahati city, the most developed city in the northeastern part of India. The city is witnessing a high urban expansion turning its eco-sensitive hilly regions into urban areas (Sarma et al. 2013, 2015). In every rainy season, the high amount of water and sediments coming from the urban-induced hilly areas are causing drainage congestion leading to devastating urban flash floods (Patowary et al. 2019). As projected by Patowary and Sarma (2020), the urban settlement in the hilly area is going to be increased by 11.12% from 2011 to 2025. This gives an idea of the worst condition of urban flash floods in Guwahati city. The location of the study area (watershed) in Japorigog hill, which is a part of the core area of Guwahati city, is shown in Fig. 3.1. Watershed has been delineated by using ArcSWAT with a Shuttle Radar Topography Mission (SRTM) DEM (digital elevation model) of a resolution of 1 arcsecond. The slope of the study area changes from 0° to 32.9° (with an average slope of 14.17°), and the elevation changes from 59 to 177 m. The total area of the hilly watershed is 74 ha, of which 30.8% was the urban settlement in 2015 (Patowary et al. 2019).

3.2.2 Estimation of Future Urban Settlement

Patowary and Sarma (2020) projected the urban settlement in the entire 15 hills of Guwahati city by applying the ASEA model. It is an indirect urban settlement assessment model, which uses the socio-economic, demographic, and geographical conditions of an area to estimate the urban settlement. The future data of the socio-economic, geographical, and demographic parameters of the model were generated concerning GMDA (2009). In this study, the projected LULC data of the study area for the year 2025 has been prepared by clipping the projected LULC map of the Japorigog hill of Guwahati city in ArcGIS 10.4.1. Figure 3.2 depicts the LULC map

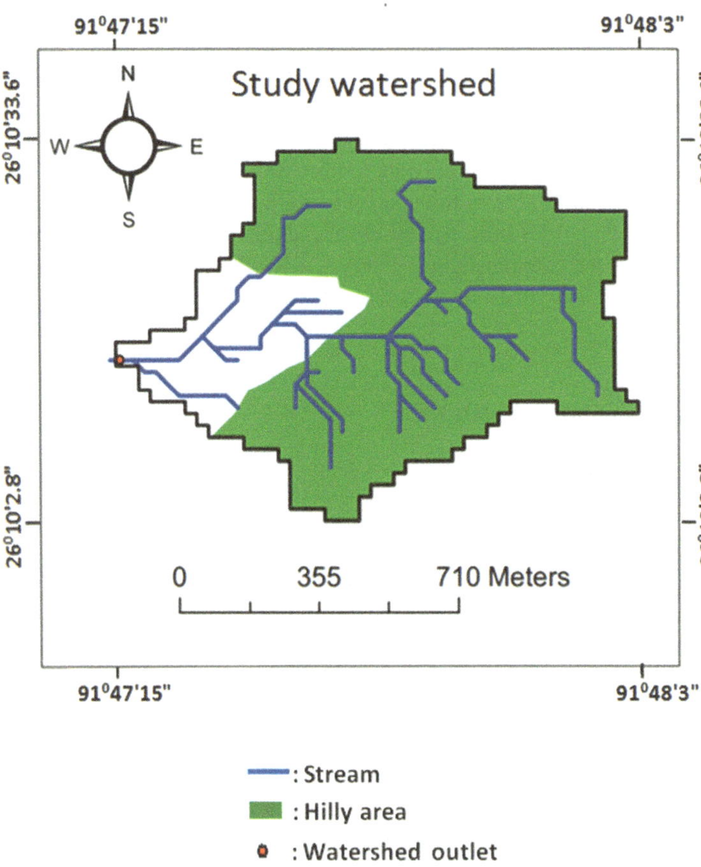

Fig. 3.1 Study area

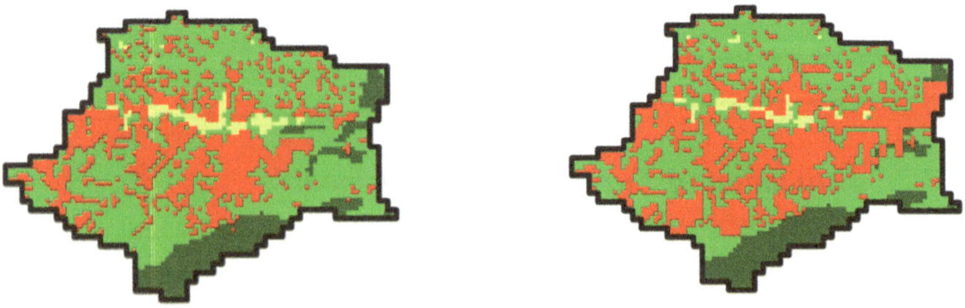

(a) (b)

Fig. 3.2 LULC maps of the watershed: (**a**) 2015 and (**b**) 2025 (light green, scrubland; dark green, forest; yellow, bare land; red, urban settlement)

of the study area in 2025. For comparison, the LULC map for the year 2015 (extracted from the LISS IV satellite image of 4 December 2015) has also been presented. It is found that from 2015 to 2025, urban settlements in the study area increase by 36.73%.

3.2.3 The Optimal Combination of EMPs in the Future

3.2.3.1 The R-OPTEMP-LS

This R-OPTEMP-LS (Patowary et al. 2019) is a linear programming-based optimization model. The objective function of the R-OPTEMP-LS is defined to minimize the sum of the costs of EMPs, essential to bring the yearly values of sediment yield and peak runoff within permissible limits. This is controlled by various conditions like existing LULC type, soil type, slope, simplicity of maintenance, and availability of land. Water and sediment yield constraints are presented with the help of rational method (Kuichling 1889) and GIS-based RUSLE (revised universal soil loss equation; Renard et al. 1991), respectively. Here, the complete watershed has been considered under a solo proprietor (in the form of private developers, society, government, etc.). The detail of the R-OPTEMP-LS has been given in Fig. 3.3.

3.2.3.2 EMP Selection

Based on the accessibility and execution convenience, the EMP selection has been done for the year 2025. For the settlements in the plain area of the watershed, the optimal values (area) of the selected EMPs—grass, garden (vegetables, shrubs, as well as bushy vegetation), forest, and detention pond (depth = 1.5 m, Sarma et al. 2015)—are represented by Xp_1, Xp_2, Xp_3, Xp_4, correspondingly. The similar EMPs for the hilly portion are symbolized by Xh_1, Xh_2, Xh_3, Xh_4, respectively. Again, for the near-vertical steep hill cuts, two EMPs—grass/creepers and retaining wall with random rubble masonry—have been selected, the optimal values of which are denoted by Yh_1 and Yh_2, respectively.

The cover management factors, the runoff coefficients, and the costs per unit area of EMPs have been used from the past literature (Sarma et al. 2005; San Diego County 2003; ODOT Highway Division 2014; Wischmeier and Smith 1961; Gelagay and Minale 2016; Patowary and Sarma 2018; Patowary et al. 2019). These studies estimated the unit costs by using the market rates of the year 2012–2013. It is worth mentioning that, although for the theoretical application purpose, here, unit costs have been taken from the past literature, it is essential to use the latest available costs for the real field application of the model.

Constraint 1: Sediment yield constraint: $Smin \leq S \leq Smax$

Smin = minimum annual sediment yield required from the watershed (t/yr), determined based on the requirement of the downstream environment.

Smax = maximum annual sediment yield allowed from the watershed (t/yr). It should not exceed the value of sediment yield from the watershed under natural land cover condition.

S = sediment yield after the application of EMPs from the watershed (t/yr)

Constraint 2: Peak runoff constraint: $Qmin \leq Q \leq Qmax$

Qmin = minimum peak runoff required from the watershed (m^3/s). It should meet the demand of downstream aquatic ecosystem for its proper ecological functioning.

Qmax = maximum peak runoff allowed from the watershed (m^3/s). It is equal to the safe carrying capacity of the drain just downstream of the watershed.

Q = peak runoff after the application of EMPs from the watershed (m^3/s)

Constraint 3: EMP area suitability constraint:

The area of an EMP cannot exceed the suitable area available in the watershed for that EMP. Again, the available suitable area for that EMP should be greater than a minimum feasible area required for the implementation of that EMP.

Objective function:

$$\text{Minimize } Z = \sum_{i=1}^{n}(Cq_i + Cm_i)Xp_i + \sum_{j=1}^{q}(Cq_j + Cm_j)Xh_j + \sum_{k=1}^{r}(Cq_k + Cm_k)Yh_k$$

Where, i= 1, 2, 3, ……, n are EMPs considered for the urban settlement area located in plain areas of the watershed.

j= 1, 2, 3, ……, q are EMPs considered for the urban settlement area located in hilly areas of the watershed.

k=1, 2, 3, ……, r are EMPs considered for the steep hill cuts associated with urban settlements in the hilly portion of the watershed.

Cq_i, Cq_j, Cq_k are construction costs of ith, j^{th} and k^{th} EMPs, respectively.

Cm_i, Cm_j, Cm_k are maintenance costs of ith, j^{th} and k^{th} EMPs, respectively.

Xp_i = Area of the ith EMP applied in urban settlements in the plain area of the watershed (m^2).

Xh_j= Area of the jth EMP applied in urban settlements in the hilly area of the watershed (m^2).

Yh_k= Area of the kth EMP applied in the steep hill cuts of the watershed (m^2).

Constraint 4: Maximum area available for EMPs: No EMP can be applied in the allowable coverage (building) area according to the applicable building bye-laws. Therefore, total EMP area should be less than or equal to the bare settlement area.

Constraint 5: Owner's choice for EMPs:

As EMPs are to be implemented and maintained within the residential area; the planned EMP area must be within the permissible amount of areas as per the owner's choice (Sarma et al. 2015). As a hilly urban watershed contains several residential plots, a general upper and lower limit of the constraint should be considered for the watershed by combining the limits of EMP areas at different plots.

Constraint 6: Non negativity constraint:

$$Xp_i \geq 0 \qquad Xh_j \geq 0 \qquad Yh_k \geq 0$$

Fig. 3.3 The detail of the R-OPTEMP-LS

3.2.3.3 Constraints

1. Constraint for sediment yield and peak runoff: As given in Fig. 3.3, the maximum and minimum limits of sediment yield and peak runoff have been calculated and obtained as follows:

$$S_{\min} = 0, \ \ S_{\max} = S_{\text{natural}} = 2608.79\,\text{t/year},$$
$$Q_{\min} = Q_{\text{natural}} = 2.979\,\text{cumec}, \ \ Q_{\max} = Q_{\text{drain}} = 4\,\text{cumec}$$

2. Constraint for suitability of EMP area: The superimposition of the soil map (collected from Assam Remote Sensing Application Centre, Guwahati), DEM, and slope map of the study area in ArcGIS gives that all the chosen EMPs are applicable to the plain bare settlement area, but to the steep hill cut area, only the grass and retaining wall are applicable.
3. Maximum area available for EMP:

   The total EMP area ≤ bare settlement area.
   As given in Patowary et al. (2019), this constraint is obtained as:

$$\sum\nolimits_{i=1}^{4} \text{Xp}_i \leq 0.4 \times U_{\text{sp}} \tag{3.1}$$

$$\sum\nolimits_{j=1}^{4} \text{Xh}_j \leq 0.4 \times U_{\text{sp}} \tag{3.2}$$

$$\sum\nolimits_{k=1}^{2} \text{Yh}_k \leq 0.6 \times \frac{\sin\theta}{\sin(\beta-\theta)} \times U_{\text{sh}} \tag{3.3}$$

where the built-up area in the plain watershed area is derived from the LULC map of the year 2025, U_{sp}= 69,188 m^2.

Similarly, settlement in the hilly watershed area, U_{sh}= 218,148 m^2.

The average hill cut angle is β = 70° (Patowary et al. 2019).

The natural slope (average) of the hilly watershed area is θ = 14.17° (derived from slope map in ArcGIS).

Again, as given in Patowary et al. (2019), owners' choices have been considered hypothetically.

3.3 Results and Discussions

The "Linear Model Solver Tool" of "Microsoft Excel" application has been used to solve the optimization problem. It is found that the execution of the chosen EMPs makes the sediment yield manageable within allowable limits. However, it is not feasible to keep the peak runoff from the watershed within the highest allowable limit due to the projected increase in urban settlements in 2025. This is because 60% of this increased urban settlement contributes only imperviousness producing a high

Table 3.1 Costs of EMP implementation obtained from R-OPTEMP-LS model

	2015 (Patowary et al. 2019)	2025
Cost ($\times 10^7$ Rs)	Total = 10.24	Total = 13.67
	Plain = 0.85	Plain = 0.85
	Hill = 1.97	Hill = 2.69
	Steep hill cut = 7.42	Steep hill cut = 10.13

Table 3.2 Comparison of sediment and water yield for the years 2015 and 2025

		Sediment yield (tonnes/year)	Peak runoff (cumec)
2025	Without EMP	41,267.98	4.75
	With EMP	2602.15	3.85
2015 (Patowary et al. 2019)	Without EMP	34,020.77	4.38
	With EMP	2602.15	4.00
	Natural land cover condition (pre-urbanized)	2608.79	2.99

amount of runoff. The EMPs implemented in the remaining 40% area of urban settlement (bare) are not sufficient to bring the peak runoff within the maximum allowable limit. Sarma et al. (2015) mentioned that the installation of the rainwater harvesting (RWH) system along with the considered EMPs is a good alternative to bring the peak discharge within the permissible limits. By adopting the rooftop rainwater harvesting, the runoff coefficient of the built-up area is becoming possible to lower by 20% in Guwahati city (Sarma et al. 2006). From the last few years, GMDA is giving utmost importance to the adoption of the rainwater harvesting system, and it is now made compulsory in building bye-laws of the city. Therefore, it has been considered that till 2025, RWH schemes will be strictly implemented in Guwahati city, and consequently, in 2025, the runoff coefficient of the impervious area has been expected to reduce by 20%. Under this consideration, a feasible solution is obtained to bring the peak discharge and soil loss within allowable limits in 2025. With the consideration of all the constraints, it is found that for a 36.73% increase in the settlement of hilly watershed areas from 2015 to 2025, there is a 33.4% increase in the overall expenditure of EMPs to maintain the soil loss and peak runoff in the permissible limits. For the ease of interpretation of the effect of future urban settlements on the management cost of the watershed, the results are also compared with that in Patowary et al. (2019) for the year 2015 (Table 3.1). Again, Table 3.2 presents the sediment yield and peak runoff for various situations. It is observed that after EMP implementation, the sediment yields from the watershed are almost the same with those from the natural land surface scenario, although peak runoff values are not the same. After the implementation of EMPs, the less peak runoff value in 2025 than that in the 2015 LULC scenario indicates that the installation of the RWH system along with the other considered EMPs is quite effective in reducing the peak discharge. Again, due to the implementation of EMPs, there is a more efficient reduction of sediment yield than the peak runoff. It is due to the fact that the impervious area produced due to the increased urban development yields no sediments but a high amount of water.

3.4 Conclusions

This study presents optimal management of an urban hilly watershed considering the future hydrological alterations to be taken place as a consequence of urban development. The study objective has been achieved with the help of two basic models—(a) ASEA model, which determines the urban settlement in an eco-sensitive hilly area based on various socio-economic, demographic, and geographic factors, and (b) R-OPTEMP-LS, an optimization model specially developed for hilly watersheds, which minimizes the cost of a combination of EMPs implemented to bring the sediment and water yields within a sustainable limit. The study has been performed for a hilly urban watershed in the central Guwahati of Northeast India. The future urban settlement data for the study area has been extracted from Patowary and Sarma (2020), who applied the ASEA model to project the urban built-up area in the entire hilly areas of Guwahati city for the year 2025. The LULC data of the study area thus obtained for the year 2025 was used in the R-OPTEMP-LS to get the optimal combination of EMPs with a minimum possible cost. The result of the R-OPTEMP-LS clearly indicates that the city expansion should be strictly prohibited in the hilly areas of Guwahati city as the ecological management of those urban-induced hilly areas will claim a huge financial investment. It is observed that for a 36.73% increase in the urban coverage in the hilly watershed area from 2015 to 2025, the overall expenditure of EMPs will increase by 33.4%. The knowledge of potential economic investment required to manage the future adverse consequences of urban developmental activities in such eco-sensitive areas can be very useful in efficient and eco-friendly urban planning.

References

Chang NB, Wen CG, Wu SL (1995) Optimal management of environmental and land resources in a reservoir watershed by multiobjective programming. J Environ Manag 44(2):144–161

Das J, Jha S, Goyal MK (2020) On the relationship of climatic and monsoon teleconnections with monthly precipitation over meteorologically homogenous regions in India: Wavelet & global coherence approaches. Atmos Res 238:104889. https://doi.org/10.1016/j.atmosres.2020.104889

Flinker P (2010) The need to reduce impervious cover to prevent flooding and protect water quality. Rhode Island Department of Environmental Management, Rhode Island. http://www.dem.ri.gov/programs/bpoladm/suswshed/pdfs/imperv.pdf

Gabriel SA, Faria JA, Moglen GE (2006) A multiobjective optimization approach to smart growth in land development. Socio Econ Plan Sci 40(3):212–248

Gelagay HS, Minale AS (2016) Soil loss estimation using GIS and remote sensing techniques: a case of Koga watershed, northwestern Ethiopia. Int Soil Water Conserv Res 4(2):126–136

GMDA (2009) Master plan of Guwahati metropolitan area 2025. Guwahati Metropolitan Development Authority, Guwahati. http://www.gmda.co.in/master_plan.php

Goyal MK, Ojha CSP (2010) Evaluation of various linear regression methods for downscaling of mean monthly precipitation in Arid Pichola Watershed. Nat Resour 1(1):11–18. https://doi.org/10.4236/nr.2010.11002

Goyal MK, Ojha CSP (2012) Downscaling of precipitation on a lake basin: evaluation of rule and decision tree induction algorithms. Hydrol Res 43(3):215–230. https://doi.org/10.2166/nh.2012.040

Goyal MK, Panchariya VK, Sharma A, Singh V (2018) Comparative assessment of SWAT model performance in two distinct catchments under various DEM scenarios of varying resolution, sources and resampling methods. Water Resour Manag 32(2):805–825. https://doi.org/10.1007/s11269-017-1840-1

Hsieh CD, Yang WF (2007) Optimal nonpoint source pollution control strategies for a reservoir watershed in Taiwan. J Environ Manag 85(4):908–917

Karterakis SM, Karatzas GP, Nikolos IK, Papadopoulou MP (2007) Application of linear programming and differential evolutionary optimization methodologies for the solution of coastal subsurface water management problems subject to environmental criteria. J Hydrol 342(3–4):270–282

Kondoh A, Nishiyama J (2000) Changes in hydrological cycle due to urbanization in the suburb of Tokyo Metropolitan area, Japan. Adv Space Res 26(7):1173–1176

Kuichling E (1889) The relation between the rainfall and the discharge of sewers in populous districts. Trans Am Soc Civ Eng 20(1):1–56

Mukheibir P (2008) Water resources management strategies for adaptation to climate-induced impacts in South Africa. Water Resour Manag 22(9):1259–1276

Ng HYF, Marsalek J (1989) Simulation of the effects of urbanization on basin streamflow 1. JAWRA J Am Water Resour Assoc 25(1):117–124

ODOT Highway Division (2014) Hydraulics design manual: appendix F-rational method. Oregon Department of Transportation, Geo-Environmental Section, Salem, pp 7-F-1–7-F-14

Ozacar BG (2013) Impacts of urbanization on flood and soil erosion hazards in Istanbul, Turkey. The University of Arizona, Tucson

Patowary S, Sarma AK (2018) GIS-based estimation of soil loss from hilly urban area incorporating hill cut factor into RUSLE. Water Resour Manag 32(10):3535–3547

Patowary S, Sarma AK (2020) Projection of urban settlement in eco-sensitive hilly areas and its impact on peak runoff. Environ Dev Sustain 22(6):5833–5848

Patowary S, Sarma B, Sarma AK (2019) A revision of OPTEMP-LS model for selecting optimal EMP combination for minimizing sediment and water yield from hilly urban watersheds. Water Resour Manag 33(4):1249–1264

Poff NL, Bledsoe BP, Cuhaciyan CO (2006) Hydrologic variation with land use across the contiguous United States: geomorphic and ecological consequences for stream ecosystems. Geomorphology 79(3–4):264–285

Poonia V, Goyal MK, Gupta BB, Gupta AK, Jha S, Das J (2021) Drought occurrence in different river basins of India and blockchain technology based framework for disaster management. J Clean Prod 312:127737. https://doi.org/10.1016/j.jclepro.2021.127737

Renard KG, Foster GR, Weesies GA, Porter JP (1991) RUSLE: revised universal soil loss equation. J Soil Water Conserv 46(1):30–33

Sadeghi SHR, Jalili K, Nikkami D (2009) Land use optimization in watershed scale. Land Use Policy 26(2):186–193

San Diego County (2003) San Diego County hydrology manual. San Diego County Department of Public Works, Flood Control Section

Sarma B (2011) Optimal ecological management practices for controlling sediment and water yield from a hilly urban system within sustainable limit (Doctoral dissertation)

Sarma AK, Chandramouli V, Singh B, Goswami P, Rajbongshi N (2005) Urban flood hazard mitigation of Guwahati City by silt monitoring and watershed modeling. Report submitted to

Ministry of Human Resources Department (MHRD) by Department of Civil Engineering, IIT Guwahati

Sarma AK, Giraud G, Baishya MD (2006) Rainwater harvesting for urban flood peak reduction. My Green Earth. J Soc Socio Eco Awareness Environ Prot 3(2):14–21

Sarma B, Sarma AK, Singh VP (2013) Optimal ecological management practices (EMPs) for minimizing the impact of climate change and watershed degradation due to urbanization. Water Resour Manag 27(11):4069–4082

Sarma B, Sarma AK, Mahanta C, Singh VP (2015) Optimal ecological management practices for controlling sediment yield and peak discharge from hilly urban areas. J Hydrol Eng 20(10): 04015005

Stutter MI, Lumsdon DG (2008) Interactions of land use and dynamic river conditions on sorption equilibria between benthic sediments and river soluble reactive phosphorus concentrations. Water Res 42(16):4249–4260

Tong ST, Chen W (2002) Modeling the relationship between land use and surface water quality. J Environ Manag 66(4):377–393

Viessman W, Lewis GL, Knapp JW, Harbaugh TE (1989) Introduction to hydrology

Wischmeier WH, Smith DD (1961) A universal equation for predicting rainfall-erosion losses: an aid to conservation farming in humid regions. Agric Res Serv ARS Special Report, pp 22–66

Chapter 4
Mainstreaming Peri-urban Ecosystems for Urban Resilience Through Policy-Planning Framework: An Opportunity Analysis for Indian Cities

Pritha Acharya, Anil Kumar Gupta, Swati Singh, Nivedita Mani, and Shiraz Wajih

Abstract Peri-urban areas are often the most challenging yet strategically important areas of the recent present. They are the primary support systems for cities as they establish active connections between rural and urban areas through dynamic interactions between socio-cultural, biophysical and economic spheres. The dependence of cities on peri-urban ecosystems is not only limited to the availability of resources, such as food, water, energy, infrastructure and development but also extends to encompass an array of ecosystem service flows and absorption of environmental and climatic shocks. India witnesses a high pace of urban growth and subsequent complex peri-urban challenges yet has significant knowledge and database (case studies, on-ground research, etc.) arguing the benefits and potentials of peri-urban ecosystems in providing urban resilience. Despite this, peri-urban landscapes and ecosystems are not recognized in urban policy planning. This paper aims to map the existing opportunities for peri-urban ecosystem services-based urban resilience into

P. Acharya
ECDRM Division, National Institute of Disaster Management (NIDM), Delhi, India

A. K. Gupta (✉)
National Institute of Disaster Management (Ministry of Home Affairs, Government of India), New Delhi, India

S. Singh
United Nations Development Programme (UNDP), Delhi, India

N. Mani
Research and Communications, Gorakhpur Environmental Action Group (GEAG), Gorakhpur, Uttar Pradesh, India
e-mail: geagdelhi@geagindia.org

S. Wajih
Gorakhpur Environmental Action Group (GEAG), Gorakhpur, Uttar Pradesh, India
e-mail: geag@geagindia.org

A. K. Gupta et al. (eds.), *Ecosystem Restoration: Towards Sustainability and Resilient Development*, Disaster Resilience and Green Growth,
https://doi.org/10.1007/978-981-99-3687-8_4

a policy-planning framework. Exploratory research has been carried out to enumerate the urban and peri-urban linkages and associate challenges, with the help of case studies in three Indian cities, namely, Shimla, Panjim and Visakhapatnam. The paper delineates suitable pathways for mainstreaming peri-urban issues into the urban resilience framework through recognizing the scope of nature-based solutions (NbS) and community involvement and by utilizing the effective policy-planning framework.

Keywords Peri-urban ecosystems · Urban resilience · Policy-planning framework · Indian cities

4.1 Introduction

Cities are the harbingers of global development and economic growth. They offer multiple benefits, facilities and opportunities which catalyses a global influx of people to cities from rural areas. Currently, more than 50% (UN Habitat 2020) of population worldwide resides in cities and urban centres, and it is estimated that by 2050, more than two-thirds of the population will inhabit the urban areas (UN 2017). As the urban areas inclusive of metros, cities and towns continue to grow, the pressure of this expansion is radiated to the urban peripheries or the peri-urban areas (Aijaz 2019). The Intergovernmental Panel on Climate Change in its special report stated climate change to be the most pressing challenge of this century; further the ramifications of climate change are experienced globally in the form of increased frequency, duration and severity of extreme events and disasters. Impacts of climate change like storm surges, urban flooding, heat islands, heat and cold waves, etc. affect urban systems globally, which account for 80% GDP (UN Habitat 2020; Poonia et al. 2021) share and pose a constant risk to business continuity. Further, socio-economic dimensions along urban peripheries or peri-urban systems are generally based out of primary production and, thus, are highly sensitive to climate change impacts. Adaptation strategies are necessary for overcoming these challenges (Rosenzweig et al. 2015).

Urban resilience is a complex process that represents an agglomerate of interdependent mechanisms. Addressing city's vulnerability to the impact of climate change and disasters through effective and proactive perspectives has been at the core of city resilience arguments (Chelleri et al. 2015; Goyal et al. 2018). Promoting the strategies and adaptation measures pertaining to city well-being has emerged as the recent focus. Though ecosystem-based approaches as adaptation measures for enhancing city resilience have found some recognition, however, it is limited to urban infrastructure and water (Goyal and Ojha 2012; Carter 2018). As the disclosure for urban resilience widened, urban biodiversity and early warning systems were included under its ambit (Bush and Doyon 2019). Cities do not operate in silos; they are inextricably linked to peri-urban spaces for their survival and smooth functioning (Goyal and Ojha 2010; Das et al. 2020; Verma 2021). Peri-urban areas are characterized by diverse land-uses including agriculture, forest, industries, residential and wetlands and water catchments, etc. The uniqueness of these ecosystems is appreciable in terms of their dynamic transitional characteristic between rural and urban areas,andtheirabilitytoabsorbthe urban stresses (Tiwari 2019). Ribeiro and Gonçalves (2019) placed urban resilience as a cumulative of five dimensions—natural, economic, social, physical and institutional. Nonetheless, the context of urban resilience remained partially effective with the negligence to peri-urban resilience.

Developed countries have well-demarcated boundaries of these transition zones, with heterogeneity in land-use types in peri-urban spaces. However, in the context of developing countries, managing peri-urban spaces and ensuring effective utilization of these spaces is challenging (Tiwari 2019). On the other hand, developing countries provide for opportunity to integrate peri-urban planning as an imperative to urban resilience (Verma 2021). India is the second most populated country in the world. Besides, it is one of the biggest economies among the developing nations. However, the country has its own sets of challenges. In the last few decades, Indian cities have outgrown into big urban centres attracting the biggest migration from rural to urbans (Addo 2010). Urban and peri-urban stresses include disaster vulnerability, environmental degradation, resources overutilization, increased consumption, health and well-being and livelihood opportunities in the backdrop of climate change scenarios.

Some studies acknowledging the role of Indian peri-urban spaces in promoting overall resilience are documented (Aijaz 2019). Furthermore, examples of existing on-ground pilots for urban and peri-urban resilience offer prominent lessons yet are unconnected (Gupta et al. 2018). In order to mainstream the benefits of peri-urban ecosystems, systemic integration is necessary. Developmental agenda like the Sustainable Development Goals (SDGs), Sendai Framework for Disaster Risk Reduction (SFDRR), Resilient Cities, Paris Climate Agreement, etc. also acknowledges the need for an integrative approach for economic, environmental and social urban growth (UN 2016). This study is an attempt to promote systemic integration of peri-urban ecosystems and urban resilience. It is an explorative study, based on insights from three cases, aimed at mainstreaming peri-urban ecosystems for urban resilience through policy-planning frameworks.

4.2 Recognizing Urban-Rural Connect and Peri-urban in Transit

Urbanization is an effective driver for poverty reduction and development, but the transformative force associated with the process reshapes rural and urban landscapes (UN 2017; Tiwari 2019). Well-being of both the urban and rural space needs are thus maintained through constant urban-rural linkage flows in the form of capacity, people, resources and services (UN 2017). Bulderberga (2014) explained urban-rural interactions as invisible and visible flow of information, goods and services,

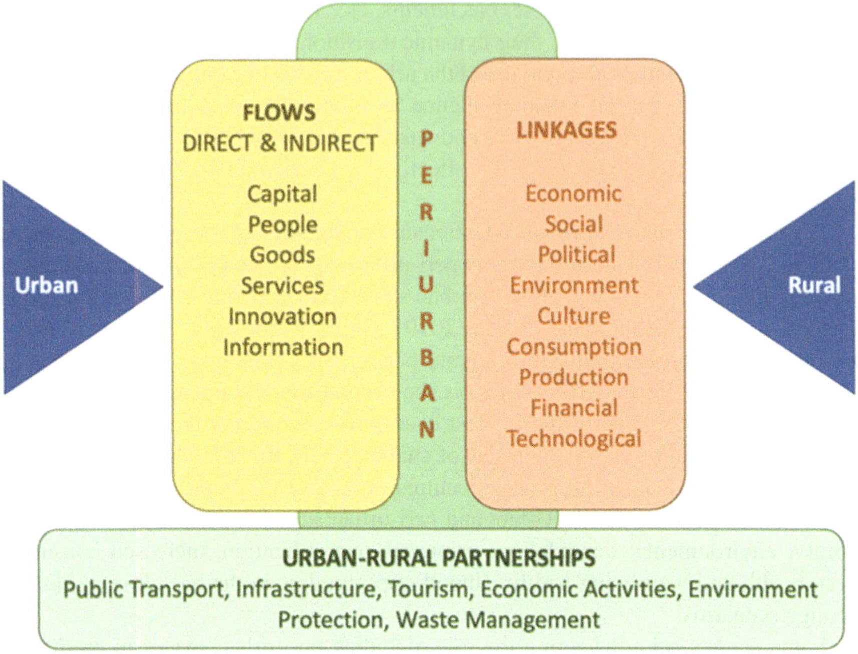

Fig. 4.1 Urban-rural flows and linkages. (Adapted from Bulderberga 2014)

technology, people and capital through structural, spatial and functional linkages (Fig. 4.1).

There exists a lot of ambiguity regarding the definition of a peri-urban area (Gomes and Hermans 2018; Shrestha 2019), but they are mainly addressed as transition zones between urban and rural areas. The peri-urbans showcase blending characteristics and mosaic of land-uses serving multi-functional attributes (Narain 2020), which, thus, extends a zone of spatial support for urban-rural flows and linkages.

Representing the overlapping belt between rural and urban, the peri-urban spaces are composed of a network of ecosystems like the croplands, wetlands, forests and agroforests, etc., which provides a plethora of ecosystem services to the urban cores. Furthermore, the associated ecosystem service flow harnesses direct and indirect value and benefits to both the urban and peri-urban habitants (Singh et al. 2013). The diversity and heterogeneity of peri-urban systems accounting for the social, economic and financial interest (Narain 2020) are the basis of synergies for urban-rural connect. These synergies guarantee food security and water security and provide effective market linkages ensuring livelihood sources (Verma 2021).

Most Indian cities depend on groundwater sources for drinking purposes. The peri-urban open spaces are crucial for collecting and harvesting rainwater and storing them—recharging the underground aquifers. In the case of Hyderabad city, rapid expansion of the city engulfing its peri-urban spaces has disturbed the city's hydrological cycle (Prakash 2014). These peri-urban ecosystems ensure water security, which reduces water scarcity and enhances the availability and quality of urban waters.

Water, energy and food are vital resources to sustain livelihoods, and, therefore, understanding the intricate relationship among them and various factors affecting the interlinkages are very critical for adaptation planning (Biggs et al. 2015; Rasul and Sharma 2016). This nexus of water-energy-food in urban systems has direct linkage with the peri-urban environment and intactness of the ecosystem services they offer (Caputo et al. 2021).

Agriculture is the most common land cover pattern seen in the peripheries of urban areas (Verma 2021). Developing as well as developed countries are dependent on peri-urban agriculture for food security (Wei et al. 2018). For example a vast proportion of all farms in the United States are situated in the peri-urbans. It also promotes an emotional and social well-being by ensuring a supply of nutritious and healthy food. Ensuring ample opportunities for marginal farmers and peri-urban dwellers, this form of agriculture provided livelihood security and continuous flow of household incomes as well as extended support to the fraction of sustenance farmers (Verma 2021).

4.2.1 Peri-urban Potentials

Peri-urbans for today are cities for tomorrow, especially in developing countries like India. Major cities in making which fail to find existence today are either in the process of development or will be developed in the peripheries of present time (Singh 2018). Thus, peri-urban areas are vehicles for taking resilience to future cities as these would bring in the challenges, strengths and opportunities with them while transforming into the part of the city.

Environmental degradation of peri-urban areas impedes and impacts the ecosystem service flows (Gupta et al. 2017). The Millennium Ecosystem Assessment Report of 2015 points out that 60% of ecosystem services have already been degraded (MEA 2005). Loss of almost three-fourths of ponds and lakes in Chennai paralysed the entire city by cutting off basic amenities to more than three million locals during the 2015 flash floods. Numerous other instances have seen cities and urban cores fall prey to natural hazards (Nel et al. 2014). Cities' reliance on peri-urban ecosystems makes it imperative to maintain the health of these ecosystems. City's resilience should inculcate overall performance of its systems which includes managing disasters, adapting to climate change and remitting the ill effects of environmental degradation. Despite the recognized need of integrating peri-urban ecosystem-based strategies to strengthen urban resilience, peri-urban ecosystems are unfortunately deteriorating in quantum and quality (Gupta et al. 2016).

Recognition of peri-urban areas by various stakeholders against the benefits they offer to reduce city vulnerability provides an important opportunity to mainstream and integrate resilience through ecosystems and ecosystem service flows in future urban development (Ribeiro and Gonçalves 2019). Assessing disasters and developmental needs and impacts on peri-urban ecosystems to better plan, manage and implement holistic actions for ecosystem integrity is essential. This can be possible if ecosystem approaches are mainstreamed in developmental planning to promote urban resilience.

4.2.2 Peri-urban Issues and Challenges

Tiwari (2019) cumulatively list all key issues of peri-urban areas and the responsible causes. Discretely demarcating boundaries of peri-urban spaces is a limitation for the present socio-political context. Reclassification of rural into cities and towns, contrasting definitions of urban and rural and co-existence of rural and urbans are the key reasons comprehensively identified by Narain (2020). These challenges in terms of spatial extent result in administrative neglect (Verma 2021). Ambiguity among the governing bodies to extend their support to integrate peri-urban spaces in their sustainable and developmental planning has been discussed broadly by many practitioners and researchers (Narain 2020). Absence of appropriate building regulations and unclear administrative planning lead to encroachment into peri-urban areas through rampant construction of illegal structures by few individuals, real estate markets and industries (Aijaz 2019; Verma 2021).

Case studies across the developing and developed nations show similar trends in rural to urban migration (UN Habitat 2020). Better job opportunity, improved living standards, lesser cost of living, lower land process and easy accessibility to multiple resources drive these peri-urban migrations (Aijaz 2019). Interestingly, amenity-led migration—driven by aesthetic factors—from cities and metros to urban fringes has gained momentum in the past couple of years (Chipeniuk 2004; Moss 2006; Perlik 2011). The locals of the small hilly town of Mukteshwar now face implications of water scarcity owing to land-use changes with urban to peri-urban migration (Narain 2020).

Expanding city limits in the backdrop of urbanization put peri-urbans as the main buffers to absorb the shocks from growing demands. Coming up of developmental projects like educational institutions, airports and commercial centres also triggers growth in the urban fringes (Tiwari 2019). After being badly hit by the cyclone Amphan, Kolkata airport faced immense economic loss due to limited functionality. The state invested in refunctioning of the Andal airport situated in the hinterlands of the Durgapur, a city in West Bengal in India. Cumulative changes including demographic changes, socio-economic conditions and physical and infrastructural factors along with changing climatic and environmental conditions are triggering the indiscriminate and rapid conversion of land-use patterns in peri-urban areas.

4.3 Methodology

Ribeiro and Gonçalves (2019), in their study on urban resilience, based urban resilience on four pillars—resisting, recovery, adaptation and transformation. They classified urban resilience across the dimensions of natural, physical, social, economic and institutional resilience. Researches that promote linkages of peri-urban with all five dimensions of urban resilience have been touched upon in the ambit of the paper. This is suggestive of the fact that knowledge products are existing in support of urban and peri-urban, but they are recluse and non-inclusive (Narain 2020). This study had been designed with an aim to map the existing opportunities for mainstream peri-urban ecosystems in policy-planning framework for urban resilience in Indian context. An exploratory research approach was used to understand policy gaps. Three case studies from India, namely, Shimla, Panjim and Visakhapatnam, had been analysed in detail for identifying opportunities for systemic integration of the peri-urbans into urban resilience framework (Nivedita and Wajih 2021). A map depicting the location and the geographic profile of the case study sites is depicted in Fig. 4.2. Drawing from the inference of analysis, this paper discusses the pathways and policy-planning protocols for effective mainstreaming.

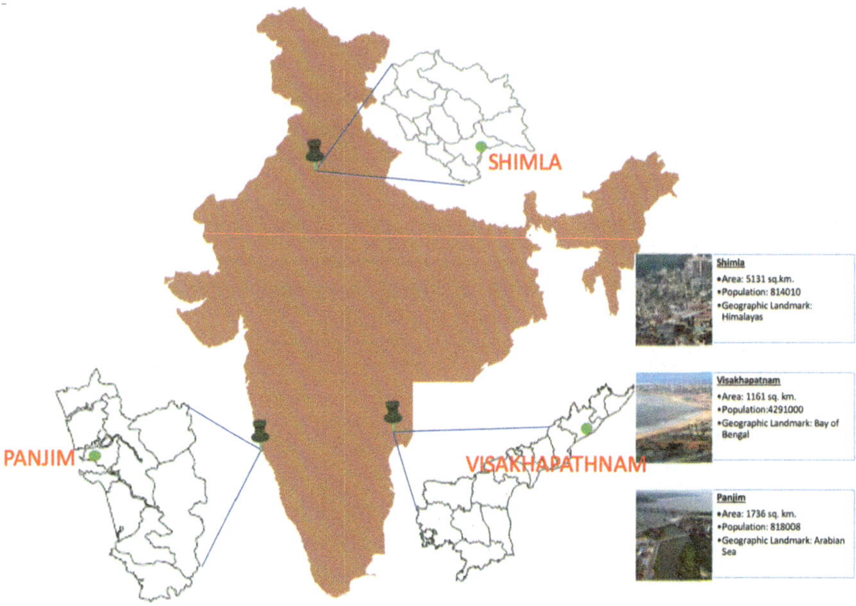

Fig. 4.2 Geographic representation of selected case study sites

4.3.1 Case Analysis: Study of Three Indian Cities

The present plight of peri-urban ecosystems around Indian cities has shaken up the planners, practitioners, researchers and conservationists to follow an inclusive approach to overcome the dichotomy of urban development at the cost of peri-urban ecosystem (Narain 2020). A number of ground interventions have been undergoing and documented in various literatures (Nivedita and Wajih 2021). Taking insights from these, the following three case studies have been selected. These studies have diverse geographic representation (Fig. 4.2) with distinct administrative, social-economic characteristics and unique climatic and disaster vulnerabilities. The geographic profile, disaster vulnerability, ecosystem dependence and challenges of the peri-urban spaces in the cities of Shimla, Visakhapatnam and Panjim have been described in Table 4.1. Table 4.2 describes the essential ecosystem services on which these three cities depend. City y Type

The observations and inferences drawn are based on the case study reported under a 'Knowledge Compendium of Case Studies', a publication by GEAG, Gorakhpur (Nivedita and Wajih 2021).

Thorough and detailed analysis of the above case studies sketches out the following deduction:

(a) Cities fail to operate as individual entities, and their dependence to peri-urban systems for flow and supply of resources—food, water and goods—is fundamental and critical. They are interwoven by delicate yet continuous linkages and cannot be replaced by installation of subsidiary systems.
(b) The peri-urban bias will only be regulated by an effective and transparent governance model. Coordination lacuna and faults often hamper with on-ground action models.
(c) Transformative changes in the mindset of peri-urban beneficiaries are needed to regulate the rapid land-use changes and degradation of peri-urban spaces.
(d) Traditional knowledge systems existing around the peri-urbans have immense potential for increasing the adaptive capacity of city cores as well as the peri-urbans. There is a need to better harness these opportunities.
(e) Among the crucial role players in mainstreaming peri-urban ecosystems in policy models and framework are local government, communities, NGOs, policymakers and most importantly individuals.

4.4 Discussions

The above documented case studies supported by global practices provide clear-cut evidence to on-ground strategies like ecosystem-based approaches (EbA) and nature-based solutions (NbS) but are primarily concerned for addressing short-term benefits (Acharya et al. 2020). Current situation demands a rather holistic and inclusive set of strategies (Fedele et al. 2019). Suitable governance models and

Table 4.1 Comparative analysis of cities under study

City name	City type	Terrain	Geographic landmark	Demographic profile	Disaster proneness	Challenges
Visakhapatnam, Andhra Pradesh	• Tyre-II city	• Coastal	• Bay of Bengal	• Area: 1161 sq. km.	• Cyclone	• Encroachment and peri-urban areas:
	• Commercial			• Population: 4,291,000	• Floods	Conversion of agricultural fields
				• State boundaries: Tamil Nadu, Karnataka, Odisha and Telangana	• Drought	Loss of open spaces for grazing
					• Heat wave	Shortage of fodder
						• Water crisis:
						Loss of water bodies
						Demand-supply gap of drinking water
						Water shortage during summer
						Groundwater contamination
						• Solid waste management: waste segregation
						Open burning a dumping ground
						Drainage shortage
						Groundwater contamination
						• Health challenges: atmospheric pollution due to improper waste disposal Deterio-rating sanitation facilities
						• Marginalization:
						Land-use changes triggered marginalization
						Violation of government rules like CRZ
						Habitat degradation in and around man-grove belts
						Loss of livelihood of informal labourers

(continued)

Table 4.1 (continued)

City name	City type	Terrain	Geographic landmark	Demographic profile	Disaster proneness	Challenges
Panjim, Goa	• Tyre-III city	• Coastal	• Arabian Sea	• Area: 1736 sq. km.	• Coastal erosion	• Marginalization:
	• Residential	• Marshy		• Population: 818,008 State borders: Maharashtra, Karnataka	• Flood	Loss of traditional livelihood aspects
					• Landslide	Loss of traditional 'Khazaans'
						• Biodiversity loss:
						Loss of habitats for local fauna in particular
						• Waste management:
						Unregulated dumping
						• Encroachment in peri-urban areas:
						Conversion of agriculture fields
Shimla, Himachal Pradesh	• Tyre-II city	• Mountainous	• Great Indian Himalayas	• Area: 5131 sq. km.	• Flood	• Water scarcity:
	• Tourism dependent			• Population: 814,010	• Cold wave	Drinking water demand-supply gap
				• State boundaries: Punjab, Jammu and Kashmir, Uttarakhand	• Heat wave	Reducing groundwater levels
					• Thunderstorm	• Health issues:
					• Air pollution	Contamination due to intermixing of sewage and drinking water leading to jaundice epidemic
						• Waste management:
						Improper and faulty facilities for managing wastewater

Source: Developed on the information and inferences drawn from Nivedita and Wajih (2021)

Table 4.2 Ecosystem services offered by the peri-urban spaces

Ecosystem services	Shimla	Visakhapatnam	Panjim
Regulatory services			
Carbon sequestration	✓	✓	✓
Groundwater recharge			
Maintaining groundwater salinity		✓	
Water cycling		✓	✓
Soil fertility			
Erosion prevention/coastal protection		✓	✓
Flood check/water level rise buffer	✓	✓	✓
Climate (local) regulation	✓	✓	✓
Air quality regulation	✓	✓	
Wastewater treatment/sanitation	✓	✓	
Water filtration			✓
Provisioning services			
Agriculture		✓	
Fodder availability		✓	
Fisheries and aquaculture		✓	
Drinking water and water for irrigation	✓	✓	
Livelihood support		✓	✓
Resource availability	✓		
Supporting services			
Grazing land			
Nutrient cycling		✓	✓
Sedimentation load			
Soil formation		✓	✓
Biomass generation			✓
Cultural services			
Recreational values			✓
Health and well-being	✓	✓	
Tourism avenues	✓		
Aesthetic value			

Source: Developed on the information and inferences drawn from Mani et al. (2021)

approaches with effective policy frameworks at its core are necessary to address long-term goals for urban resilience harnessing the potentials that peri-urban spaces and systems provide to urban resilience (Acharya et al. 2020).

4.4.1 Opportunity Analysis for Policy-Planning Framework

Howlett and Cashore (2014) quoted policies as actions containing goals at the same time and providing ways or means to achieve them. Implementation of these policies relies on roots and pathways. Pathways are thus crucial in implementing these policies (Bhardwaj et al. 2020). Previous studies have highlighted various types of pathways of successful implementation of policies: some being participatory and others being technical. Participatory pathways involve community as the key stakeholders and are rather successful in achieving on-ground pilots and actions (Acharya et al. 2020). However, when peri-urbans are in a central role, the regulatory pathways are the most appropriate ones (Narain 2020; Parvez 2020; Verma 2021). The results from the above described three case studies are concomitant to this. The ambiguity in spatial extent, interference by land mafia and marginalization of local inhabitants and immediate beneficiaries challenge the uptake and reliance singularly on participatory pathways. Besides, the ambiguity of administrative status is rather suggestive of the faults in the planning process.

4.4.2 Mainstreaming Pathways

Taking insights from Schröter-Schlaack et al. (2016) and Berghöfer et al. (2011), Bhardwaj et al. (2020) in their paper shared a policy framework which identified key entry points for mainstreaming into policies. On a similar note, Gupta et al. (2017) identified some essential approaches and pathways for subject mainstreaming in developmental policies. However, the context of utilizing these approaches differs. A comparative description of both of these can be seen in Fig. 4.3.

4.4.3 Mainstreaming Challenges

Transformative adaptations address the root causes of vulnerabilities and, thus, are evident in ensuring full proof adaptation strategies (Karki 2017; Acharya et al. 2020). The number of examples highlighted in this paper, are suggestive of present on-ground actions taken up to conserve peri-urban spaces, but misses the prospects of transformative adaptations. This is complementary to the willingness of people to take up strategic actions. The peri-urban socio-economic and ecological diversity and their conflicting niches make community participation difficult. A unique challenge in mainstreaming ecosystem-inclusive action plans is the administrative status of peri-urban areas. The ambiguity that exists in the governance framework makes these spaces easy victims of the ill effects of rapid urbanization (Narain 2020).

Thus, the constant threat posed to these ecosystems from degradation and overexploitation of resources adds on to their fragility and vulnerability. As peri-urban spaces are inhabited by diverse group of people, often there is a need for transformative changes in the mindset of the local (Acharya et al. 2020). Another prominent institutional challenge to mainstream on-ground action as well as developing appropriate planning mechanisms is the lack of coordination between various

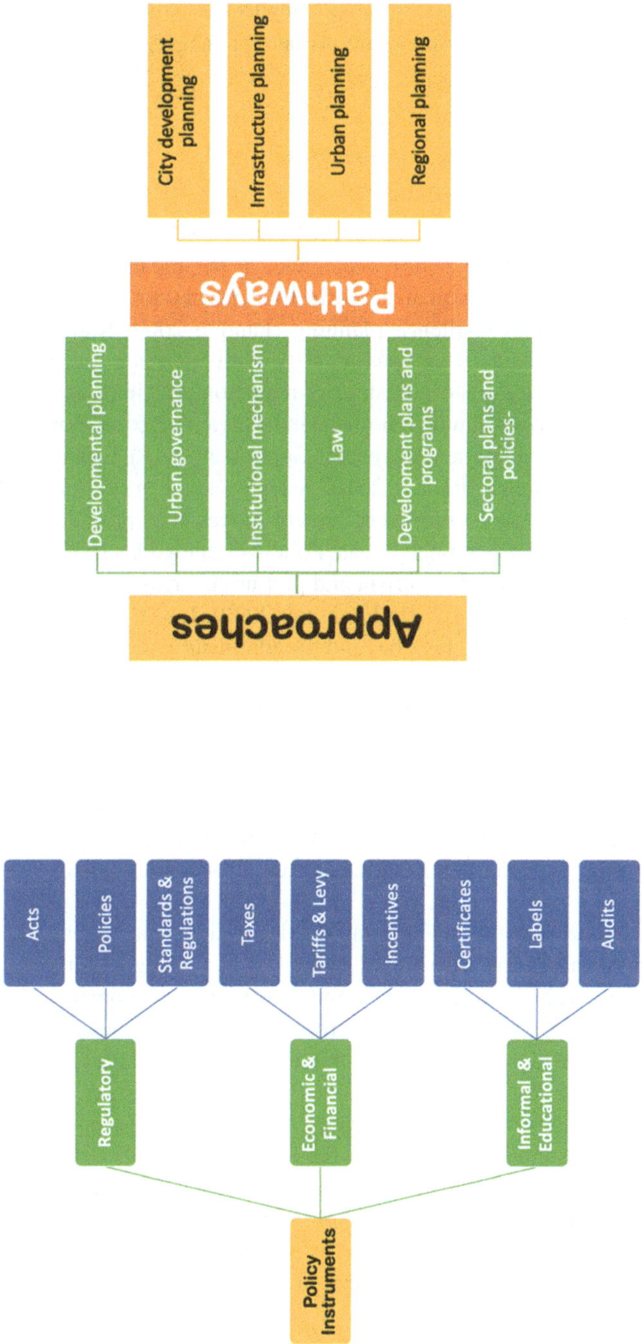

Fig. 4.3 Pathways and approaches to mainstream peri-urban ecosystems in policy-planning framework for urban resilience. Source: (Adapted from Bhardwaj et al. 2020)

line departments. Lacking capacity and knowledge clarity, limited participatory procedures and institutional support are further challenging.

4.4.4 Mainstreaming Opportunities

India's National Action Plan on Climate Change (NAPCC) is an umbrella framework under which multiple policy strategies can be implemented. The missions, thus, formulated under the plan are effective drivers for uptake of peri-urban considerations in urban resilience planning (Pandve 2009; Rattani et al. 2018). Prime Minister's Agenda 10 on Disaster Risk Reduction is a holistic guiding tool for promoting the integration of disaster risk reduction principles in the development sector, and, thus, peri-urban integration and mainstreaming could potentially harness access in the urban developmental process (Gupta et al. 2018). Furthermore, the draft National Land Utilisation Policy, 2013, recognizes the need for developmental projects to preserve the environment and resources that provide the ecosystem services. Similarly, the National Urban Policy Framework (2020) acknowledges the greater risk to ecosystems in urban and peri-urban areas if the business-as-usual models continue (GoI 2018; Circle Economy 2020). It also provides guidelines to states and asks institutions to look beyond the administrative boundaries considering the urban and peri-urban interdependence. The Rurban Mission of GoI (2020) talks about preserving and nurturing rural communities focusing on inclusiveness without compromising the urban-like facilities (Verma 2021). However, India's National Urban Policy Framework 2018 gives states the authority to plan for urban and peri-urban development (Parvez 2020). The State Action Plans on Climate Change (SAPCC) engages the stakeholder departments to come on a common platform in making the State Actions Plans conserving the prominent challenges (Chaturvedi et al. 2019).

India has taken big strides in enabling institutional and legal framework on disaster management and also in implementing the Hyogo Framework of DRR and climate change actions. Policies and law in related domains envisaged for designated planning protocols and a tiered approach at national, sub-national and local levels. It is evident from Fig. 4.4 that district- and local-level policy-planning frameworks offer real-ground opportunities for translating science-policy-planning interface into actions. State action plan for climate change and district disaster management plans have recognized co-benefits (Gupta et al. 2016) and the recent development in the form of district-level environmental action plan (DEAP) on the directive of the National Green Tribunal offer strategic support system and resources for promoting ecosystem protection in peri-urban systems. However, the policy-planning framework for this purpose needs to ensure forward and backward linkages and transformational shifts in approach towards city resilience against climate change disasters.

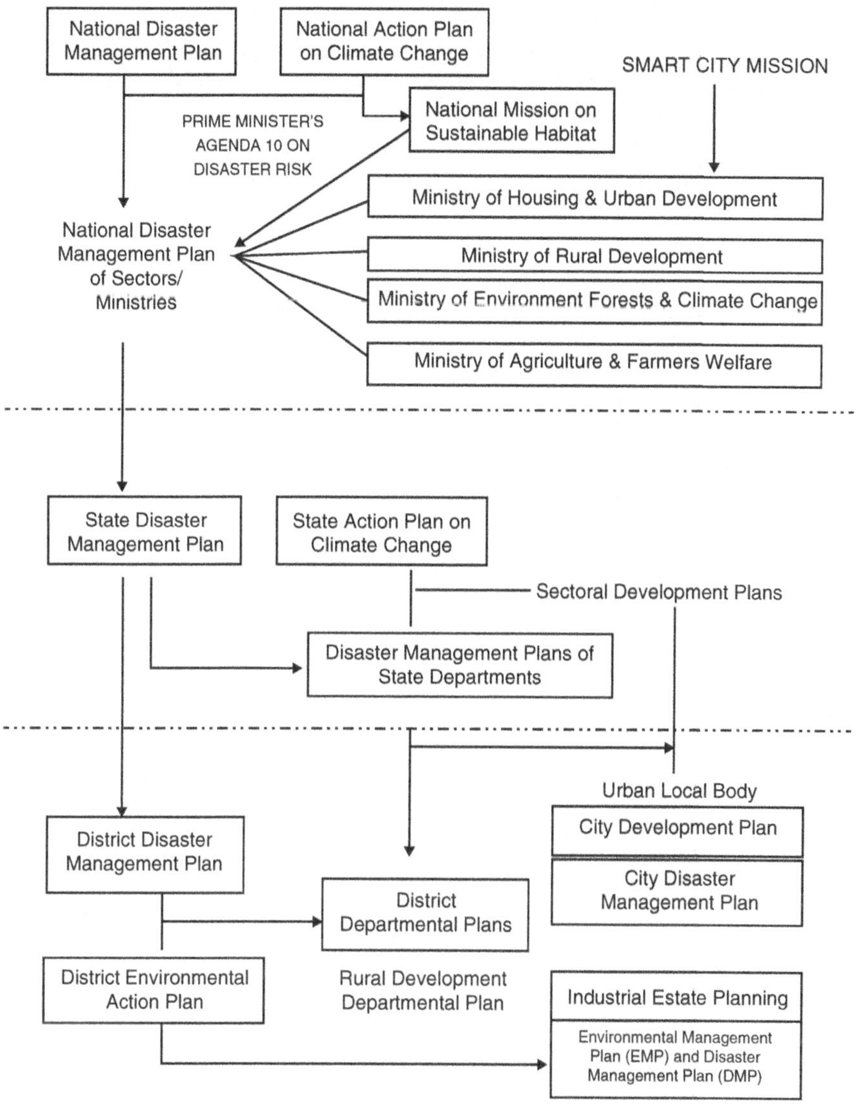

Fig. 4.4 Policy-planning framework for disaster risk reduction, climate change and city resilience at different levels in India, showing opportunities for mainstreaming peri-urban ecosystems

4.5 Conclusion

Despite the simulative discussions on the peri-urbans, the administrative, spatial, socio-economic dynamics pose a tricky environment for imperatives, appropriate policy interventions and further mainstreaming them (Narain 2020; Nivedita and

Wajih 2021). The existing policies at district, sub-national and national level have limited mention of peri-urbans and are only on context basis (Parvez 2020). Framework supporting direct linkages to peri-urban spaces is definitely lacking (Verma 2021). However, it is evident that across India the value of peri-urban systems and in particular the ecosystem services, associated challenges and need for integrated approach to increase urban resilience needs due attention (Nivedita and Wajih 2021).

Witnessing the fast pace of urbanization and growing complexity of peri-urban challenges, it is imperative to consider peri-urban ecosystems in developmental planning for the services, flows and interlinkages they provide (Narain 2020). Recognizing the scope of NbS integration and promotion through peri-urban systems, peri-urban plans and policies should be developed to address the climate change needs. Knowledge-based mainstreaming is easy, but the challenge comes from value-based mainstreaming. For this, a two-pronged approach is required—developing suitable national policies and plans and ground-level implementation of successful mainstreaming products. The transformative approach of mainstreaming peri-urban ecosystem benefits for urban resilience would, therefore, need to utilize the existing policy-planning framework. It is further recommended to look at peri-urban systems as ecological imperatives for urban resilience building (UN Habitat 2020).

References

Acharya P, Gupta AK, Dhyani S, Karki M (2020) New pathways for NbS to realise and achieve SDGs and post 2015 targets: transformative approaches in resilience building. In: Nature-based solutions for resilient ecosystems and societies. Springer, Singapore, pp 435–455

Aijaz RUMI (2019) India's peri-urban regions: the need for policy and the challenges of governance. ORF issue brief (285)

Appeaning Addo K (2010) Urban and peri-urban agriculture in developing countries studied using remote sensing and in situ methods. Remote Sens 2(2):497–513

Berghöfer A, Mader A, Patrickson S, Calcaterra E, Smit J, Blignaut JN, Wit M, Zyl HV (2011) TEEB manual for cities: ecosystem services in urban management. In: The economics of ecosystems and biodiversity. ICLEI/IUCN, New York/Gland

Bhardwaj S, Gupta AK, Dhyani S, Thummarukudy M (2020) Nature-based solution entry points through sectoral policies, strategic instruments and business continuity. In: Nature-based solutions for resilient ecosystems and societies. Springer, Singapore, pp 409–433

Biggs EM, Bruce E, Boruff B, Duncan JM, Horsley J, Pauli N, McNeill K, Neef A, Van Ogtrop F, Curnow J, Haworth B (2015) Sustainable development and the water–energy–food nexus: a perspective on livelihoods. Environ Sci Policy 54:389–397

Bulderberga Z (2014) Rural–urban partnership for balanced spatial development in Latvia. Int J Arch Environ Eng 8(4):925–932

Bush J, Doyon A (2019) Building urban resilience with nature-based solutions: how can urban planning contribute? Cities 95(102):483

Caputo S, Schoen V, Specht K, Grard B, Blythe C, Cohen N, Fox-Kämper R, Hawes J, Newell J, Poniży L (2021) Applying the food-energy-water nexus approach to urban agriculture: from FEW to FEWP (food-energy-water-people). Urban For Urban Green 58(126):934

Carter JG (2018) Urban climate change adaptation: exploring the implications of future land cover scenarios. Cities 77:73–80

Chaturvedi A, Rattani V, Awasthi K (2019) State action plans on climate change need upscaling and capacity enhancement. https://www.cprindia.org/projects/state-action-plans-climate-change-india

Chelleri L, Waters JJ, Olazabal M, Minucci G (2015) Resilience trade-offs: addressing multiple scales and temporal aspects of urban resilience. Environ Urban 27(1):181–198

Chipeniuk R (2004) Planning for amenity migration in Canada. Mt Res Dev 24(4):327–335

Circle Economy (2020) The urban policy framework. https://www.circle-economy.com/resources/the-urban-policy-framework

Das J, Jha S, Goyal MK (2020) On the relationship of climatic and monsoon teleconnections with monthly precipitation over meteorologically homogenous regions in India: wavelet & global coherence approaches. Atmos Res 238(104):889. https://doi.org/10.1016/j.atmosres.2020.104889

Fedele G, Donatti C, Corwin E, Pangilinan M (2019) Nature-based transformative adaptation: a practical handbook. Conservation Union, Arlington

GoI (2018) National urban policy framework. https://smartnet.niua.org/nupf

GoI (2020) The Shyama Prasad Mukherji Rurban Mission (SPMRM). https://rurban.gov.in/index.php/Public_home/about_us

Gomes SL, Hermans LM (2018) Institutional function and urbanization in Bangladesh: how peri-urban communities respond to changing environments. Land Use Policy 79:932–941

Goyal MK, Ojha CSP (2010) Evaluation of various linear regression methods for downscaling of mean monthly precipitation in Arid Pichola Watershed. Nat Resour 1(1):11–18. https://doi.org/10.4236/nr.2010.11002

Goyal MK, Ojha CSP (2012) Downscaling of precipitation on a lake basin: evaluation of rule and decision tree induction algorithms. Hydrol Res 43(3):215–230. https://doi.org/10.2166/nh.2012.040

Goyal MK, Panchariya VK, Sharma A, Singh V (2018) Comparative assessment of SWAT model performance in two distinct catchments under various DEM scenarios of varying resolution, sources and resampling methods. Water Resour Manag 32(2):805–825. https://doi.org/10.1007/s11269-017-1840-1

Gupta AK, Katyal S, Chopde S, Wajih SA, Kumar A (2016) Climate-smart district disaster management plan as effective tool for implementing state action plan on climate change: lessons from three states in India. Thematic paper. GEAG India and ISET Colorado USA (India of CE), published under CDKN. https://doi.org/10.13140/RG.2.2.28547.91686

Gupta AK, Singh S, Wajih SA, Mani N, Singh AK (2017) Urban resilience and sustainability through peri-urban ecosystems: integrating climate change adaptation and disaster risk reduction. Gorakhpur Environmental Action Group, Gorakhpur

Gupta A, Chopde S, Singh S, Wajih S, Katyal S (2018) Prime minister's agenda 10. India's disaster risk management roadmap to Climate resilient and Sustainable Development, New Delhi

Howlett M, Cashore B (2014) Conceptualizing public policy. In: Engeli I, Allison CR (eds) Comparative policy studies. Research methods series. Palgrave Macmillan, London

Karki M (2017) Need for transformative adaptation in South Asia. Int J Multidiscip Stud 4(2):1–7

Mani N, Wajih SA, Gupta AK, Narain V, Prakash A (2021) Peri-urban ecosystems and urban resilience: knowledge compendium of case studies. Gorakhpur Environmental Action Group, Climate Development and Knowledge Network, Gorakhpur (U.P.), Cape Town, p 46. https://nidm.gov.in/PDF/pubs/Peri-urban_Knowledge_Compendium2021.pdf

MEA (2005) Overview of the Millennium Ecosystem Assessment. https://www.millenniumassessment.org/en/About.html

Moss LA (ed) (2006) The amenity migrants: seeking and sustaining mountains and their cultures. CABI, Wallingford

Narain V (2020) Building the resilience of periurban communities to the impacts of climate change and urbanization. Int J Disaster Resilience Built Environ 12:115–125

Nel JL, Le Maitre DC, Nel DC, Reyers B, Archibald S, Van Wilgen BW et al (2014) Natural hazards in a changing world: a case for ecosystem-based management. PLoS One 9(5):e95942

Nivedita M, Wajih SA (2021) Peri-urban ecosystems and urban resilience: knowledge compendium of case studies. Gorakhpur Environmental Action Group, Gorakhpur. Climate Development and Knowledge Network, Cape Town, South Africa, p 46

Pandve HT (2009) India's national action plan on climate change. Indian J Occup Environ Med 13(1):17

Parvez S (2020) Policy brief suggests integrated approach for inclusive development of peri-urban areas. https://talkofthecities.iclei.org/policy-brief-suggests-integrated-approach-for-inclusive-development-of-peri-urban-areas/

Perlik M (2011) Alpine gentrification: the mountain village as a metropolitan neighbourhood. New inhabitants between landscape adulation and positional good. Journal of Alpine Research | Revue de géographie alpine 99(1):1–13

Poonia V, Goyal MK, Gupta BB, Gupta AK, Jha S, Das J (2021) Drought occurrence in Different River Basins of India and blockchain technology based framework for disaster management. J Clean Prod 312(127):737. https://doi.org/10.1016/j.jclepro.2021.127737

Prakash A (2014) The periurban water security problem: a case study of Hyderabad in Southern India. Water Policy 16(3):454–469

Rasul G, Sharma B (2016) The nexus approach to water–energy–food security: an option for adaptation to climate change. Clim Policy 16(6):682–702

Rattani V, Venkatesh S, Pandey K, Jitendra, Kukreti I, Somvanshi A, Sangomla A (2018) India's national action plan on climate change needs desperate repair. https://www.downtoearth.org.in/news/climate-change/india-s-national-action-plan-on-climate-change-needs-desperate-repair-61884

Ribeiro PJG, Gonçalves LAPJ (2019) Urban resilience: a conceptual framework. Sustain Cities Soc 50(101):625

Rosenzweig C, Solecki W, Romero-Lankao P, Mehrothra S, Shakal S, Bowman T, Ali Ibrahim S (2015) ARC3. 2 summary for city leaders. Climate change and cities. In: Second assessment report of the urban climate change research network, urban climate change research network. Columbia University, New York

Schröter-Schlaack C, Albert C, Christina VH, Hansjürgens B, Krätzig S, Albert I (2016) Ecosystem Services in rural areas - basis for human wellbeing and sustainable economic development. Naturkapital Deutschland - TEEB DE, Leipzig

Shrestha A (2019) Urbanizing flows: growing water insecurity in peri-urban Kathmandu Valley, Nepal (Doctoral dissertation, Wageningen University)

Singh B (2018) Peri urban reality in India. http://bwsmartcities.businessworld.in/article/Peri-Urban-Reality-in-India/03-01-2018-136246/

Singh S, Nair SS, Gupta AK (2013) Ecosystem Services for disaster risk reduction: a case study of wetland in East Delhi region, India. Global J Hum Soc Sci 13(4):1–12

Tiwari P (2019) Dynamics of peri urban areas of Indian Cities

UN (2016) Department of Economic and Social Affairs Sustainable Development. https://sdgs.un.org

UN (2017) World population projected to reach 9.8 billion in 2050, and 11.2 billion in 2100. https://www.un.org/development/desa/en/news/population/world-population-prospects-2017.html

UN Habitat (2020) World cities report 2020; the value of sustainable urbanization. https://unhabitat.org/sites/default/files/2020/10/wcr_2020_report.pdf

Verma S (2021) Peri-urban ecosystems: the potential for a planned approach in India. CDKN, New Delhi

Wei S, Bai ZH, Chadwick D, Hou Y, Qin W, Zhao ZQ et al (2018) Greenhouse gas and ammonia emissions and mitigation options from livestock production in peri-urban agriculture: Beijing—a case study. J Clean Prod 178:515–525

Chapter 5
Urban Forest and Ecosystem Services Intercede Urban Habitat in Delhi

Vartika Awasthi, Pritha Acharya, and Anil Kumar Gupta

Abstract Delhi is viewed as the second most crowded city in India, with a populace of 11 million individuals as per 2011 statistics. This unplanned or planned urbanized expansion has taken a toll on the green infrastructure and caused climatic changes over a period of 20 years. In this study, we used Landsat imagery of the years 2000, 2010, and 2020 to do a comparative study, by mapping the LULC, LST, and NDVI. This study is also supplemented with field-based surveys and questionnaires done with random field selection in the area of study. This research helps us in understanding the importance of urban forestry, the effect of urbanization on urban habitat, and the biodiversity in India's largest city.

Keywords Urban green · Ecosystem services

5.1 Introduction

Over the twentieth hundred years, the total world populace has expanded by more than 400% (Cavender and Donnelly 2019). In 2019, the total populace was recorded to be 767.35 crores, and it was assessed that by 2025, 66% of the total populace will be gathered in metropolitan regions, which would all in all increases the need to build the significance of giving decent natural quality and spaces as well as food security and versatile food frameworks (Haberman et al. 2014). The exponential increase in urbanization rate is joined by natural debasement (prompting air contamination, heat island effects, soil disintegration, territory and untamed life

V. Awasthi (✉)
Forest Research Institute (FRI), Dehradun, Uttarakhand, India

P. Acharya
ECDRM Division, National Institute of Disaster Management (NIDM), Delhi, India

A. K. Gupta
National Institute of Disaster Management (Ministry of Home Affairs, Government of India), New Delhi, India

A. K. Gupta et al. (eds.), *Ecosystem Restoration: Towards Sustainability and Resilient Development*, Disaster Resilience and Green Growth,
https://doi.org/10.1007/978-981-99-3687-8_5

misfortune, fossil fuel by-products, clamor levels, and so on) accordingly affecting human well-being and personal satisfaction (Cavender and Donnelly 2019; Gurjar et al. 2008; Roy et al. 2012). Especially in the tropics where the impacts are more severe as most developing countries are located in that region and it was predicted that almost 90% of urbanization would occur in coming decades in these areas (Emmanuel 2005; Vailshery et al. 2013). The escalating global population front-line the imperative to re-calibrate how metropolitan areas are constructed and populated (Russo et al. 2017).

Therefore, the worldwide dispersal of the unconstrained vegetation that rules deserted metropolitan land is as much a social peculiarity as it is a natural one (Mack and Erneberg 2002; Rees and Wackernagel 2008). This urban vegetation comprises all herbs, shrubs, lawns, trees, and other vegetation in cities and is characterized by the inclusion of natural resources with human developments (Baró et al. 2014; Nowak 1994; Poonia et al. 2021). They look after the source to ameliorate environmental quality, enhance the well-being of individuals and communities, provide a large range of services to individuals and communities, and determine a more healthy environment (Nowak et al. 2001).

The word "urban forestry" elucidates an image of the concrete jungle of large spread-out metropolitan cities all over the world (Goyal et al. 2022). However, it has a pleasant insinuation with the addition of all trees on public and private underused land, as well as those in utility and traffic corridors, as well as all street trees, residential trees, park trees, and greenbelt vegetation (Chaturvedi et al. 2013). Consequently, it is defined as the art, science, and technology of administrating trees and forest resources in and around urban ecosystems for the physiological, sociological, economic, and aesthetic benefits trees provide for society (Chaudhry et al. 2011). Therefore, urban forestry provides a direct relationship between humans and nature focusing on preserving and improving urban forest ecosystems and the range of positive ecological services they provide (Ahmad and Goparaju 2016; Paul and Nagendra 2015; Steenberg et al. 2017).

The government of India committed to the Bonn Challenge, which calls for the restoration of 13 million hectares of degraded land by 2020 and a further 8 million hectares by 2030, at the 2015 Conference of the Parties (CoP) of the United Nations Framework Convention on Climate Change (UNFCCC) in Paris; therefore, urban forestry is not only helping to mitigate degraded environment but also helping the country to fulfill the pledge made on the international stage and acts as a barrier to various natural disasters.

5.2 Methodology

5.2.1 Study Area

Delhi lies under geographical coordinates of 28°24′17″ N and 28°53′00″ N longitude and 76°50′24″ E and 77°20′37″ E latitude (Paul and Nagendra 2015) (Fig. 5.1).

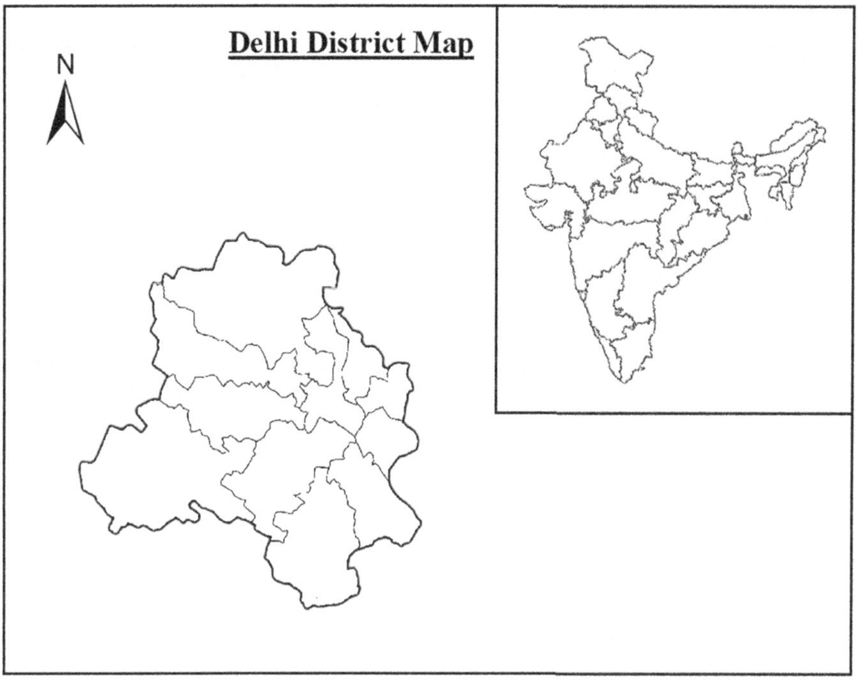

Fig. 5.1 Delhi administrative map

NCTD is spread out to approximately 1484 km (i.e., 573 square mi), with an elevation of 200–250 m (650–820 ft.), and lies at the transverse of river Yamuna from north to south. Delhi with a population of 11 million is stated as India's second most populated metropolitan city (*Census 2011.Pdf* 2016). The city supports xerophytic vegetation, with thorny deciduous and fleshy evergreen species, in humid subtropical climatic conditions, with lengthy, sweltering summers and brief, dreary winters (Paul and Nagendra 2015). The yearly temperature differs from around 3 °C in the colder time of year to around 45 °C in the late spring. The storm season goes from July to September, with a yearly precipitation of 400–600 nm.

Delhi has a continuously expanding urban infrastructure with increasing population density, which is supported by a broad spectrum of city forests like the Asola Bhatti wildlife sanctuary area and Delhi ridge which are considered as lungs of the city to remnants of forest and pockets of highly modified artificial landscaped parks containing diverse biodiversity in them (Das et al. 2020; Goyal et al. 2018; Goyal and Ojha 2010, 2012).

The city has approximately about 195.44 sq. km of area under tropical dry deciduous and tropical thorny forest cover, with about 129 sq. km area under tree cover supporting 63 different plants specified (Forest Survey of India 2011), there fore administratively divided into 9 districts (Paul and Nagendra 2015) and out of which 2 locales were chosen as the area of study are; Northwest Delhi and South Delhi, Northwest Delhi is primarily thought of as an agricultural plain area, covering 442.84 km^2, while South Delhi, covering 250 km^2, is the foothills of the Aravalli range, supporting a significant amount of biodiversity.

5.2.2 Data and Material

This study is carried out using a ground-based survey and questionnaire along with satellite image data.

Satellite image data of Landsat 7 + ETM for the years 2000 and 2010 and Landsat 8 OLI + TIRS 2020 was downloaded from the USGS Earth Explorer (EarthExplorer (usgs.gov)) (Hegazy and Kaloop 2015; Kumar et al. 2012; Kumari et al. 2018) (Table. 5.2).

5.2.3 Method

5.2.3.1 Survey Assessment

According to Clark et al. (1997), urban forest provides important services rather than goods (e.g., reducing environmental contamination, providing social and mental well-being, improving water quality, etc.), which indirectly or directly benefits the community residents in urban habitat, and therefore, urban forests are majorly growing on private lands with human intervention. Therefore, to execute comparative research in the area of study, a formal survey questionnaire is formed with questions regarding people's perception of the vegetation and biodiversity in their locality. The survey includes open- and close-ended questions in Likert scale format, which address the awareness of people toward the aesthetic and positive as well as negative views of having a plantation. A field survey is used to gather data, which is then processed by basic, intricate, and statistical procedures to produce results that are shown in the form of a table or chart (Bhalla and Bhattacharya 2015; Clark et al. 1997) (Table 5.1).

The data collected on the perception of people about the greenness of their locality through a linear scale represents the difference in opinion and contrasting perceptions about people of different localities in the area of study through a bar chart (Fig. 5.2).

The percentage of response was measured by quantifying the responses by the respondents, and these four statements are used to assess the desired information. These statements were based on literature reviews and on-ground surveys of the areas, and each of these statements is correspondingly provided with values as they were given in the survey with the response 1 indicating "strong agreement," 2 indicating "mild agreement," 3 indicating "neutral response," and 4 indicating

Table 5.1 Relative importance of statement by all the respondents conducted in both study areas

	Responses				Weightage
Statement	Strongly agreed (1)	Agreed (2)	Neutral (3)	Disagreed (4)	Mean
Trees in urban areas are important for human health	0.8	0.4	0.3	0	1.3
Trees in urban areas reduce air pollution and assist in odor control	0.7	0.6	0.3	0	1.5
Trees in urban areas increase their aesthetic value	0.7	0.6	1.2	0.4	2.3
Roadside plantations are important in urban areas	0.7	0.4	0.3	0	1.4

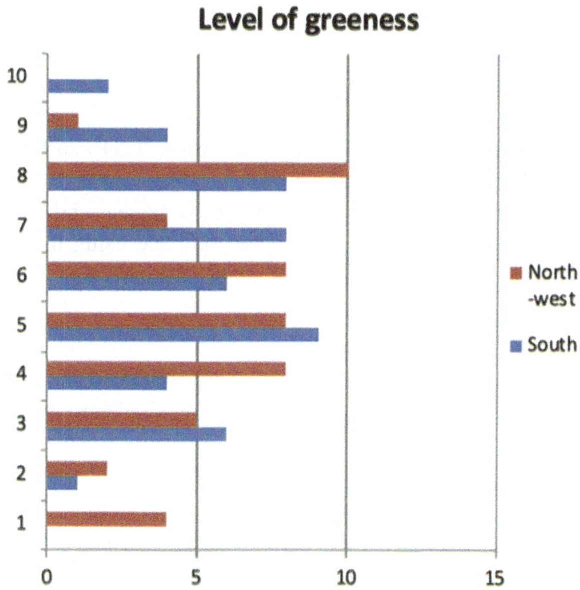

Fig. 5.2 People's perception in South and Northwest (study area)

"disagreement" to the statements, out of which respondents are free to choose one according to their perception of the area corresponding to the respective statement. As each statement carries a different value to each individual surveyed, the weighted mean ($G\mathcal{D}$) was calculated to the considered relative importance of all four statements (Bhalla and Bhattacharya 2015).

$$G\mathcal{D} = \frac{\sum_{i=1}^{n} \omega_i X_i}{\sum_{i=1}^{n} \omega_i}$$

where:
GD = weighted average
n = no. of terms to be averaged
ω_i = no. of weight applied to x values
X_i = data values to be averaged

5.2.3.2 Land Surface Temperature (LST)

LST is a major parameter to study environmental processes and to monitor responses such as the interpretation of change in climate, surface energy, and water balance through satellite imagery data (Fig. 5.3). The main data source concerns Landsat 7 and Landsat 8 acquisitions, provided by USGS (Parastatidis et al. 2017) (Table 5.2).

Retrieval of LST from Landsat 7

According to the Landsat data user book, LST can be retrieved in four steps. In the first step, the image pixels are converted from digital number (DN) of thermal infrared band (band 6) into spectral radiance, and in the second step, the spectral radiance is converted into sensor radiance using the equations given below (Ihlen and USGS 2019; Kumar et al. 2012; Kumari et al. 2018; Parastatidis et al. 2017):

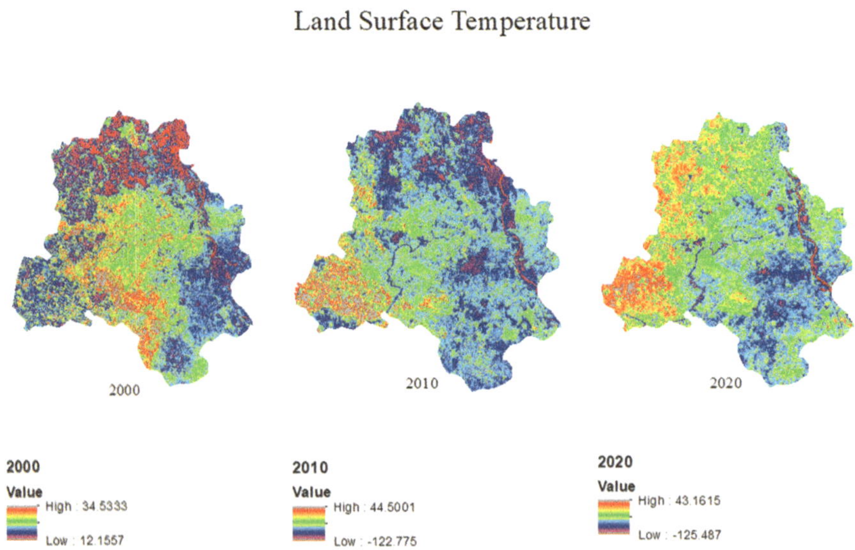

Fig. 5.3 Land surface temperature of Delhi from the years 2000, 2010, and 2020

Table 5.2 Satellite information

S. no	Satellite	Sensors	Year	No of spectral bands	Spatial resolution (m)	Source
1	Landsat 7	Enhanced Thematic Mapper Plus (ETM)	2000	8	30	USGS
2	Landsat 7	Enhanced Thematic Mapper Plus (ETM)	2010	8	30	USGS
3	Landsat 8	Operational Land Imager (OLI)	2020	11	15	USGS

$$L_\lambda = \min + (\max - \min)^* \mathrm{DN}/255$$

where:
L_λ = spectral radiance
min = 1.238 (spectral radiance of DN value 1)
max = 15.600 (spectral radiance of DN value 255)
DN = digital number

$$L_\lambda = \frac{(L_{MAX\lambda} - L_{MIN\lambda}) * (Q_{CAL} - Q_{CALMIN}) + L_{MIN\lambda}}{(Q_{CALMAX} - Q_{CALMIN})}$$

where:
L_λ = spectral radiance
Q_{CAL} = quantized calibrated pixel value in DN
$L_{MAX\lambda}$ = spectral radiance scale to Q_{CALMAX} in (Watts/(m^{2} * sr * μm))
$L_{MIN\lambda}$ = spectral radiance scale to Q_{CALMIN} in (Watts/(m^{2} * sr * μm))

Q_{CALMAX} = the maximum quantized calibrated pixel value (corresponding to L_{MAX}) in DN = 255

Q_{CALMIN} = the minimum quantized calibrated pixel value (corresponding to L_{MIN}) in DN = 1 (Table 5.3)

In the third step, sensor effective radiance is converted into effective sensor brightness, which is also known as black body temperature. This is done using Planck's inverse function.

Table 5.3 Radiance maximum and minimum obtained from metadata file from satellite imagery

Band no.	Satellite/sensor	L_{MAX}	L_{MIN}
6.1	Landsat 7/ETM + high gain	12.650	3.200
6.2	Landsat 7/ETM + low gain	17.040	0.000

$$T_B = \frac{K_2}{\ln(K_1/L_\lambda + 1)}$$

Where,
T_B = Effective at satellite temperature in Kelvin
K_2 = Calibration constant 2
K_1 = Calibration constant 1
L_λ = Spectral radiance
ln = natural logarithm
In this last step of the calculation of LST, the temperature is converted from Kelvin to Celsius from the equation given below:

$$T\,(^\circ\mathrm{C}) = T\,(\mathrm{K}) - 273:15$$

Retrieval of LST for Landsat 8 OLI/TIRS

Conversion to Sensor Spectral Radiance

Landsat 8 Handbook describes the steps for retrieval of LST from Landsat 8 OLI/TIRS (Kumari et al. 2018; Zanter 2015). In step one, the band data is converted into spectral radiance using the radiance rescaling factor provided in the metafile of satellite imagery. Using the equation given below:

$$L_\lambda = M_L * Q_{CAL} + A_L$$

Where,
L_λ = Spectral radiance (W/(m^2 * sr * μm))
M_L = Band-specific multiplicative rescaling factor from the metadata (RADIANCE_MULTI_BAND_*x*)
A_L = Band-specific additive rescaling factor from the metadata (RADIANCE_ADD_BAND_*x*)
where *x* is the band number
Q_{CAL} = Quantized and calibrated standard product pixel value (DN)

Conversion to Top-of-Atmosphere (TOA) Reflectance

The band data is converted into reflectance using the reflectance coefficient provided in the metadata file. The following equation is used to convert 1 DN value into TOA reflectance:

$$\rho\lambda' = M_\rho Q_{\mathrm{CAL}} + A_\rho$$

where:

$\rho\lambda'$ = top-of-atmosphere planetary spectral reflectance, without correction for solar angle (unitless)

M_ρ = reflectance multiplicative scaling factor for the band (REFLECTANCE_MULT_BAND_*x* from the metadata)

A_ρ = reflectance additive scaling factor for the band (REFLECTANCE_ADD_BAND_*x* from the metadata), where *x* is the band number

Q_{CAL} = quantized and calibrated standard product pixel values (DN)

Conversion to TOA Brightness Temperature

In the third step, band data is converted from spectral radiance to TOA brightness temperature using the thermal constant provided in the metadata file (.MTL file).

$$T_{\mathrm{B}} = K_2 / \ln(K_1/L_\lambda + 1) - 273.15$$

where:

T_{B} = TOA brightness temperature (Kelvin)

L_λ = TOA spectral radiance (Watts/(m^2 * sr * μm))

K_1 = band-specific thermal conversion constant from the metadata (K_1_CONSTANT_BAND_*x*)

K_2 = band-specific thermal conversion constant from the metadata (K_2_CONSTANT_BAND_*x*)

Subtraction of 273.15 converts T_{B} from Kelvin to Celsius.

Satellite/sensor	K_1 constant	K_2 constant
Landsat-7/ETM + SLC	666.09	1282.71
Landsat-8 OLI/TIRS	774.8853	1321.0789

In LST estimation, emissivity is a critical variable, as a small uncertainty in it could lead to major errors in the LST (up to 1 k) depending on the setting of the sensor, the climatologically conditions, and the geographical setting of the area (Parastatidis et al. 2017).

NDVI Method for Emissivity Correction

$$\mathrm{NDVI} = (\mathrm{NIR} - \mathrm{RED})/(\mathrm{NIR} + \mathrm{RED})$$

NDVI-based emissivity is calculated from the Landsat visible and near-infrared bands and typical emissivity values. The amount of vegetation present is an important factor for LST estimation; therefore, NDVI is an essential component in LST calculation (Kumari et al. 2018; Parastatidis et al. 2017).

The proportion of the vegetation P_V is calculated by:

$$P_V = (\mathrm{NDVI} - \mathrm{NDVI}_{\mathrm{MIN}})/(\mathrm{NDVI}_{\mathrm{MAX}} + \mathrm{NDVI}_{\mathrm{MIN}})^2$$

where,
P_V = proportion of vegetation
NDVI = DN values from NDVI Image
$\mathrm{NDVI}_{\mathrm{MIN}}$ = Minimum DN values from NDVI Image
$\mathrm{NDVI}_{\mathrm{MAX}}$ = Minimum DN values from NDVI Image

Land surface emissivity is a proportionality factor that scales blackbody radiance to predict emitted radiance, and it is the efficiency of transmitting thermal energy across the surface into the atmosphere; therefore, it is important to calculate LSE to estimate LST (Jimenez-Munoz et al. 2009; Kumari et al. 2018).

$$\varepsilon = 0.004^{*}\ P_v + 0.986$$

where,
ε = Land surface emissivity
P_V = Proportion of vegetation

Table 5.4 Wavelength of the emitted radiance of different satellites/sensors

Satellite	Band	λ (μm)
Landsat 4, 5 and 7	6	11.45
Landsat 8	10	10.8
Landsat 8	11	12

LST Estimation

This is the final step, when the TOA brightness temperature is converted to LST, using the following equation (Table 5.4):

$$\text{LST} = T_{\text{B}}/(1 + (\lambda^{*} T_{\text{B}}/c^{2})) * \ln\varepsilon$$

where,
λ = wavelength of emitted radiance
$c^2 = h*c/s$ = 104,388*10^{-2} mk
h = plank's constant = 6.626 * 10^{-34} Js
s = Boltzmann constant = 1.38 * 10^{-23} J/k
c = velocity of light = 2.998 * 10^{8} m/s

5.2.3.3 Land Use and Land Cover

For the dynamic assessment of the study area, multi-temporal data from three different years (2000, 2010, and 2020) have been interpreted, to understand the changes that occurred over the years in the urban structure, vegetation index, and overall LULC changes in the area (Mukherjee and Sarma 2016; Rahman et al. 2011) (Figs. 5.4 and 5.5).

The study is based on data collection which includes the topographical sheets of scale 1:50,000 and multispectral Landsat satellite imagery data of the past three decades collected from the Survey of India and USGS, respectively. Table 5.5 shows the source of data collected from satellite imagery (Gour et al. 2014; Raju et al. 2018).

Satellite bands are layer stacked to obtain false color composite (FCC) image which on sub-setting extract our area of study. Similarly, the topography sheets of Delhi with different parts were scanned, and it is mosaics in ERDAS Imagine 2015. Then the area of interest (AOI) is subset through that mosaic toposheet image which is geo-referenced (Landsat image) and geometrically to a common Universal Transverse Mercator (UTM) WGS84 coordinate system and was resampled to its spatial resolution using the nearest neighborhood algorithm (Javed Mallick and Bharath 2008; Rahman et al. 2012).

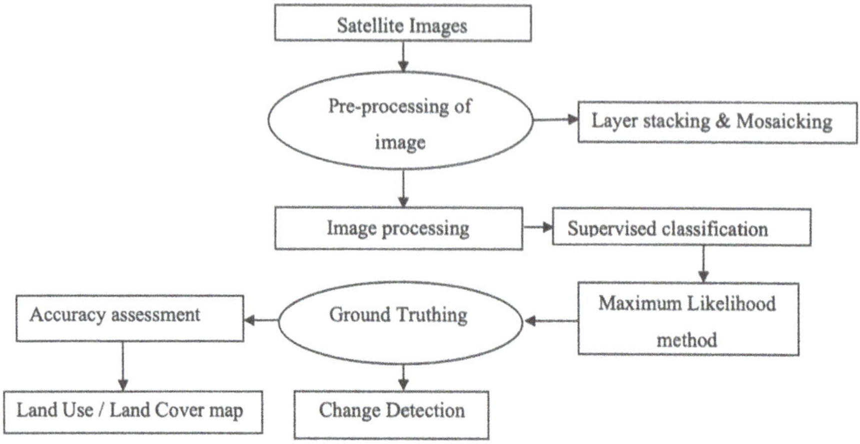

Fig. 5.4 Flowchart methodology for land use/land cover map

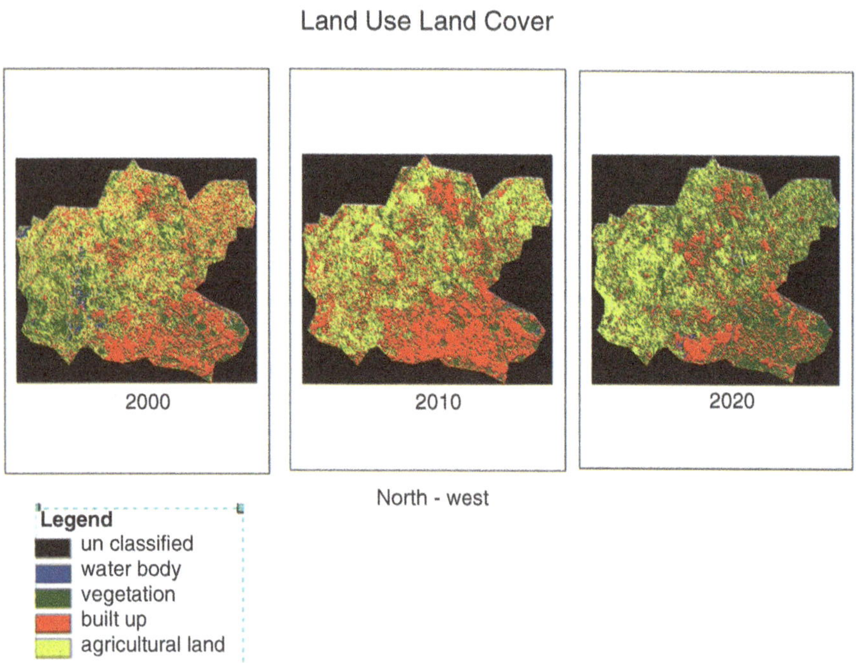

Fig. 5.5 Land use/land cover map of Northwest Delhi

The classification method depends on the resolution of the image and the availability of classification software. In this study, supervised as well as unsupervised classification approaches are applied to classify the pre-processed images.

Table 5.5 Detail of Landsat data collected from USGS

S. no	Satellites/sensor	Date	Path row
1	Landsat 7 + ETM	2000-02-19	146/040
2	Landsat 7 + ETM	2010-01-29	146/040
3	Landsat 8 + OLI	2020-05-09	146/40

Unsupervised Classification

The ISODATA clustering approach with the minimum spectral distance formula is used by the ERDAS envision software to create clusters for unsupervised classifications. The means of these clusters are shifted every time the clustering is repeated, starting with either arbitrary cluster means or the means of an existing signature set. The process continues until the total number of iterations or the number of pixels in each class changing by less than the chosen pixel change threshold is reached. The new cluster means are utilized for the following iteration (Gour et al. 2014; Xiuwan et al. 1999).

Supervised Classification

The satellite image has undergone extensive analysis in supervised classification to determine the likely land use classes and their associated range of reflectance values (DN values). The land cover has been divided into five classes for each time period: unclassified, agricultural land, vegetation, built-up, and water body (Fig. 5.6). Spectral profiles have been created to determine the separability and relative difference in pixel values of different land use classes in different spectral bands. Therefore, we use the ERDAS Imagine software to do supervised classification using the MLC (maximum likelihood) methodology. For better map accuracy, locations with questions were ground verified using Google Earth imagery, and incorrectly classified places were fixed. The quality of maps would not be met without accuracy evaluation. Consequently, an accuracy evaluation has been done (Gour et al. 2014; Jat et al. 2008; Paul and Nagendra 2015; Raju et al. 2018).

5.2.3.4 Accuracy Assessment

The final step in the classification process is accuracy assessment (Foody 2002; Rwanga and Ndambuki 2017). In remotely sensed data, the term accuracy is used to express the "degree of correctness or righteousness." A classified thematic map is considered accurate if it provides an unbiased representation of the land cover of the studied area; therefore, classification accuracy is typically taken to mean the degree to which the derived image classification agrees with reality or conforms to the "truth." A classification error is, thus, some discrepancy between the situation depicted on the thematic map and reality (Foody 2002).

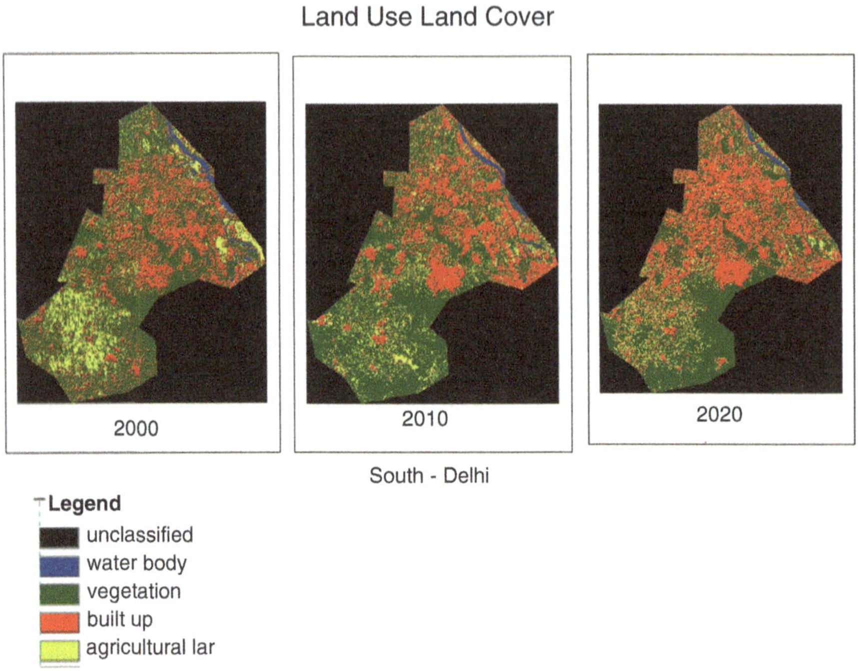

Fig. 5.6 Land use/land cover map of South Delhi

The ideal way to perform error evaluation is to locate reference test pixels in the study area. Both stratified and random sampling provide statistical and practical application validity. Henceforth, the mostly stratified random sampling method is preferred to select a minimum number of samples from each stratum, and therefore these locations of the reference test sites are determined through ground truthing or by Google Earth Engine. This reference test information is collected from the random sites and compared on a pixel-by-pixel basis with the information presented in thematic maps and summarized in the cells of the error matrix (Xiuwan et al. 1999) (Tables 5.6 and 5.7).

Table 5.6 Accuracy assessment of Northwest Delhi (2000, 2010, and 2020)

	Water body	Vegetation	Built up	Agricultural land	User (total)	User accuracy
Accuracy assessment of Northwest Delhi (2000)						
Water body	4	0	0	1	5	80
Vegetation	0	10	0	0	10	100
Built up	0	1	44	3	48	91.6
Agricultural land	0	1	0	36	37	97.2
Producer (total)	4	12	44	40	100	
Producer accuracy	100	83.3	100	90		
Overall accuracy	94					
Kappa accuracy	90.4					
Accuracy assessment of Northwest Delhi (2010)						
Water body	2	1	0	2	5	40
Vegetation	0	10	0	0	10	100
Built up	1	0	56	4	61	91.8
Agricultural land	0	1	0	23	24	95.8
Producer (total)	3	12	56	29	100	
Producer accuracy	66.7	83.3	100	79.3		
Overall accuracy	91					
Kappa accuracy	84.3					
Accuracy assessment of Northwest (2020)						
Water body	5	1	2	0	8	62.5
Vegetation	0	12	1	1	14	85.7
Built up	0	1	50	1	52	96.2
Agricultural land	0	9	0	17	26	65.4
Producer (total)	5	23	53	19	100	
Producer accuracy	100	52.2	94.3	89.5		
Overall accuracy	84					
Kappa accuracy	74.9					

Table 5.7 Accuracy assessment of South Delhi (2000, 2010, and 2020)

	Water body	Vegetation	Built up	Agricultural land	User (total)	User accuracy
Accuracy assessment of South Delhi (2000)						
Water body	12	2	1	0	15	80
Vegetation	0	30	1	0	31	96.8
Built up	0	6	33	1	40	82.5
Agricultural land	0	2	1	11	14	78.6
Producer (total)	12	40	36	12	100	
Producer accuracy	100	75	91.7	91.7		
Overall accuracy	86					
Kappa accuracy	79.9					
Accuracy assessment South Delhi (2010)						
Water body	7	0	0	1	8	87.5
Vegetation	0	38	3	1	42	90.5
Built up	0	0	42	0	42	100
Agricultural land	0	0	1	7	8	87.5
Producer (total)	7	38	46	9	100	
Producer accuracy	100	100	91.3	77.8		
Overall accuracy	94					
Kappa accuracy	90.5					
Accuracy assessment of South Delhi (2020)						
Water body	7	0	0	1	8	87.5
Vegetation	0	38	1	0	39	97.4
Built up	0	0	39	0	39	100
Agricultural land	0	0	0	14	14	100
Producer (total)	7	38	40	15	100	
Producer accuracy	100	100	97.5	93.3		
Overall accuracy	98					
Kappa accuracy	97.01					

$$\text{User Accuracy} = \frac{\text{Number of correctly classified pixel in each category}^{*}100}{\text{Total number of classified pixels}}$$

$$\text{Producer Accuracy} = \frac{\text{Number of correctly classified pixels in each category}^{*}100}{\text{Total number of reference pixels in the category}}$$

$$\text{Overall Accuracy} = \frac{\text{Total number of currently classified pixels}^{*}100}{\text{Total number of reference pixels}}$$

$$\text{Kappa Coefficient} = \frac{(\text{TS}^{*}\text{TCS}) - \Sigma(\text{Column total}^{*}\text{Row total}) * 100}{\text{TS}^{2} - \Sigma(\text{Column total} - \text{Row total})}$$

5.2.3.5 Change Detection

Landsat multi-spectral satellite images of the years 2000 and 2020, respectively, have been used for generating LULC maps, which undergoes accuracy assessment based on ground truthing or by using Google Earth Engine (Rahman and Netzband 2007) (Fig. 5.7). Change detection represents and assesses temporal differences between two satellite images of different periods. Therefore, change analysis is an important technique to identify various changes occurring in different classes of land use like an increase in built-up structure, a decrease in vegetation land, etc. (Gour et al. 2014; Hegazy and Kaloop 2015) (Table 5.8).

5.2.3.6 Normalized Difference Vegetation Index (NDVI)

NDVI is a measure of the amount and ruggedness of vegetation at the surface. The spatial and temporal distribution of vegetation communities, the vegetation biomass, and the extent of land degradation in various ecosystems are determined using NDVI (Kumar et al. 2012; Pettorelli et al. 2005).

In Landsat satellite image data, NDVI is generated from the red (third) and near-infrared (fourth) bands; therefore, the ratios of these two bands are used for assimilation of NDVI because absorption by chlorophyll of these two bands of the electromagnetic spectrum is the highest. However, green leaves have a reflectance of 20% or less in the 0.5–0.7 μm range (green to red) and about 60% in the 0.7–1.3 μm range (near infrared), and the value is then normalized to $-1 <= \text{NDVI} <= 1$ to partially account for differences in illumination and surface

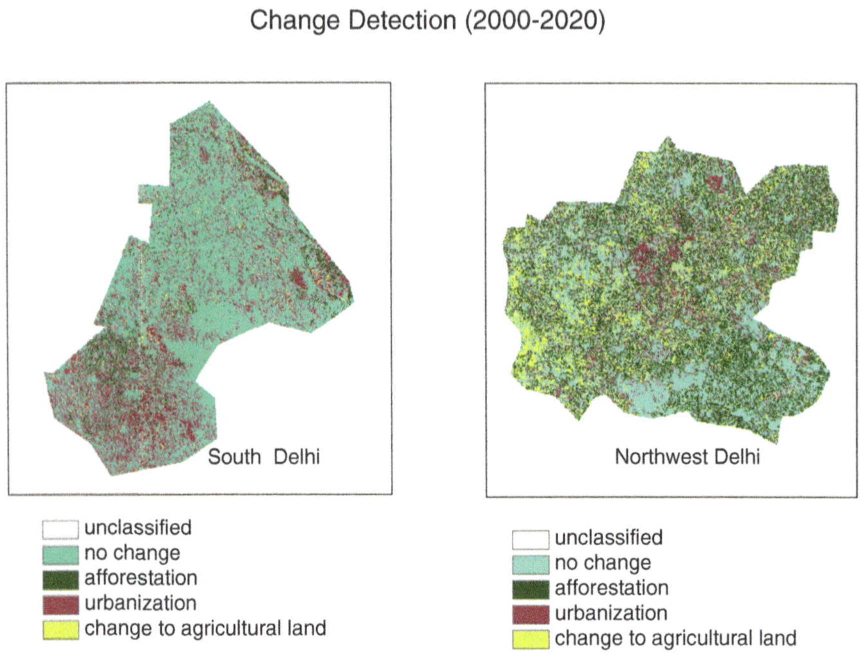

Fig. 5.7 Change matrix that occurred from 2000 to 2020 in Northwest and South Delhi

Table 5.8 The change in both areas of study from 2000 to 2020

Change in the area (South)	Area (in ha)	Percentage
No change	17,214.3	33%
Afforestation	1169.15	2%
Urbanization	5400.09	10%
Change to agricultural land	572.13	1%
Change in the area (Northwest)	**Area (in ha)**	**Percentage**
No change	20,939.9	31%
Afforestation	12,774.2	19%
Urbanization	4252.16	6%
Change to agricultural land	6453.07	10%

slope (Gour et al. 2014; Grover and Singh 2015; Kumar et al. 2012; Kumari et al. 2018) (Figs. 5.8 and 5.9). The index is defined by the following equation:

$$\mathrm{NDVI} = (\mathrm{Band4} - \mathrm{Band3})/(\mathrm{Band4} + \mathrm{Band3})$$

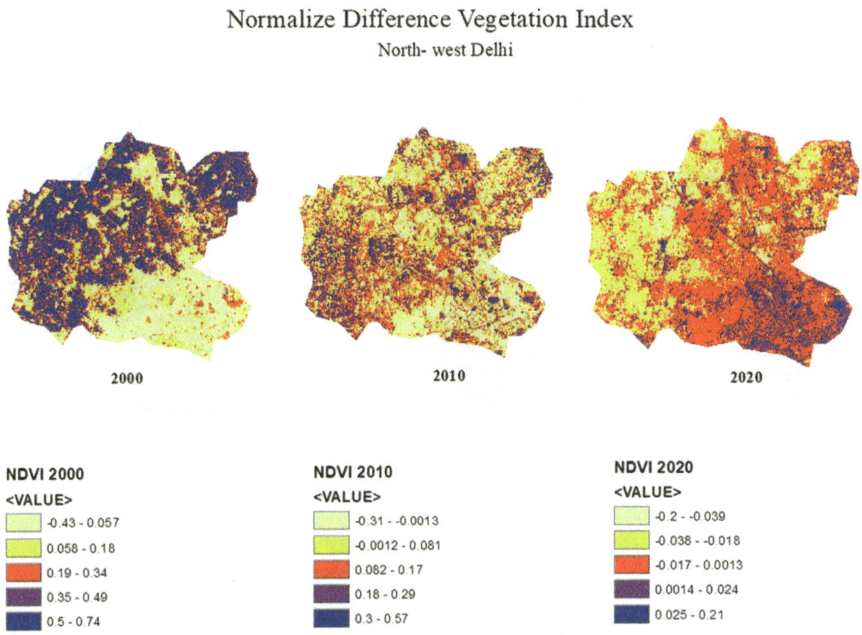

Fig. 5.8 NDVI map of Northwest Delhi

5.3 Result

5.3.1 Survey Result

People were surveyed in their respective colonies, monumental places, and famous parks of the study area, respectively; the general socioeconomic profile of respondents in both studied areas is in a 2:1 ratio of male and female. Respondents interviewed include natives and non-natives of Delhi above 18 years of age.

The aesthetic importance of trees and their ability to reduce pollution in the area are two important benefits most commonly viewed from the conversation with the people from the studied areas. About 60% of individuals from both studied areas mention the decrease in the tree canopy in their area, while more than 50% of individuals from Northwest Delhi claim the decrease in biodiversity in their area, and about 18% of the individuals in South Delhi claim for increase in the biodiversity in the area over the following years of their stay. The majority of individuals from both areas view pollution as a major environmental problem and believed trees in their proximity have a positive effect on them and help with various health issues. Therefore, about 0.2% of people mentioned the disadvantages like pollen allergy, lack of parking spaces, insect infestation, and shedding of leaves and fruits on the path making it sticky and slippery as a negative view of having trees near urban structures.

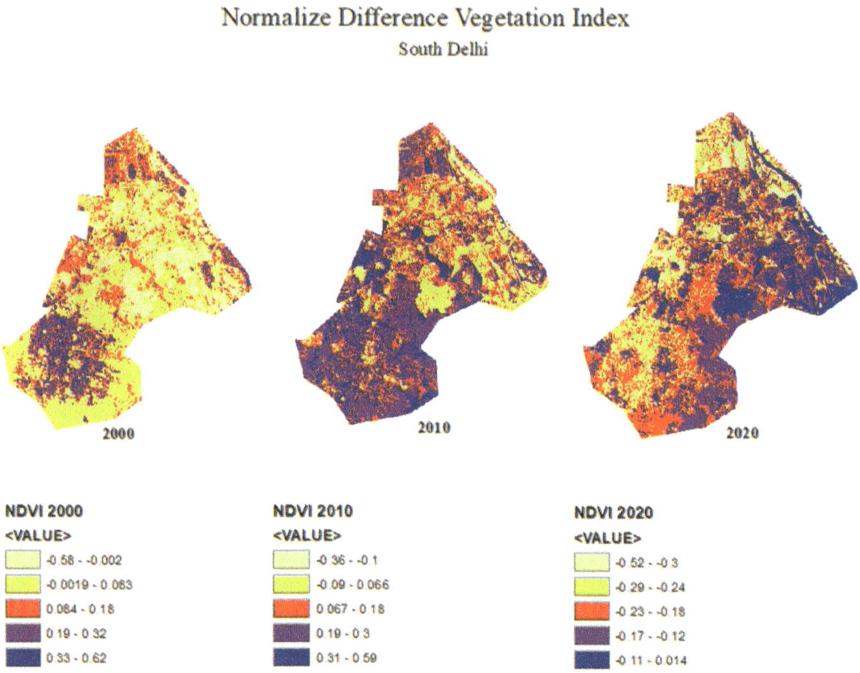

Fig. 5.9 NDVI map of South Delhi

Each statement determined the relative importance and expressed the level of agreement for all 100 responders (Bhalla and Bhattacharya 2015) (Table 5.9). The weighted mean analysis in (Table 5.1) indicates that all the respondents agreed with all four statements (Bhalla and Bhattacharya 2015).

5.3.2 Land Surface Temperature (LST)

Delhi has witnessed a deep socio-spatial transformation and rapid population growth in the last few decades as it has gained political, administrative, and economic importance (Véron 2006). The rapid urbanization of the city has led to an increase in an impermeable concrete surface structures like industrial units, transport systems, and more built-up structures resulting in a decline of green infrastructure, which therefore increases anthropogenic heat in the city district. As a result, the core of the urban area becomes warmer than its peripheral parts and forms the urban heat island (UHI) effect. The increase in concrete surface area is linked with the increased UHI, which indirectly describes the relationship between urbanization and increased temperature (Ahmad and Goparaju 2016; Javed Mallick and Bharath 2008; Kumari et al. 2018) (Figs. 5.10 and 5.11).

Table 5.9 Ground survey questionnaire results from the Northwest and South Delhi area

1	Survey questions	options given	Response %	
			Northwest	South
2	Are you native of Delhi?	Yes	77%	60%
		No	22%	40%
3	How many years you stayed in delhi	0–5	28%	42%
		5–10	6%	12%
		10–15	4%	12%
		>15	56%	34%
4	How many male and female members reside in your family			
		Male		
	0		6%	18%
	1		12%	18%
	2		40%	36%
	3		20%	22%
	4		16%	2%
	>5		2%	4%
		Female		
	0		6%	12%
	1		20%	20%
	2		44%	38%
	3		16%	14%
	4		6%	2%
	>5		2%	12%
5	Is your area well managed and organized?			
	Yes		54%	62%
	No		46%	36%
6	Shade of tree canopies is enough in your locality			
	Yes		34%	36%
	No		64%	62%
7	What difference you have noticed regarding environment of area	Less clean	50%	34%
		Clean	38%	44%
		Very clean	4%	16%
		Dirty	8%	4%
8	What environmental problems you encounter in your locality			
		Pollution	32%	50%
		Garbage issue	24%	20%
		Less green space	16%	12%
		All of the above	26%	18%
9	Faunal biodiversity?			
		Pigeon	60%	36%

(continued)

Table 5.9 (continued)

		Squirrel	28%	34%
		Peacock	0%	4%
		Butterflies	6%	8%
		All above	0%	8%
		Other (monkey, birds, moth etc.)	6%	16%
10	Faunal biodiversity change observed?			
		Decreased	40%	36%
		Neutral	44%	44%
		Increased	14%	18%
11	Do you see any disadvantages of having tree in your locality?			
		Yes	12%	14%
		no	80%	86%

The spatial distribution of LST is done by using the spatial radiance model (SRM) method for Landsat satellite image data, and it is observed that LST is highest over built-up areas, fallow land, and open areas in comparison to land under green cover and water bodies (Chakraborty_2013.Pdf n.d.; Gour et al. 2014; Javed Mallick and Bharath 2008).

The spatial distribution of land surface temperature of Northwest Delhi in 2000 is estimated to ranges from 18 to 34 °C; in 2010, it ranges from 20 to 30 °C, and in 2020, it is estimated to ranges from 21 to 36 °C with a mean and standard deviation of 31.90 and 5.25 °C, respectively. Similarly, the spatial distribution of 1 LST values of South Delhi in 2000 is estimated to ranges from 12 to 32 °C; in 2010, it ranges from 23 to 34 °C, and in 2020, it is estimated to ranges from 22 to 39 °C with a mean and standard deviation of 32.21 to 7.56 °C, respectively. In the analysis of LST data, it can be said that there is a slight increase in the temperature in both the areas of study in Delhi, although the temperature in South Delhi is slightly higher as a large part of the district is covered in lattice soil with sparse vegetation and concrete jungle of urban land area than that of Northwest Delhi having a large part under agricultural land.

5.3.3 *Land Use/Land Cover (LU/LC)*

The FCC (false color composite) image of the satellite is obtained from USGS EarthExplorer and clipped out the area of study with approx. 99% accuracy. Based on satellite image analysis, the images are classified into four major LULC classes, agricultural land (including non-cultivated fertile land, fallow land, etc.), built-up area (land covered with the concrete surface), vegetation (roadside trees, gardens,

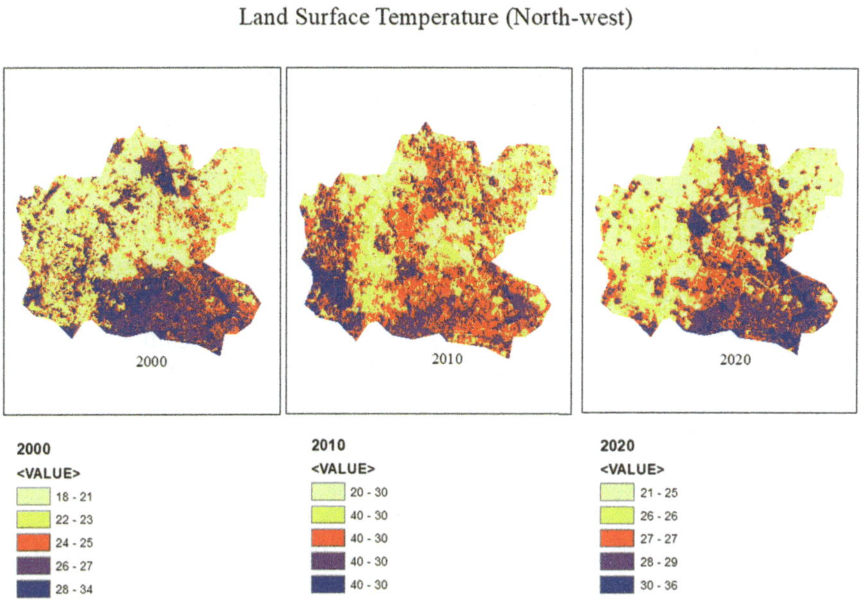

Fig. 5.10 Land surface temperature of Northwest Delhi (2000, 2010, and 2020)

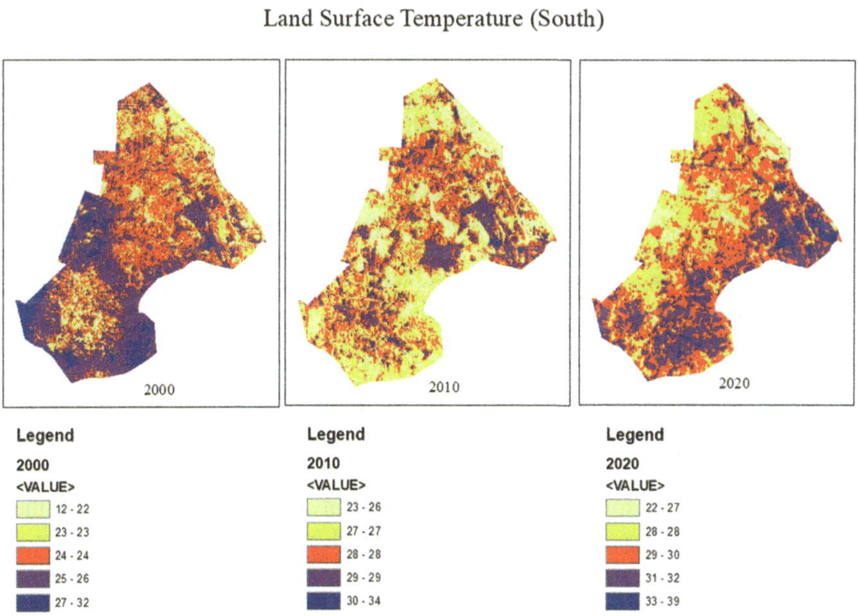

Fig. 5.11 Land surface temperature of South Delhi (2000, 2010, and 2020)

Table 5.10 Northwest Delhi land use/land cover data

Northwest Delhi			
Land use	2000	2010	2020
Agricultural land	27%	25%	24%
Built-up	22%	26%	34%
Vegetation	12%	15%	21%
Water body	1%	1%	1%

Table 5.11 South Delhi land use/land cover data

South Delhi			
Land use	2000	2010	2020
Agricultural land	13%	12%	9%
Built-up	37%	34%	46%
Vegetation	48%	53%	44%
Water body	2%	1%	1%

community park, and street trees), and water body (river, lake, and pond) (Kumari et al. 2018; Rahman et al. 2012; Tilahun 2015).

The Landsat images were classified for the years 2000, 2010, and 2020 on the bases of data obtained from Google Earth and ground truthing at the area of study, and the area for each class is calculated by taking into account the pixel count and total area. Therefore, the percentage of each classified area from all three years and for both the study area is mentioned in Tables 5.10 and 5.11 (Rwanga and Ndambuki 2017).

In Northwest Delhi, agriculture is seemed to be a dominant practice in comparison to the South Delhi district; in the time snap of 20 years, vegetation in the Northwest district has increased with huge tree canopy above built-up areas. Therefore, it is tough to classify built-up areas using Google Earth. However, in South Delhi vegetation is comparatively high in comparison to Northwest Delhi, and within 20 years of period, agricultural land in the South has decreased with an overall increase in built-up areas and a comparative decline in green infrastructure and water bodies (Figs. 5.12 and 5.13).

5.3.4 Accuracy Assessment

It is important to perform an accuracy assessment for the individually classified image. Henceforth, accuracy assessment is carried out for both study areas for all three years (2000, 2010, and 2020).

In the classified image of research areas, 100 random sites were marked. and a confusion matrix is formed on the bases of a comparative study between these points and the corresponding pixels of LU/LC of the classified image to compute the user's

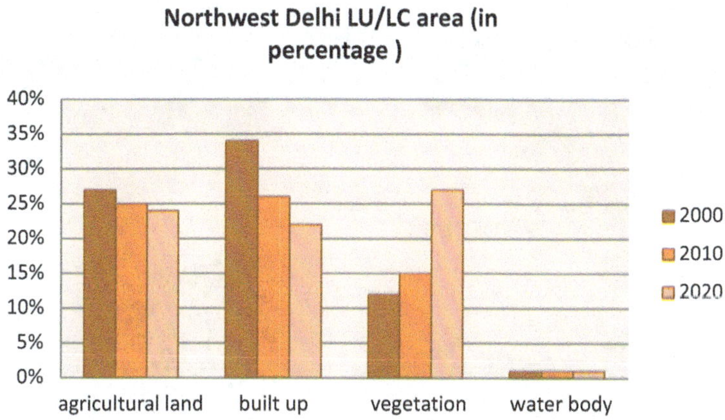

Fig. 5.12 Chart representing Northwest Delhi LU/LC area

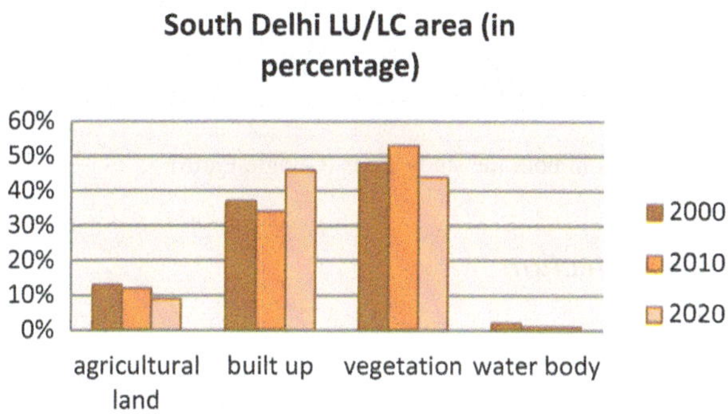

Fig. 5.13 Chart representing South Delhi LU/LC area

accuracy, producer's accuracy, and kappa coefficient (Raju et al. 2018; Tilahun 2015) (Fig. 5.14 and Tables 5.6 and 5.7).

Using the formulae mentioned in Sect. 5.2.3.4, various accuracy parameters are calculated and tabulated in Tables 5.6 and 5.7. The result shows the overall accuracy obtained from a random sampling of the classified image of the years 2000, 2010, and 2020 for Northwest Delhi is 94%, 91%, and 84%, although for South Delhi, it is 86%, 84%, and 98%, respectively. The broad range of accuracy indicates the severe confusion of vegetation with other LU/LC classes. The producer's accuracy gives the accuracy of prediction of a particular category, while the user's accuracy gives the reliability of the classification to the user. Therefore, all the individualized parameter apart from overall accuracy gives a detailed description of the model performance of a particular class or category of interest or study (Rwanga and Ndambuki 2017).

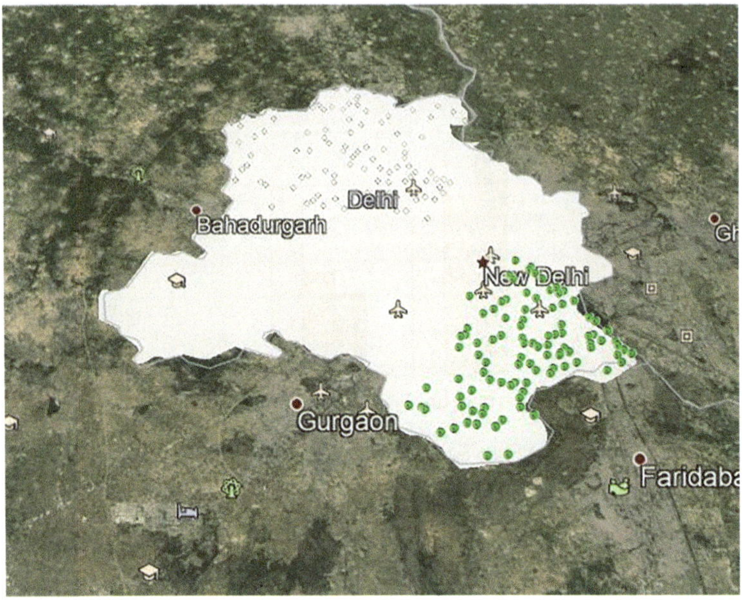

Fig. 5.14 Reference points in both the study areas (Google Earth)

5.3.5 Change Detection

LULC change is a continuous process which resulted from natural and human intervention. The study of LU/LC requires understanding and monitoring various factors that have led to these changes. In this study, we observed that there is an interchange of land between certain LU/LC classes (Bijender and Joginder 2014; Rahman et al. 2012).

Both study areas have undergone a significant transformation during the past 20 years (Fig. 5.7). To meet the needs of the expanding population, the majority of LU/LC changes in Northwest Delhi are seen in agriculture and vegetation, with roughly 6% of the growth occurring in built-up areas that encroach on agricultural land. Similar trends are seen in South Delhi, where there has been a significant growth in the built-up area and a 2% increase in green infrastructure (Table 5.8). The loss of harmony between green infrastructure and concrete land surface is explained by all the changes, emphasizing the value of urban forests inside urban environments.

5.3.6 *Normalize Difference Vegetation Index*

The most common use of NDVI in satellite imaging is to track and assess the prevalence of sparse vegetation, such as shrubs or grassland. The value of NDVI varies from −1 to +1. Therefore, a higher value of NDVI indicates richer and healthier vegetation in the area (Kumar et al. 2012; Kumari et al. 2018).

The NDVI values change in both areas with the preceding year. In Northwest Delhi, the NDVI value from 2000 to 2020 changed from (−0.43 to 0.74) to (−0.2 to 0.21) which indicates the decrease in vegetation area (Fig. 5.8). Consequently, in South Delhi, the values changed from (−0.53 to 0.62) to (−0.52 to 0.014) for the years 2000 to 2020 which indicates a decrease in vegetation area and an increase in concrete surfaces (Fig. 5.9). Therefore, on comparison of both areas, the vegetation in South Delhi is comparatively more than that of Northwest Delhi. The differences occurred due to the large area under the ridge and wildlife sanctuary with the number of gardens in the South Delhi area, while in Northwest Delhi, large areas under agriculture and built-up area affect the NDVI.

5.4 Conclusion

At the present, India's urban policies lack the potential for green spaces to contribute to the sustainability of social, economic, and environmental conditions in urban areas (Bolund and Hunhammar 1999; Govindarajulu 2014). The megacity like Delhi had undergone various land use changes over the year, witnessing increasing urbanization and unplanned development of the city which had increased the environmental issues and decreased the green infrastructure (Paul and Nagendra 2015). Through the above study, we could draw out the major environmental issues faced by the public in urban areas and the various ecosystem services they drive off from urban greening. The study majorly highlights the LULC changes that occurred in 20 years and their effect on the urban habitat (Subramoniam et al. 2022). Consequently, the study concludes that Northwest Delhi has a low vegetation index with subsided LST values as a large part is under agricultural cultivation and at the convergence of the Yamuna River. This area lacks biodiversity and an adequate amount of green canopy near urban structures. In contrast South Delhi has a slightly high vegetation index with a considerable amount of biodiversity and an effective tree canopy as it consists of Asola wildlife sanctuary, but LST values are slightly high due to the presence of rocky lattice soil in the region. The study also demonstrates the positive effects of urban forestry on the urban environment, which directly and indirectly affect urban life.

References

Ahmad F, Goparaju L (2016) Geospatial technology in urban forest suitability: analysis for Ranchi, Jharkhand, India. Ecol Questions 24:45–57. https://doi.org/10.12775/EQ.2016.011

Baró F, Chaparro L, Gómez-Baggethun E, Langemeyer J, Nowak DJ, Terradas J (2014) Contribution of ecosystem services to air quality and climate change mitigation policies: the case of urban forests in Barcelona, Spain. Ambio 43(4):466–479. https://doi.org/10.1007/s13280-014-0507-x

Bhalla P, Bhattacharya P (2015) Urban biodiversity and green spaces in Delhi: a case study of new settlement and Lutyens' Delhi. J Hum Ecol 52(1–2):83–96. https://doi.org/10.1080/09709274.2015.11906933

Bijender S, Joginder S (2014) Land use/land cover change of Delhi: a study using remote sensing and GIS techniques. Int Res J Earth Sci 2(1):15–20

Bolund P, Hunhammar S (1999) Ecosystem services in urban areas. Ecol Econ 29(2):293–301. https://doi.org/10.1016/S0921-8009(99)00013-0

Cavender N, Donnelly G (2019) Intersecting urban forestry and botanical gardens to address big challenges for healthier trees, people, and cities. Plants People Planet 1(4):315–322. https://doi.org/10.1002/ppp3.38

Census 2011.Pdf (2016)

Chakraborty_2013.pdf (n.d.)

Chaturvedi A, Kamble R, Patil NG, Chaturvedi A (2013) City-forest relationship in Nagpur: one of the greenest cities of India. Urban For Urban Green 12(1):79–87. https://doi.org/10.1016/j.ufug.2012.09.003

Chaudhry P, Bagra K, Singh B (2011) Urban greenery status of some Indian Cities: a short communication. Int J Environ Sci Dev 2(2):98–101. https://doi.org/10.7763/ijesd.2011.v2.104

Clark JR, Matheny NP, Cross G, Wake V (1997) A model of urban forest sustainability. J Arboricult 23(1):17–30

Das J, Jha S, Goyal MK (2020) On the relationship of climatic and monsoon teleconnections with monthly precipitation over meteorologically homogenous regions in India: wavelet & global coherence approaches. Atmos Res 238:104889. https://doi.org/10.1016/j.atmosres.2020.104889

Emmanuel R (2005) Thermal comfort implications of urbanization in a warm-humid city: the Colombo Metropolitan Region (CMR), Sri Lanka. Build Environ 40(12):1591–1601. https://doi.org/10.1016/j.buildenv.2004.12.004

Foody GM (2002) Status of land cover classification accuracy assessment. Rem Sens Environ 80(1):185–201. https://doi.org/10.1016/S0034-4257(01)00295-4

Forest Survey of India (2011) India state of forest report 2011. In: India state of forest report 2011: vol I

Gour D, Soumen D, Sujan S (2014) An application of remote sensing and GIS to analyze urban expansion and land use land cover change of Midnapore Municipality, WB, India. Int Res J Earth Sci 2(5):8–20

Govindarajulu D (2014) Urban green space planning for climate adaptation in Indian cities. Urban Clim 10(P1):35–41. https://doi.org/10.1016/j.uclim.2014.09.006

Goyal MK, Ojha CSP (2010) Evaluation of various linear regression methods for downscaling of mean monthly precipitation in Arid Pichola Watershed. Nat Resour 1(1):11–18. https://doi.org/10.4236/nr.2010.11002

Goyal MK, Ojha CSP (2012) Downscaling of precipitation on a lake basin: evaluation of rule and decision tree induction algorithms. Hydrol Res 43(3):215–230. https://doi.org/10.2166/nh.2012.040

Goyal MK, Panchariya VK, Sharma A, Singh V (2018) Comparative assessment of SWAT model performance in two distinct catchments under various DEM scenarios of varying resolution, sources and resampling methods. Water Resour Manag 32(2):805–825. https://doi.org/10.1007/s11269-017-1840-1

Goyal MK, Gupta AK, Jha S, Rakkasagi S, Jain V (2022) Climate change impact on precipitation extremes over Indian cities: non-stationary analysis. Technol Forecast Social Change 180: 121685. https://doi.org/10.1016/j.techfore.2022.121685

Grover A, Singh RB (2015) Analysis of urban heat island (UHI) in relation to normalized difference vegetation index (NDVI): a comparative study of Delhi and Mumbai. Environ - MDPI 2(2): 125–138. https://doi.org/10.3390/environments2020125

Gurjar BR, Butler TM, Lawrence MG, Lelieveld J (2008) Evaluation of emissions and air quality in megacities. Atmos Environ 42(7):1593–1606. https://doi.org/10.1016/j.atmosenv.2007.10.048

Haberman D, Gillies L, Canter A, Rinner V, Pancrazi L, Martellozzo F (2014) The potential of urban agriculture in Montréal: a quantitative assessment. ISPRS Int J Geo-Inf 3(3):1101–1117. https://doi.org/10.3390/ijgi3031101

Hegazy IR, Kaloop MR (2015) Monitoring urban growth and land use change detection with GIS and remote sensing techniques in Daqahlia governorate Egypt. Int J Sustain Built Environ 4(1): 117–124. https://doi.org/10.1016/j.ijsbe.2015.02.005

Ihlen V, USGS (2019) Landsat 7 (L7) data users handbook. USGS Landsat User Serv 7:151

Jat MK, Garg PK, Khare D (2008) Monitoring and modelling of urban sprawl using remote sensing and GIS techniques. Int J Appl Earth Observ Geoinform 10(1):26–43. https://doi.org/10.1016/j.jag.2007.04.002

Javed Mallick YK, Bharath BD (2008) Estimation of land surface temperature over Delhi using Landsat-7 ETM+. J Ind Geophys Union 12(3):131–140

Jimenez-Munoz JC, Cristobal J, Sobrino JA, Sòria G, Ninyerola M, Pons X (2009) Revision of the single-channel algorithm for land surface temperature retrieval from landsat thermal-infrared data. IEEE Trans Geosci Rem Sens 47(1):339–349. https://doi.org/10.1109/TGRS.2008.2007125

Kumar KS, Bhaskar PU, Padmakumari K (2012) Estimation of land surface temperature to study urban heat island effect using Landsat Etm+ Image. Int J Eng Sci Technol 4(2):771–778

Kumari B, Tayyab M, Shahfahad S, Mallick J, Khan MF, Rahman A (2018) Satellite-driven land surface temperature (LST) using Landsat 5, 7 (TM/ETM+ SLC) and Landsat 8 (OLI/TIRS) data and its association with built-up and green cover over Urban Delhi, India. Rem Sens Earth Syst Sci 1(3–4):63–78. https://doi.org/10.1007/s41976-018-0004-2

Mack RN, Erneberg M (2002) The United States naturalized flora: largely the product of deliberate introductions. Ann Missouri Bot Garden 89(2):176–189. https://doi.org/10.2307/3298562

Mukherjee A, Sarma K (2016) Assessment of land use dynamics of Okhla Bird Sanctuary, Delhi using geospatial technology. J Biodivers Manag For 5:1. https://doi.org/10.4172/2327-4417.1000151

Nowak DJ (1994) Understanding the structure. J For 92(10):42–46

Nowak J, Noble FDMH, Sisinni M (2001) Urban areas. J For 99(3):37–42

Parastatidis D, Mitraka Z, Chrysoulakis N, Abrams M (2017) Online global land surface temperature estimation from landsat. Rem Sens 9(12):1–16. https://doi.org/10.3390/rs9121208

Paul S, Nagendra H (2015) Vegetation change and fragmentation in the mega city of Delhi: mapping 25 years of change. Appl Geogr 58:153–166. https://doi.org/10.1016/j.apgeog.2015.02.001

Pettorelli N, Vik JO, Mysterud A, Gaillard JM, Tucker CJ, Stenseth NC (2005) Using the satellite-derived NDVI to assess ecological responses to environmental change. Trends Ecol Evol 20(9): 503–510. https://doi.org/10.1016/j.tree.2005.05.011

Poonia V, Goyal MK, Gupta BB, Gupta AK, Jha S, Das J (2021) Drought occurrence in Different River Basins of India and blockchain technology based framework for disaster management. J Clean Prod 312:127737. https://doi.org/10.1016/j.jclepro.2021.127737

Rahman A, Netzband M (2007). An assessment of urban environmental issues using remote sensing and GIS techniques an integrated approach: a case study: Delhi, India Atiqur. In: PRIPODE workshop on urban population, development, and environment dynamics in developing countries, June, 1–25.

Rahman A, Aggarwal SP, Netzband M, Fazal S (2011) Monitoring urban sprawl using remote sensing and GIS techniques of a fast growing urban centre, India. IEEE J Sel Top Appl Earth Observ Rem Sens 4(1):56–64. https://doi.org/10.1109/JSTARS.2010.2084072

Rahman A, Kumar S, Fazal S, Siddiqui MA (2012) Assessment of land use/land cover change in the North-West District of Delhi using remote sensing and GIS techniques. J Indian Soc Rem Sens 40(4):689–697. https://doi.org/10.1007/s12524-011-0165-4

Raju RS, Sudarsana Raju G, Rajasekhar M (2018) Land use/land cover change detection analysis using supervised classification, remote sensing and GIS in Mandavi River Basin, YSR Kadapa District, Andhra Pradesh, India. J Rem Sens GIS 9(3):46–54

Rees W, Wackernagel M (2008) Urban ecological footprints: why cities cannot be sustainable- and why they are a key to sustainability. In: Urban ecology. Springer, Boston. https://doi.org/10.1007/978-0-387-73412-5_35

Roy S, Byrne J, Pickering C (2012) A systematic quantitative review of urban tree benefits, costs, and assessment methods across cities in different climatic zones. Urban For Urban Green 11(4): 351–363. https://doi.org/10.1016/j.ufug.2012.06.006

Russo A, Escobedo FJ, Cirella GT, Zerbe S (2017) Edible green infrastructure: an approach and review of provisioning ecosystem services and disservices in urban environments. Agric Ecosyst Environ 242:53–66. https://doi.org/10.1016/j.agee.2017.03.026

Rwanga SS, Ndambuki JM (2017) Accuracy assessment of land use/land cover classification using remote sensing and GIS. Int J Geosci 8(4):611–622. https://doi.org/10.4236/ijg.2017.84033

Steenberg JWN, Millward AA, Nowak DJ, Robinson PJ (2017) A conceptual framework of urban forest ecosystem vulnerability. Environ Rev 25(1):115–126. https://doi.org/10.1139/er-2016-0022

Subramoniam SR, Ravindranath S, Rakkasagi S, Ram H (2022) Water resource management studies at micro level using geospatial technologies. Springer, Berlin, pp 49–74. https://doi.org/10.1007/978-3-030-98981-1_2

Tilahun A (2015) Accuracy assessment of land use land cover classification using Google Earth. Am J Environ Prot 4(4):193. https://doi.org/10.11648/j.ajep.20150404.14

Vailshery LS, Jaganmohan M, Nagendra H (2013) Effect of street trees on microclimate and air pollution in a tropical city. Urban For Urban Green 12(3):408–415. https://doi.org/10.1016/j.ufug.2013.03.002

Véron R (2006) Remaking urban environments: the political ecology of air pollution in Delhi. Environ Plan A 38(11):2093–2109. https://doi.org/10.1068/a37449

Xiuwan C, Heping H, Tateishi R, Ahn CH (1999) A method on land cover classification by combining unsupervised algorithm and training data. Geocarto Int 14(4):15–20. https://doi.org/10.1080/10106049908542124

Zanter K (2015) Landsat 8 (L8) data users handbook version 1.0 June 2015. United States Geol Surv 1:1–106

Chapter 6
Restoration and Rejuvenation of Rivers, Streams and Wetlands: Challenges and Way Forward

Urvashi Sharma, Dhananjay Kumar, Sanjeev Kumar, Venkatesh Dutta, and Narendra Kumar

Abstract It is vital for the existence and survival of humans to have access to the satisfactory quality and adequate quantity of water. Inland water bodies were considered as community asset or resource over the centuries. In recent years, several big cities have witnessed unprecedented flood and droughts. This can be attributed to depleting inland water storage systems such as ponds, lakes and streams. A five-phase planning and execution system has been discussed here for the restoration and rejuvenation of the water bodies. The five phases are recognition, restoration, protection, improvement and sustenance. These phases begin with preparing the inventory of the water bodies and identification of the water bodies that need to be restored, then the best designated use of the identified water body is decided, and then the water body is prevented from any type of pollution and unnatural changes, and later on, to improve the overall catchment of the water body, in situ treatment techniques, green buffer development and proper management of the drainage basin are being done. The last phase is to sustain the desired results in the restored water body, and it is the toughest of all the phases. To achieve the desired results in restoring the water bodies, public and stakeholder participation and government's initiatives are the key.

Keywords River · Ponds/lakes · Restoration · Rejuvenation · Wetlands

---

U. Sharma · D. Kumar · V. Dutta · N. Kumar (✉)
Department of Environmental Science, Babasaheb Bhimrao Ambedkar University, Lucknow, Uttar Pradesh, India

S. Kumar
Department of Environmental Science, Central University of Jharkhand, Ranchi, India

A. K. Gupta et al. (eds.), *Ecosystem Restoration: Towards Sustainability and Resilient Development*, Disaster Resilience and Green Growth,
https://doi.org/10.1007/978-981-99-3687-8_6

6.1 Introduction

Access to satisfactory water quality and adequate quantity is the key for existence and excellence of all the living creatures. Ponds, lakes, streams, wetlands and rivers were considered as community assets or resources over the centuries. The observed structural shift in the state of mind of individuals towards their water bodies can be attributed not only to urbanization, industrialization, and the advancement in surface water supply chains but also to the increase in the number of tube wells and hand pumps used to abstract groundwater, leading to the initiation of pollution (Goyal et al. 2022). As a result, the locals and the government both had stopped caring, fostering and preserving these community resources and asset (Poonia et al. 2021; Paerl et al. 2014). Bourgeoning urbanization, industrialization and infrastructural expansion has altered the precious water bodies from precious natural resources of the community to a measly dumping ground for all types of domestic sewage, industrial effluents, solid wastes, religious offering, construction debris, etc.; as a result, these water bodies have degraded (Lyu et al. 2020; Dang et al. 2019; Goyal et al. 2018).

In the recent years, several big cities have witnessed unprecedented flood and droughts (NASA 2019). The recent studies have shown that worldwide there is a decline in the number of water bodies due to several factors such as ramped urbanization, developmental activities, increasing population, population density, land use/land cover change, increase in cash crop cultivation, decline in groundwater table and increasing unplanned urban outskirts. To restore and rejuvenate water bodies, the planners must keep the following points in view (Søndergaard et al. 2007):

1. The water bodies should be pollution-free, and they should meet the desired criteria of water quality.
2. The excess water available during the monsoon should be preserved.
3. The storage capacity of the water bodies should be restored and augmented.
4. The groundwater recharge should be increased.
5. The water availability should be enhanced for different intended purposes.

6.2 Restoration and Rejuvenation of the Water Bodies

The planners while planning the long-term management plans for the urban areas so that the water bodies may serve as a good resource require understanding on the following aspects:

(a) What is the current condition of the water bodies?
(b) For what purpose the water body is best suitable?
(c) Why the water bodies should be managed and conserved?

So the water body post restoration and rejuvenation will be able to serve for the future potential needs of the nearby urban communities facing problems due to water scarcity. Its planning and execution have the following five phases (CPCB 2019):

1. Recognition phase
2. Restoration phase
3. Protection phase
4. Improvement phase
5. Sustenance phase

6.2.1 Recognition Phase

The initial step in the restoration of water bodies is the recognition phase, wherein the existing or lost water bodies' problems are recognized and identified. The probable causes of the problems and their effects on the water bodies will be analysed. Suitable solutions to the identified problems will be identified. To do this exercise, the following information will be needed:

(a) **Collection and maintenance of historic information in relation to ponds or lakes:** It includes collection of geographical information of the water body using remote sensing and GIS such as GPS location, elevation above mean sea level, area, dimensions, address, ownership and boundary conditions and mapping the water body for its interaction with the indigenous people of that area. Additionally, information about its hydrology, including whether it is a natural or man-made water body, the description of its catchment area as the source of water, water depth during monsoon and non-monsoon periods, total storage capacity, and its classification based on water logging (permanent or intermittent) and bank overflows are essential aspects to be considered in this phase. Furthermore, understanding the purposes for which the community uses the water body, the status of open water, and the percentage of aquatic vegetation within the water body are crucial factors to be documented. Moreover, detailed data regarding the major towns surrounding the water body, including their total population, total sewage generation, and their contributions to water pollution, should be gathered to facilitate effective restoration efforts. Major industries, industrial clusters and estates discharging the pollutants in the water body. Existing STPs and common effluent treatment plants (CETPs) and their treatment capacities in the vicinity will also be needed. Total solid waste being generated and provisions for its scientific handling and management will also be needed. Other pertinent information such as biodiversity details and wetland Ramsar sites will also be needed.

(b) **Collection and maintenance of historic information in relation to river or streams:** The recognition phase of the stream or river under consideration will include preparation of digital maps of river or stream showing the GPS location of tributaries and their salient features such as its origin, confluences, length of the river, average cross-sectional area (in m^2), average depth (in m), velocity (in m/s), volume or discharge (in m^3/s) and drains contributing to river pollution. The catchment details of the river will also be required, which include the purpose the river or stream is serving or was serving in the past. Details of major towns, industrial clusters, solid waste, STPs, CETPs and biodiversity will also be required the same as required for the ponds or lakes. Along with it, data on the status of the groundwater, its consumption and the quality will also be required.

(c) **Digital Mapping:** The information will be placed together on the maps, which also needs to be updated on a regular basis.

6.2.2 Restoration Phase

The restoration phase includes defining the designated best use of the water body in order to articulate strategies with regard to the level of treatment required for achieving the desired level of restoration of selected water body. The user who requires the highest level of water quality and purity among all the users is labelled as the "designated best use" for that area the water body is located or flowing. Restoration needs to be done on seven principles for sustainable use of lakes. Firstly, humans should develop a harmonious relationship with the nature; secondly, the starting point for planning and management of lake sustainably should be the drainage basin; thirdly, the long-term preventive approach should be directed towards prevention of the prime causes of degradation of water body; fourthly, scientific and best available information should be the base for decision-making and policy development; fifthly, while resolving the conflicts of the competing users, the needs of the nature should be taken into consideration along with the present and future generations; sixthly, public participation should be encouraged in identification and resolving problems specific to their local water body; and seventhly, the governance should be as such that it must work fairly and maintain a transparency in its actions and decisions and it should be targeted to empower its stakeholders and local community.

Detailed Gap Analysis During the restoration process of the water body in the catchment area, it is crucial to address the gaps in data related to the total generation of industrial effluent, municipal sewage, and waste generation. These projections should be made for a minimum of 15–20 years to anticipate future challenges accurately. Additionally, thorough assessment of the volumetric flow of all drains contributing to the pollution in the water body must be conducted to develop effective strategies for mitigating and managing the sources of pollution.

Additional site-specific measures that may be taken are as follows:

1. Development of buffer zone and maintaining the existing activities within the buffer zones.
2. Where there is an adequate land available close to the water body, feasibility to develop biodiversity park should be checked.

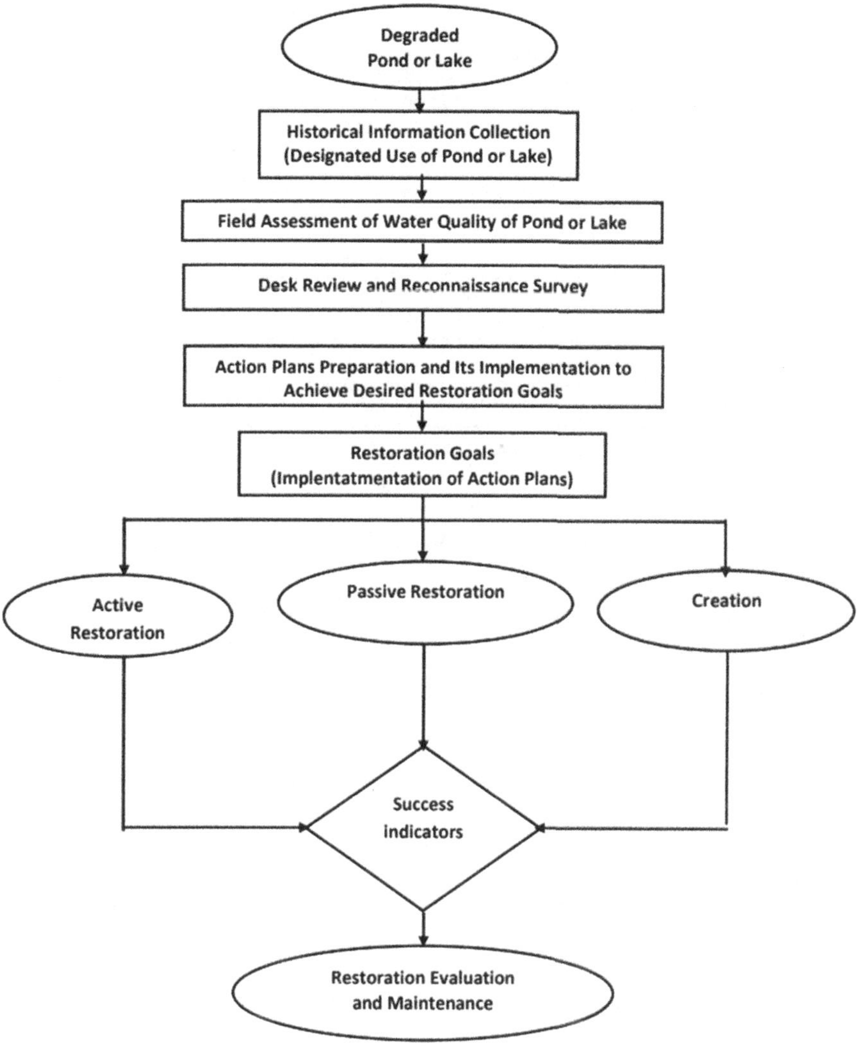

Fig. 6.1 Model flow chart for restoration of lake and ponds (CPCB 2019)

3. Develop greenery in the vicinity as well as introduce recreational facilities such as building jetty and paddle boats.
4. Arrangement of the man power and machinery for maintaining the water body after its restoration has been done.
5. Arrangement for the disposal of the de-siltation and de-weeding wastes.
6. Awareness and training programmes should be done regularly.
7. While preparing the action plans, room for inclusion of any other action required in the future should be analysed (Figs. 6.1 and 6.2).

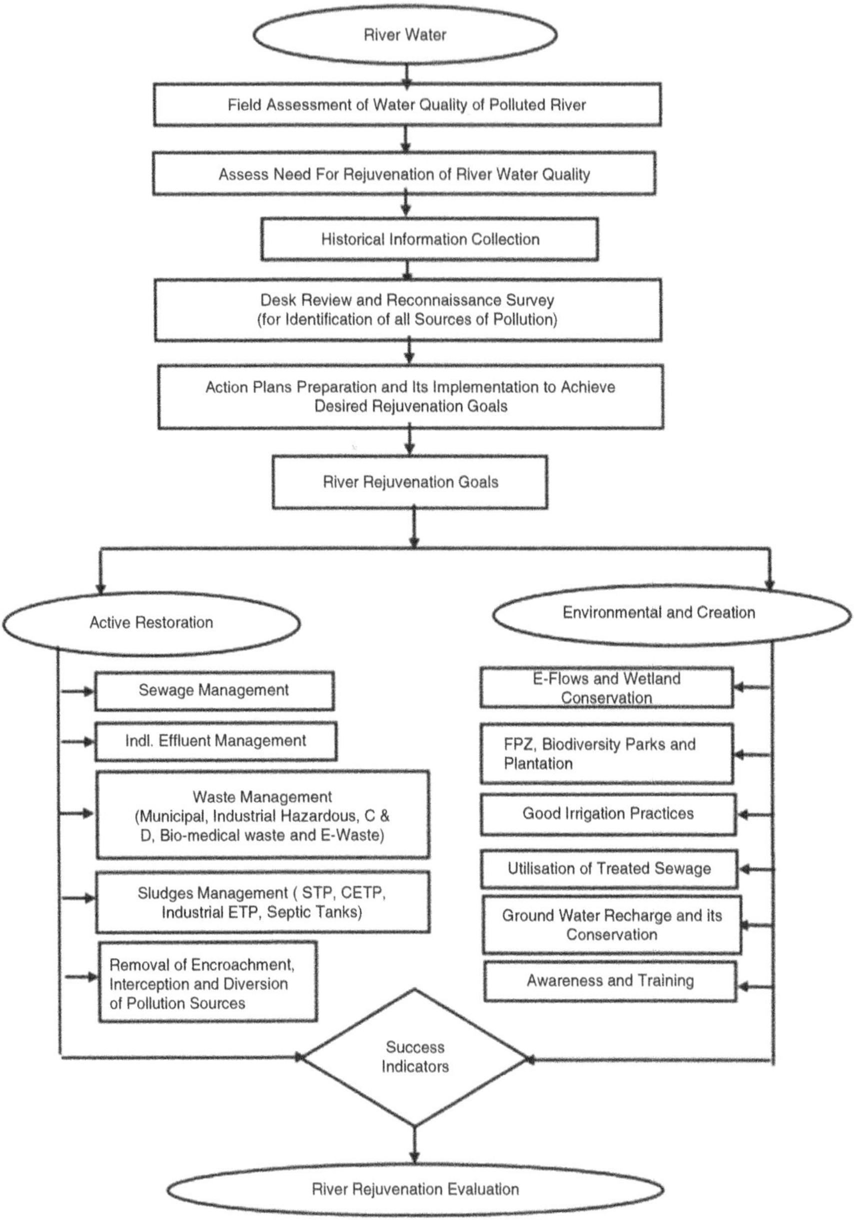

Fig. 6.2 Model flow chart for rejuvenation of polluted rivers (CPCB 2019)

6.2.3 Protection Phase

In the protection phase, emphasis is given on the general health and natural functioning of the water body. To prevent the causes of water body degradation, a long-term preventive approach is required. The action plan thus prepared must take into consideration the precise time targets, the budget estimates and the government body who will be implementing the action plans. The action plan should be sound enough to deal problems related to sewage management, industrial effluent management and waste management. To protect water body, periodic de-siltation will also be needed (Subramoniam et al. 2022). De-siltation is helpful in the removal of nutrient-enriched accumulated sludges, which promotes the elimination of contaminated sediments, increases the storage capacity of the water body as well as enhances the groundwater recharge potential (Zhong et al. 2018). Periodic dredging of the weeds is also required. Dredging 80% of condense and gruffly aquatic plant cover quarterly should be done, and regular de-weeding improves the overall water quality and water body's aesthetics, which can be done using manual or physical control methods, mechanical control measures and biological control methods (Zhang et al. 2018).

Controlling soil erosion is also essential for protecting the water bodies, which is under consideration for restoration and rejuvenation. As part of the restoration efforts various measures can be implemented to improve the stability and protection of the water body's surrounding areas. These measures may involve stabilizing earthen bunds, embankments, and shorelines. To achieve this, techniques such as vegetative cover, rock riprap, stone revetment, or pitching can be employed. These methods will help prevent erosion and provide effective shoreline protection for the drainage channels, contributing to the overall restoration and sustainability of the water body. All the inflow drainage channels should have silt barriers, sediment traps and sediment detention basins to trap the silt. The inflow of floating material should also be prevented using strains or traps.

Protection of a lake, pond or the river should also include protection of its drainage basin. To protect the drainage basin, all the drainage channels carrying water to the water body need to be restored. Restoration efforts can be undertaken using the following interventions: firstly, the inflow of untreated industrial effluent and municipal sewage should be stopped; secondly, based on the historical data, identification of major channels should be done, and to protect them, suitable buffer land needs to be kept free from any impervious cover; thirdly, any encroachment and blockage in its pathway should be removed (PPSCR 2016); fourthly, to naturally enhance the aeration in the water body, encroachments should be removed periodically from the drainage channels (PPSCR 2016); fifthly, to avoid unauthorized entry in the ponds or lakes, fencing should be done; and sixthly, for smooth passway of excess flow and monsoon runoff, well-designed spill ways along with control gates should be constructed, which will control the excess flooding from the drainage basin (Fig. 6.3).

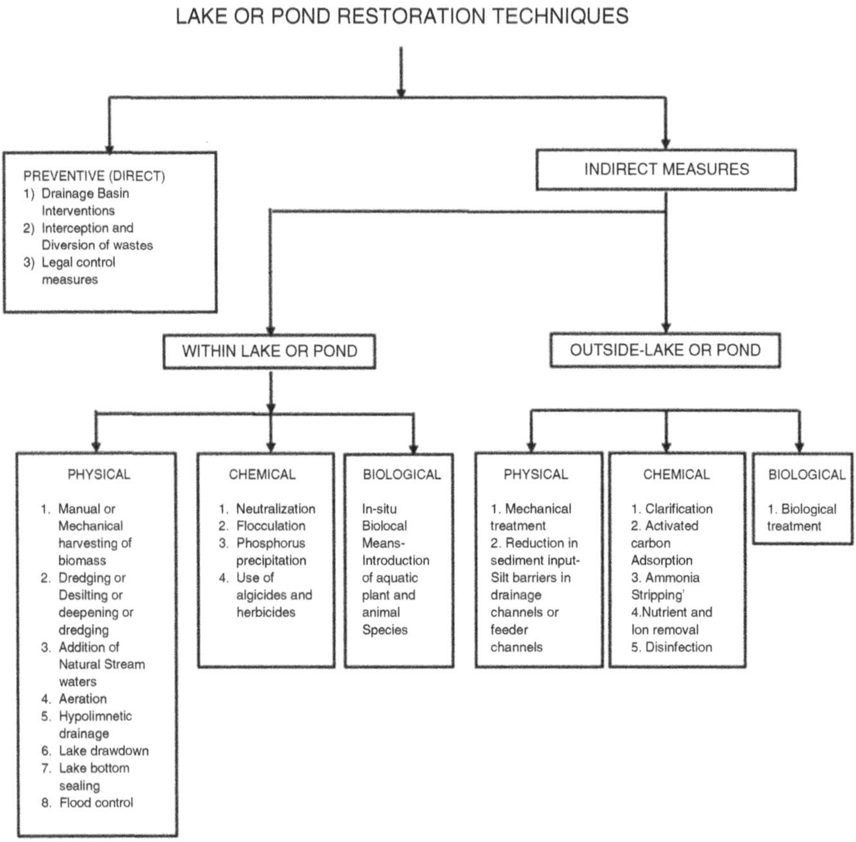

Fig. 6.3 Restoration technique for lake and ponds (CPCB 2019)

6.2.4 *Improvement Phase*

This phase is dedicated to the comprehensive improvement of the water body, with a focus on considering the needs of nature, present and future generations, and resolving conflicts among competing users. Various approaches can be employed in this phase, such as physical and chemical treatments, the use of aquatic macrophytes and animals, hydroponic techniques, floating treatment wetlands (FTW), and nutrient mixture to encourage algae growth, promoting oxygen release into the water. This, in turn, facilitates aerobic bacteria in reducing organic matter load and minimizing odors.

To manage the drainage basin effectively, efforts should be made to prevent non-point source pollution from entering the water body. Structural and land treatment measures, nutrient interception and diversion, sediment control, and the adoption of good irrigation and crop management practices are essential. Proper residue management, creation of shelter belts, and runoff control provisions for highly contaminated agriculture runoff (containing fertilizers and pesticides) should also be implemented. Maintaining green buffer zones of at least 50–100 meters around the water body as green belt zones or no activity zones is vital. These zones should not have any impervious cover and should be planted with deep-rooted plants, trees, shrubs, and grasses to absorb nutrients from anthropogenic activities. Furthermore, creating a biodiversity-friendly environment that supports migratory bird species with adequate shelter and suitable conditions for laying eggs and successful propagation should be a key consideration in the restoration efforts.

Regular monitoring of implementation of action plans for restoration of water bodies should be done at least once in 3 months. For this purpose, the action plans should have clear action points for each activity along with specific timelines to complete the work, details of the organization responsible for implementation of work, estimated budget and Program Evaluation and Review Technique (PERT) chart.

6.2.5 Sustenance Phase

The sustenance phase includes good governance based on fairness, transparency and empowerment of the stakeholders. The ownership of the lakes and ponds should be determined, and the water body should be treated as "natural resource" which will be a potential activator for better community health, providing recreational facilities and improving tourism and source to meet the probable water needs. To sustain a restored and rejuvenated water body, it is important to have well-aware citizens, resident welfare associations, local organizations, activist groups and education institutes to protect the water bodies (Goyal and Ojha 2010, 2012; Das et al. 2020). Periodic trainings of all the stakeholders should be organized on aspects related to maintenance during post-restoration phase of water bodies. The public participation should be promoted at each and every level for identification of critical problems of the water bodies as well as for its maintenance. Dissemination of information related to the water body should be displayed publicly such as water quality and dos and don'ts.

6.3 Conclusion

The restoration and rejuvenation of a water body is a non-stop progressing process where sustaining the desired results is the toughest of all the phases. The five-phase water body management plan scheme discussed in this chapter can be of use to the planners and implementers for conservation and maintaining the lost as well as polluted water bodies. The industrial clusters and local communities can play a key role in maintaining their local water bodies and can not just get aesthetic benefits but can also get monitory benefits from these healthy natural water sources. Further, in order to make any plan successful, there must be a room for strong feedback system as well as for improving the existing plan.

References

CPCB (2019) Indicative guidelines for restoration of water bodies. CPCB, MoEF&CC, GOI

Dang C, Lu M, Mu Z, Li Y, Chen C, Zhao F, Yan L, Cheng Y (2019) Phosphorus fractions in the sediments of Yuecheng Reservoir, China. Water 11:2646

Das J, Jha S, Goyal MK (2020) On the relationship of climatic and monsoon teleconnections with monthly precipitation over meteorologically homogenous regions in India: wavelet & global coherence approaches. Atmos Res 238:104889. https://doi.org/10.1016/j.atmosres.2020.104889

Goyal MK, Ojha CSP (2010) Evaluation of various linear regression methods for downscaling of mean monthly precipitation in Arid Pichola Watershed. Nat Resour 1(1):11–18. https://doi.org/10.4236/nr.2010.11002

Goyal MK, Ojha CSP (2012) Downscaling of precipitation on a lake basin: evaluation of rule and decision tree induction algorithms. Hydrol Res 43(3):215–230. https://doi.org/10.2166/nh.2012.040

Goyal MK, Panchariya VK, Sharma A, Singh V (2018) Comparative assessment of SWAT model performance in two distinct catchments under various DEM scenarios of varying resolution, sources and resampling methods. Water Resour Manag 32(2):805–825. https://doi.org/10.1007/s11269-017-1840-1

Goyal MK, Gupta AK, Jha S et al (2022) Climate change impact on precipitation extremes over Indian cities: non-stationary analysis. Technol Forecast Soc Change 180:121685. https://doi.org/10.1016/j.techfore.2022.121685

Lyu T, Song L, Chen Q, Pan G (2020) Lake and river restoration: method, evaluation and management. Water 12:977. https://doi.org/10.3390/w12040977

NASA (2019) Earth's freshwater future: extremes of flood and drought. Global climate change. Link: Earth's Freshwater Future: Extremes of Flood and Drought – Climate Change: Vital Signs of the Planet (nasa.gov)

Paerl H, Gardner W, McCarthy M, Peierls B, Wihelm S (2014) Algal blooms: noteworthy nitrogen. Science 346:175

Parliamentary Standing Committee Report (2016) Repair, renovation and restoration of water bodies- encroachment on water bodies and steps required to remove the encroachment and restore the water bodies. https://eparlib.nic.in/handle/123456789/65926?view_type=browse

Poonia V, Goyal MK, Gupta BB, Gupta AK, Jha S, Das J (2021) Drought occurrence in Different River Basins of India and blockchain technology based framework for disaster management. J Clean Prod 312:127737. https://doi.org/10.1016/j.jclepro.2021.127737

Søndergaard M, Jeppesen E, Lauridsen T, Skov C, Van Nes E, Roijackers R, Lammens E, Portielje R (2007) Lake restoration: successes, failures and long-term effects. J Appl Ecol 44:1095–1105

Subramoniam SR, Ravindranath S, Rakkasagi S, Ram H (2022) Water resource management studies at micro level using geospatial technologies. Springer, Cham, pp 49–74

Zhang H, Shang Y, Lyu T, Chen J, Pan G (2018) Switching harmful algal blooms to submerged macrophytes in shallow waters using geo-engineering methods: evidence from a 15N tracing study. Environ Sci Technol 52:11778–11785

Zhong J-C, Yu J-H, Zheng X-L, Wen S-L, Liu D-H, Fan C-X (2018) Effects of dredging season on sediment properties and nutrient fluxes across the sediment–water interface in Meiliang Bay of Lake Taihu, China. Water 10:1606

Chapter 7
Ecosystem Services for Water Management: A Review of Global Approaches and Experiences

Jayshree Shukla, Sunidhi Singh, Kavita Bramhanwade, Shalini Dhyani, Paras Pujari, and Parikshit Verma

Abstract The interlinkages between humans and nature can be effectively assessed through the ecosystem service concepts, which argue for the restoration and management of critical ecosystems. The lack of practical approaches and methodologies for these concepts in water resource management has hampered their global applications despite the increasing awareness of water conservation worldwide. This study reviews the present global and regional challenges in water conservation and ecosystem-based approaches for water management, along with case studies where service-based approaches are used for sustainable water use. The study of interconnections between multiple pressures, their impacts, the status of ecosystems, and the delivery of ecosystem services is necessary for the biophysical assessment of ecosystem services. To gain a better understanding of the interface between human rights and ecological service-based approaches, more multidisciplinary cooperation between ecologists, economists, and stakeholders should be developed. There is currently no worldwide governance mechanism in place to protect ecosystem services. According to studies, such a system is feasible in the medium- to long-term future.

Keywords Ecosystem-based approaches · Water conservation · Payment of ecosystem services · Demand mitigation

7.1 Introduction

Ecosystem services are the profits which humans derive from ecosystems, as well as ecosystems' direct and indirect contributions to human well-being. The notion of ecosystem services is important for bridging the gap between humans and nature. It emphasizes the importance of ecological functioning and biodiversity in sustaining

J. Shukla · S. Singh · K. Bramhanwade · S. Dhyani (✉) · P. Pujari · P. Verma
National Environmental Engineering Research Institute (CSIR-NEERI), Nagpur, Maharashtra, India

A. K. Gupta et al. (eds.), *Ecosystem Restoration: Towards Sustainability and Resilient Development*, Disaster Resilience and Green Growth,
https://doi.org/10.1007/978-981-99-3687-8_7

numerous human benefits (Grizzetti et al. 2016). Following the Millennium Ecosystem comprehensive assessment of global ecosystems, the idea of "ecosystem services" received attention. Later, in 2010, ecosystem services were added in the Convention on Biological Diversity (CBD) as part of the Aichi Targets as an essential indicator for assessing human well-being. The parties to the Convention on Biological Diversity approved a new Strategic Plan for Biodiversity in 2010, which included the Aichi Biodiversity Targets, which are a strengthened action to stop biodiversity loss and ensure ecosystems remain resilient and continue to offer important services. The close connection between human well-being and ecosystem health was the driving force behind this development (Carrasco 2021). In accordance with this international framework, the European Union (EU) announced the European Biodiversity Strategy to 2020 in 2011, emphasizing the importance of ecosystem services and establishing a particular objective for conserving and restoring ecosystems and their functions. Following that, the Intergovernmental Science-Policy Platform on Biodiversity and Ecosystem Services (IPBES) was established in 2012, and the notion was included in the Sustainable Development Goals (SDGs) (Pascual et al. 2017; Díaz et al. 2018; IPBES 2019).

The water cycle is directly related to a number of essential ecological services, including supply of safe drinking water, controlling flood flow, and offering opportunities for water-based recreation and cultural traditions (Hackbart et al. 2017). Aquatic ecosystems (rivers, lakes, groundwater coastal waters, and seas) can provide important ecosystem services including fish production, water provisioning, and recreation. The hydrological cycle in the river basin is also linked to key ecosystem functions such as water purification, water retention, and climate control (Goyal et al. 2018; Poonia et al. 2021; Kandel et al. 2021). Water ecosystem services are among the most beneficial ecosystem services provided by the forests and are known as even more essential because of their influences on native livelihoods, as major resources for command area users, and also as broad-ranged climate controllers through transpiration and cloud formation (Ellison et al. 2017; Netzer et al. 2019) (Fig. 7.1). Agriculture, aquaculture, industry, energy generation, human health, and ecological equilibrium all rely on water yield ecosystem services. Water availability, both in quantity and quality, is essential for socioeconomic growth, healthy ecosystems, and human life (Das et al. 2020; Tortajada 2020). Water yield assessment is increasingly being utilized as a method to transfer the values derived from natural environments to policymakers, therefore assisting in conservation prioritization and proclaiming nature's free contribution to civilization (Subramoniam et al. 2022). Globally, the ever-increasing need for water, food, and energy is putting an unsustainable strain on natural resources, frequently resulting in environmental deterioration, which has an impact on water, food, and energy security (Goyal and Ojha 2010a, b, 2012; Pérez-Sánchez et al. 2019).

The recognition of complex interconnections between multiple sectors has led to the development of holistic environmental decision-making approaches such as the Integrated Natural Resource Management (INRM), Integrated Water Resources Management (IWRM), Virtual Water (VW), Water Footprint (WF), and, more recently, Water-Environment-Energy-Food Nexus (WE2F). All of these strategies aim to strengthen collaboration between historically dissimilar sectors in order to

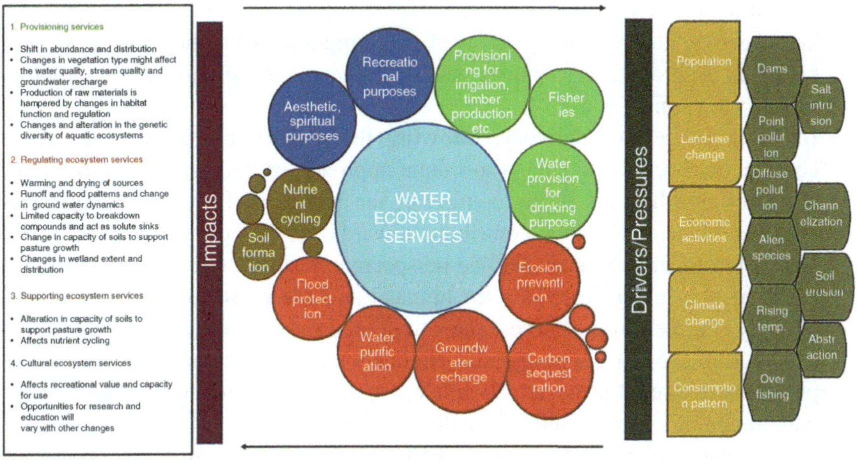

Fig. 7.1 Water ecosystem services and the drivers that affect the supply of these services along with the impact of the pressures on different services

improve resource efficiency and promote sustainability, and the only difference between them is the number and weightage of the fields included in their structure. They confront the same challenges and obstacles to implementation, some of which may never be entirely addressed (Blackstock et al. 2015). Given the knowledge base about how to apply ecosystem service theory and the fact that the Millennium Ecosystem Assessment was published 15 years ago, this article offers a brief perspective on some of the major challenges in water resource management on a global as well as national level. In addition, the chapter also briefly discusses the challenges ahead and considers how we might enhance the application of ecosystem service-based techniques to water ecosystem management in the future.

7.2 Ecosystem Service-Based Approach in Water Management

An ecosystem service-based approach is a means for comprehending the intricate interactions amid nature and people for informed policymaking with the goal of reversing ecosystem decline and guaranteeing sustainable resource use, management, and conservation. As outlined by Martin-Ortega et al. (2015), an ecosystem service-based strategy involves the following essential aspects:

(a) **The emphasis on ecological health and understanding of its implications for human well-being.** An ecosystem service-based approach adopts an anthropocentric instrumentalism perspective, emphasizing the advantages that mankind derives from environment and acknowledging that people are the ones providing worth to ecosystem components. That's in contrast to other interpretations of human-nature connections, which regard the people to be a component of a larger ecological system and deny the assumption that policymaking is only based on anthropocentric values, such as inherent value and bio- or environmental viewpoints.

(b) **In terms of service delivery, knowing the biophysical underpinnings of ecosystems is critical.** This is a unique approach of understanding about and characterizing habitats in the context of the biophysical components, mechanisms, and activities that contribute towards service provision to humans. According to an ecosystem service perspective, research should be reframed in terms of how nature provides for humans and the roles humans play in that delivery. Furthermore, it necessitates the definition and accurate assessment of the interlinkages of environmental elements and their impacts on a particular provision or a set of provisions (while recognizing intricate interlinkages) over a wide range of temporal and geographical levels.

(c) **For a profound comprehension of the service delivery process, ecological and cultural sciences, as well as other strands of knowledge, must be integrated.** By definition, an ecosystem service-based approach is transdisciplinary; it necessitates the implementation of multiple academic disciplines, for example, through codeveloped concepts, which invariably trade off accuracy in theoretical approaches in order to achieve results that are useful for decision-making. Non-academic strands of information, such as stakeholder perspectives and opinions at relevant scales, must also be included in an ecosystem service-based approach.

(d) **The evaluation of ecosystem services in order to incorporate them in policymaking.** Inherently, an ecosystem service-based approach entails a qualitative or quantitative assessment of ecosystem services, as well as the determination of monetary and non-monetary social/individual values of services. The necessity to include these principles into decision-making processes motivates this approach (Fig. 7.2)

In a water context, an ecosystem service-based approach:

- recognizes that erosion rates, water chemistry, peak flow levels, sediment load, total flow, groundwater recharge, or base flow, i.e. all the forest structural changes, may have a variety of effects on watershed processes and that these effects have a variety of effects on human well-being (Lele 2009) (core element 1).
- needs an understanding of the biophysical processes which govern how soil-vegetation dynamics, forest structure, forest cover, and other factors influence the quantity and quality of water available to a level that it affects human well-being (via use or non-use) by benefactors (core element 2).
- integrates natural science-based insights of service delivery processes (e.g. plant physiology, biodiversity, and hydrogeology) with data from sociology

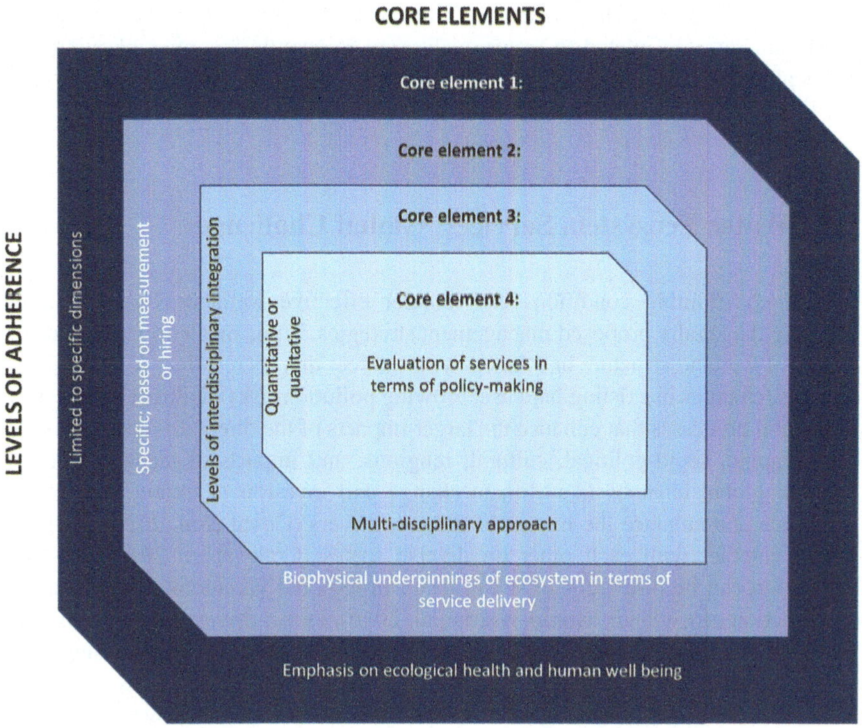

Fig. 7.2 Levels of ecosystem service-based approaches and their core elements

(e.g. economics, cognitive science, and politics) and (native) stakeholder understanding (e.g. farmers, floodplain residents, hydropower companies, and regulators) to understand, for instance, where ecosystem change takes place in relation to where benefits arise (core element 3).

- necessitates a bare minimum quantification of variations in the concluding services delivered (e.g. forest cover related with increased water flow), as well as a qualitative explanation of the impacts on human health, or the estimation of benefits related through, for instance, willing to pay for higher water accessibility, so that these incentives could be integrated into a policy (for instance, on protected forest creation or afforestation) (core element 4).

Recognizing the dependence of human well-being on natural capital is vital to overcome contemporary sustainability concerns. Natural components can be included in the system analysis using integrative frameworks like the ecosystem service method (Liu et al. 2013). The ecosystem service approach's notion of a human-ecological system is significant in combining biophysical processes and social benefits, and it allows ecosystem services to be evaluated and integrated into water resource policy planning. Moreover, economic systems for valuing ecosystem services related to water quality are frequently incompatible with biophysical models that describe the natural processes that underpin them (Keeler et al. 2012).

7.3 Water Ecosystem Services: Global Challenges

Several constraining conditions can hamper effective wetland conservation by affecting the legally proposed management strategies. Three mega drivers that result in loss of wetland structure and functions are direct drivers resulting in biophysical changes (changes in existing land use, growing pollution risks), indirect drivers such as societal processes that enhance the larger impacts of the direct drivers (economic, demographic, socio-political, cultural, religious, and impacts of technology), and ongoing global changes like climate change and invasion of exotic species that accelerate and enhance the impact of indirect drivers (Goyal et al. 2022). Multiple drivers interact with each other on diverse scales (Craig et al. 2017), and this interaction can be disastrous as it can result in regional variations in the structure and function of wetlands (Sordo-Ward et al. 2016). Direct drivers (altered course of water supply, loss of vegetation, invasion by exotic species, and influx of nutrients) that alter the natural wetlands are considered fundamental components to manage man-made wetlands. Growing urban expansion and urban solid waste will also enhance degradation of coastal wetlands (Poeta et al. 2014). The capacity of water ecosystems to provide ecosystem services is based on their ecological functions and components which are highly susceptible to changes due to direct or indirect drivers. The impacts of the combination of all direct and indirect drivers on water ecosystem services are listed in Table 7.1. Adaptation options and the fundamental aim of adaptation to various ecosystem threats are not particularly associated with specific ecosystem services and can thus be associated with overall water ecosystem services.

7.3.1 Climate Change

Water habitats are commonly exposed to a wide range of climate stresses. Wetlands are particularly vulnerable to changes in rainfall patterns, but also the effects of these on additional aspects of the water cycle (e.g. groundwater replenishment), in addition to CO_2 enrichment and warming (Martin-Ortega et al. 2015). Although the direction and magnitude of estimated changes in rainfall vary significantly by location and therefore are subject to significant uncertainty, the global shift is generally towards intensification, with moist areas becoming wetter and drier areas becoming drier, and more precipitation is expected to fall in intense rainstorms in most areas (IPBES 2019). Freshwater ecosystems are frequently exposed to severe events (e.g. floods, droughts, and intense storms) as a result of their topographic

Table 7.1 Key water ecosystem goods and services along with the pressures faced by them and the long-term impacts of those pressures along with their adaptation options

Water ecosystem services (WES)	Example of WES	Pressure on ecosystem services	Impacts	Adaptation options
1. Provisioning services				
Food	Fishing	**(a) Water quantity alteration:** Flow modifications (hydrological alterations): • Quantity and frequency (dams, water abstractions, irrigation, transfers) • Groundwater abstractions • Changes in precipitation and temperature • Changes in runoff	Shift in abundance and distribution	Protection of water ecosystem services: • Management of existing, non-climatic stressors • Expansion of protected area network • Restriction of development
Fresh water	Retention and storage		Changes in vegetation type might affect the water quality, stream quality, and groundwater recharge	
Fiber and fuel	Timber production		Production of raw materials is hampered by changes in habitat function and regulation	
Biochemical services	Material extraction from wetlands		Changes and alteration in the genetic diversity of aquatic ecosystems	
Genetic supplies	Extraction of genes from aquatic biota			
2. Regulating ecosystem services				
Climate regulation	Acts as sink for greenhouse gases	**(b) Water quality alterations** Diffuse and point pollution: nutrients, chemicals (pesticides, endocrine-disrupting compounds, nanoparticles, etc.), metals, pathogens, litter, groundwater salinization, sediments, increased turbidity, and brownification	Warming and drying of sources	Restoration of water ecosystem services: • Riparian vegetation restoration • Flow regime restoration • Removal of river bank armoring • Restore connectivity, e.g. fish ladders
Water regulation	Groundwater recharge		Topography and vegetation change will affect runoff and flood patterns and groundwater dynamics	
Water purification and waste treatment	Removal of pollutants (water filtration)		Aquatic and riparian vegetation, soils, and biogeochemistry changes may limit capacity to breakdown compounds and act as solute sinks	
Erosion regulation	Sediment retention		Hydrology and vegetation change will alter capacity of soils to support pasture growth	

(continued)

Table 7.1 (continued)

Water ecosystem services (WES)	Example of WES	Pressure on ecosystem services	Impacts	Adaptation options
Natural hazard regulation	Flood mitigation		Changes in wetland extent and distribution, aquatic vegetation, and topography will influence patterns of flooding	
Pollination and propagule dispersion	Provision of habitat for pollinators and animal dispersal vectors		Pollination and dispersal will be affected by changes in aquatic habitat and biota	
3. Supporting ecosystem services				
Soil formation	Accumulation of organic matter	(c) **Habitat alteration:** hydromorphological alterations (physical alteration of channels, bed disruption, dams)	Changes in hydrology and vegetation will alter capacity of soils to support pasture growth	Enhancement of water ecosystem services: • Planting of fast-growing, high-shade riparian trees • Storage and delivery of dilution flows to address water quality problems • Species translocations, e.g. fish stocking
Nutrient cycling	Nutrient storage		Changes to aquatic and riparian soils and biota Will affect nutrient cycling	
4. Cultural ecosystem services				
Spiritual and inspirational	Source of inspiration	(d) **Alteration in biota and biological communities:** alien species, other changes in biological communities		Replacement of water ecosystem services: • Construction of new water storage infrastructure, e.g. dams • Construction of artificial ecosystems, e.g. wetlands • Construction of sea walls
Recreational	Recreational opportunities		Changes in climate, topography, soil, water, and biota will affect recreational value and capacity for use	
Aesthetic	Appreciation for aesthetic beauty		Changes in regulating and habitat functions will affect scenery	
Educational			Opportunities for research and	

(continued)

Table 7.1 (continued)

Water ecosystem services (WES)	Example of WES	Pressure on ecosystem services	Impacts	Adaptation options
	Opportunities for education and research		education will vary with other changes	

Adapted from Capon et al. 2013; Martin-Ortega et al. 2015; Grizzetti et al. 2016

location, and the frequency and intensity of these occurrences are expected to rise in the future in many locations (Capon et al. 2013; IPBES 2019).

Water quality and quantity, flow regimes, temperature, influx of invasive species, and nutrient balance can be influenced by ongoing global climate change (Finlayson et al. 2017). Climate change has been observed to impact decision-making, as clearly evidenced for developing and expanding hydropower projects. Climate change as a direct driver can lead to biophysical changes to the wetlands by having a significant impact on the local climate, warming of wetland water, and changing the water levels and the hydroperiods (Renton et al. 2015). Climate change coupled with indirect and direct drivers, e.g. in hydropower projects, can additionally strain the natural wetlands. Natural drivers of conversion of wetlands are likely to promote man-made wetlands, directly as well indirectly. However, natural wetlands have evolved over a long period of time and are ingrained in the existing natural terrain and have been fulfilling diverse ecological functions to ensure human well-being.

7.3.2 *Agriculture Intensification*

Agriculture is the world's largest water consumer, accounting for about 70% of all water consumed annually. Agriculture can consume more than 90% of the water utilized annually in developing nations, where agriculture contributes more towards the country's economy than in developed ones (Piesse 2020). As the agricultural industry grows increasingly resource-efficient, worldwide agricultural water usage is anticipated to drop over the next three decades. Industries are the world's second-largest water user, accounting for around 19% of all water usage (Richey et al. 2015). Wastewater is also becoming a more major threat to global water security. Over 80% of the world's wastewater is believed to be discharged into the environment without being recovered or purified, damaging the ecosystem and squandering a renewable resource. Agricultural runoff frequently contains high amounts of nitrogen, pesticides, and other pollutants, contributing to the development of coastal "dead zones." Overgrowth of algae, aided by nutrient-rich wastewater, results in these zones, which break down and decrease oxygen levels. Marine life perishes or drifts away as the amount of accessible oxygen dissolved in the water decreases (Piesse 2020).

7.3.3 Extensive Chemical Use

Global fertilizer use is projected to cross 200 million tonnes by 2018 that will be a drastically high increase of 25% from 2008 levels (FAO 2015). Pesticides lead to loss of biodiversity, population structure, and also productivity of natural wetlands (Zhang et al. 2011). Eutrophication because of excessive nutrient influx from sewage wastewater, effluent discharge from industries, and fertilizer- and pesticide-laden wastewater from agriculture as well as aquaculture can alter the native biodiversity, water quality, productivity, and biological and chemical oxygen demand of the wetlands. Atmospheric nitrogen deposition is on rapid rise in developing countries, which is affecting the natural wetlands (Liu et al. 2011). Increasing nutrition due to nitrogen and phosphorus in wetlands enhances growth of algae and aquatic plants, and their health and decomposition significantly reduces the oxygen concentrations (Smith et al. 2006; Paerl and Otten 2013). Hypoxia is increasing in the coastal ecosystems (Rabalais et al. 2010), and more than 500 "dead zones" in coastal areas are identified by UNEP (UNEP 2014).

7.3.4 Increasing Water Demand

The 2018 World Water Development Report (WWDR) from the United Nations (UN) provides an update on current trends in clean water availability and future projections. Water security, defined as a population's ability to ensure long-term access to sufficient amounts of water of adequate standard, seems to be in jeopardy for many, and the issue is only going to get worse in the coming decades (Burek et al. 2016). In today's world of 7.7 billion people, clean water shortage is a serious concern. By 2050, the world population will have risen by 22–34% to 9.4 to 10.2 billion people,[1] putting a pressure on the water infrastructure. Uneven rise in population in numerous locations, which is unconnected to local resources, will exacerbate the burden. The majority of this population expansion will occur in developing nations, first in Africa and subsequently in Asia, where clean water shortage is already a significant concern. The demand for water must not exceed the supply (Gude 2017). While water demand is rising, water supply is decreasing due to dwindling supplies and pollution. At the continental level, accessible surface water supplies are expected to stay relatively stable, albeit quality will degrade and spatial and temporal distribution might shift. Aquifers will most likely decline, and salt intrusion in coastal regions will be severe. Population, gross domestic product (GDP), and water demand, on the other hand, will rise worldwide and unequally (Wada et al. 2016).

The global usage of groundwater in the 2010s was 800 km^3 per year. Sixty-seven percent of worldwide extractions were made in India, the United States, China, Iran,

[1] www.unwater.org/publications/world-water-development-report-2018/

and Pakistan (Mekonnen and Hoekstra 2016). Groundwater depletion is mostly caused by agricultural water withdrawals across the world. By 2050, groundwater extractions will have increased by 39%, to 1100 km^3 per year. Enhancing agricultural water use might result in an overall acceleration of water abstraction at the watershed level. Current worldwide withdrawals, at over 4600 km^3 per year, are already approaching optimum sustainable levels (Boretti and Rosa 2019a, b).

7.3.5 Governance

To ensure that the fundamental reasons behind loss and degradation of the wetlands are appropriately addressed, effective policy for managing and rational use of wetland requires a comprehensive understanding about the drivers. Effective governance approaches at the global, regional, national, and local levels are required to reduce, slow down, and reverse the loss of existing wetlands. Effective wetland management requires interactive governance approaches. Governance approaches and instruments (legal, economic, right based, etc.) used in wetland conservation are expected to be adaptable and transparent, having an all-inclusive approach, should be responsible, and also consider power dynamics (Mauerhofer et al. 2015). Successful wetland conservation policies should include learning, assimilation of new knowledge, formal as well as informal cooperation, assessment, and modification properties. Good governance can ensure and address effective wetland conservation (Amano et al. 2018), whereas poor governance that is built on short-term options and opportunities ignores concerns of minor stakeholder groups and reduces the effectiveness that can address long-term wetland conservation (Amano et al. 2018).

7.4 Water Ecosystem Services: Indian Perspective

About 50% of the world's urban population (1.693–2.373 billion) is anticipated to reside in water-scarce areas by 2050, with roughly a quarter concentrated in India, and 63% (19) of the world's megacities would suffer water shortages (He et al. 2021). In a backdrop of rapid industrialization, urbanization, and economic reform, India presents an essential case study for understanding the organizational and bureaucratic elements of water governance along with several other factors that pose as critical threats for present water ecosystem services.

7.4.1 Rapid Land-Use Change

Diverse wetlands located across length and breadth of the country provide diverse ecosystem services. Wetlands across India ensure nature's contributions through diverse ecosystem functions that ensure human well-being. Aquifer recharge and control of outflow followed by reducing the severity of flooding and other disaster risks are few of the key contributions.[2] In the last few decades, encroachment of natural wetlands, rapid land-use changes, emerging pollution risks, indiscriminate expansion of aquaculture in coastal areas, siltation, and infestation of wetlands by invasive alien species are constricting the conservation of wetland ecosystems in India. Accelerated extreme climate events and disaster risks coupled with climate change are further worsening the situation. More than 1/3 of that natural wetlands in the country[3] are already lost or are in severely degraded state. Around 26 wetlands notified under Ramsar convention are facing loss due to rapid shrinkage[4] from rampant infrastructure build-up. In the last few years, urban areas are new hotspots of disaster. Urban floods across Indian urban sprawls have been due to uncontrolled infrastructure development by reclaiming natural wetlands, river flood plains, or low-lying areas (Tiwari Dubey et al. 2020).

7.4.2 *Urban-Rural Water Conflicts*

The political economics of urban water governance in India is part of a larger social and political remodeling trend, in which the rise and expansion of large and expanding cities has left an enormous ecological impact on the rural areas. Growing urban water demand, combined with a failure of conservation and sewage treatment, has resulted in massive urban-rural disputes (Swyngedouw 2004; Ahluwalia 2014; Shah 2016).

On the surface, it appears that national policy conversations regarding the problems of enhancing both access and urban water quality are gaining traction. On keen investigation, archaic water regulations and the lack of a nationwide framework for water legislation have hampered the capability of Indian governments at all levels to enhance urban water management and division. The current govt's signature Smart Towns initiative has prioritized upgrades in urban water quality and hygiene but has been severely criticized for providing insufficient financing, increasing competition among cities in a competitive quest for public funds (Hoelscher 2016; Kaika 2017). The "fragmented" form of water appropriation in India is explained by the political and fiscal limits of India's government rules and allocations. Urban-rural water transfers have been marked by interstate conflict in courts, legislature, Gram

---

[2] https://www.downtoearth.org.in/blog/natural-disasters/urban-flooding-the-case-of-drowning-cities-and-rising-vulnerability-67203

[3] https://www.dailypioneer.com/2016/columnists/wetlands-are-not-wastelands.html

[4] https://www.livemint.com/Politics/rOFyi3baM1800Zif2DDrkK/The-vanishing-wetlands-in-India.html

panchayat, and city corporations, as well as in the neighborhoods and the countryside, rather than policy integration (Shah 2016; Punjabi and Johnson 2019).

7.4.3 Lack of Transparency

Lack of transparency in the utilization of funds by multi-stakeholder groups is a threat to the conservation and management of wetland in the country. One of the recent reports by the Parliamentary Standing Committee on Science and Technology, Environment, Forests and Climate Change highlighted the matter of dearth of transparency and "misuse" of capitals in some orders under the MoEF&CC.[5] Numerous such cases of misappropriation of funds have been reported in conservation of wetlands[6], while most of them have been repeatedly left unaddressed.

The institution of the Right to Information Act has resulted in noteworthy result for civil society participation and stresses for administrative transparency and responsibility, as well as responsiveness of states. However, these processes require serious monitoring from civil society by promoting democratic conversation to address power inequalities that are not extreme in nature. To realize justifiable outcomes in conservation, rational use of wetlands includes multi-stakeholder involvement and transparency for assigning ecosystem trade-offs associated with diverse wetland uses (Finlayson et al. 2017). By operating as a societal response scheme, economic assessments enhance probability of taking smart decisions for wetland use and conservation.

7.5 Water Ecosystem Services Management: Case Studies

When ecosystems are used to satisfy human requirements (e.g. water supply and food production), all ecosystems are influenced in some manner. The question is whether these effects are significant enough to overwhelm an ecosystem's ability to supply such services in a sustainable and equitable manner or to provide various ecosystem services as people and countries evolve. The following are some case studies from across the world where ecosystem-based approaches for water management have been implemented and the lessons learnt from those practices for further improvements:

[5] https://theprint.in/india/governance/misappropriation-of-funds-lack-of-transparency-in-environment-ministry-schemes-parliamentary-panel/618163/

[6] https://tunza.eco-generation.org/ambassadorReportView.jsp?viewID=44546

7.5.1 Chilika Lake: India (Sahu et al. 2014)

The largest coastal brackish water lagoon lake situated along the eastern coast in India, the Chilika lake, is famous for its rich aquatic biodiversity (presence of Irrawaddy dolphins (*Orcaella brevirostris*) that have made the lake a major tourist attraction) and provides a wide range of ecosystem services such as enormous fisheries sources (sustaining numerous economic livelihoods), resting site for over one million winter migratory avian species, and various vegetation sources (aquatic weeds, etc.), along with several recreational, socioeconomic, and religious values. The lake was added in the Montreux Record (threatened list of Ramsar sites) in 1993 due to constructions of hydraulic structures upstream altering the water flow and sediment transport causing loss of hydraulic connection between lake and ocean leading to increased turbidity, decreased salinity, and introduction of alien invasive species with reduced surface area of the lake. Introduction of aquaculture and native-corporate conflicts were also among the major reasons for ecosystem degradation and loss of many livelihoods that depended on the ecosystem services provided by the lake.

The creation of the Chilika Development Authority (CDA) in 1992 was a significant step in stopping the deterioration of the lake environment. In September 2000, the lake mouth was opened, and a canal was created through the barrier beach at Satpara, resulting in the biological regeneration and restoration of the coastal lake environment. The interchange of marine and brackish waters was secured by a decreased canal length of 18 km and the resulting de-siltation, improved water quality, restoration of micro- and macro-habitats, enhanced fishery sources, and recovery of threatened fish and prawn species along with enhanced ecotourism owing to the return of dolphins, supporting several financial livelihoods.

7.5.1.1 Important Lessons Learnt

1. Restoration initiatives are accelerated by the application of ecosystem-based approaches by relieving the pressure on the ecosystem services along with stabilization of energy and matter cycles.
2. Ecosystem approach-based management and integration of community-based management also facilitates restoration of fragmented habitats improving the ecosystem productivity which is the basis of all ecosystem services.
3. Ecotourism generates numerous economic benefits to the surrounding communities only if practiced within ecological limits.

7.5.2 Payment of Ecosystem Services (PES): Nepal[7]

Kulekhani is a 12,500-ha watershed in Makwanpur District situated about 50 km southeast of Kathmandu in Nepal. The catchment area is home to more than 46,000 people living in 8 villages, which practice sloping land agriculture, use of forest products, and livestock rearing. The Kulekhani reservoir was built on the site of the Indra-Sarobar Lake in 1982 for the collection of monsoon rain and channel water from the reservoir to the hydropower plant downstream. Later, another hydropower plant was added just below it. These plants generate about 17% of the total hydroelectricity in Nepal. A continuous rainfall (542 mm) in a 24-h period in July 1993 led to massive sedimentation and many landslides in the reservoir. A partially excavated hill above the reservoir was washed into the reservoir which led to the reduction of life expectancy of the reservoir from 100-year life span to a third of it in that 1 day. Owing to the commercial value of the hydropower plants and the necessity for their management, several participatory conservation programs were initiated by the government by employing the native people to build sediment trap dams and adopt measures to control gullies. Large pine monocultures were planted on both state forest and village lands by the government, and the farmers were provided seedlings of pine trees to plant them on their agriculture terraces.

A PES scheme was established in 2003 through a collaboration between RUPES program of ICRAF and Winrock International, for the payment of services scheme between the hydropower plant and upland communities in Kulekhani watershed. Government receives royalties from all the hydropower plants and channels it to various levels of developmental activities, as established by law. However, 12% of these royalties was earlier used for the district where the plant is situated, 38% for other district, and the remaining 50% for other development regions of the country.

Within this regulatory framework, the new PES scheme was proposed in the following ways:

1. A portion of its revenue could be paid by the hydropower company directly to the upland people for their ecosystem services.
2. A portion of the royalties received by the government could be paid directly to the upland communities.

Based on these, in 2006, the Makwanpur DDC enacted a rule that outlined how the government's 12% royalty would be distributed. The district development center (DDC) should now spend 12% royalty in the hydropower plant-affected region, while the other half can be spent in other parts of the district, according to the Hydropower Royalty Distribution and Use Directive 2062. The rule further states that 20% of the 50% allocation to the affected region will go to the upstream watershed area (catchment), 15% to the area affected by power plant, and the last 15% to the downstream area. As a result, the upstream catchment community

[7] https://www.indiawaterportal.org/sites/default/files/iwp2/Payments_for_ecosystem_services_and_food_security_Food_and_Agriculture_Organisation_2011.pdf

receives a larger royalty share as compared to other regions, and the amount is placed in the DDC's Environmental Management Special Fund (EMSF). The funds will be utilized to help watershed communities' conservation and development initiatives. The EMSF is a payment made to upland watershed societies in compensation for ecological services provided by them.

7.5.2.1 Important Lessons Learnt

1. Individual preferences and land tenure concerns are not necessary constraints for a PES system to be implemented at the community level. This sort of implementation benefited greatly from the long tradition of forest management at the community level.
2. The project's primary flaw was the indirect payment method, which was handled by a government agency (Makwanpur DDC), making it subject to local disputes and political instability. As a result, despite the fact that the local bodies were given full management responsibility of local resources by the 1999 Local Self-Government Act and the 1992 Decentralization Act, the current budgetary conflict is impeding the effective ongoing implementation of the PES scheme.

7.5.3 *Lakes Osmansagar and Himayatsagar: India*[8],

The twin drinking supply reservoirs of Osmansagar and Himayatsagar are roughly 100 km northwest of Hyderabad. The lakes were built in 1908 as a result of disaster relief operations following the catastrophic Musi River floods. The two reservoirs together supply more than 6000 m^3 of drinking water or approximately 5–10% of Hyderabad's total water consumption. The lakes harbor several species of algae, zooplankton, and fishes which support a booming fishery and are used for recreational purposes. The fact that these two ecosystems maintain such a high degree of biodiversity demonstrates their outstanding ecological health.

Because of the construction of large-scale rainwater collection assemblies, the watersheds of the two reservoirs have changed dramatically in recent years. According to a recent research, every 2 ha of the watershed has some form of intervention, such as check dams, infiltration tanks, or contour trenching. The constructions have effectively impeded water flows to the point that water bodies have not touched full tank level (FTL) in recent years and in some cases have gone totally dry. Observable fluctuations in the monsoon pattern over the previous 25 years have aggravated the problem. The major problem in putting the comprehensive Government Order (GO 111) into action is figuring out how to fulfill the basin's developmental demands and economic goals while also protecting the

[8] http://cdn.cseindia.org/userfiles/hyderabad_portraits.pdf

reservoirs' ecological integrity. In this regard, a significant people's movement has emerged in Hyderabad. The demands of the civil movement include removal of check dams and other water harvesting structures that obstruct water inflows to ensure natural water flows and maintain reservoir hydrology. It also includes the promotion of organic farming and farmers receiving all possible incentives to reduce pesticide and fertilizer loads on water resources. Forest reserves in the catchment areas of both reservoirs must be revived with funding. A properly designed subsidy policy should encourage decentralized and ecologically sustainable sewage treatment systems in basin habitations, and modern technologies such as GPS and remote sensing should be used to monitor the 22 water inlets feeding both reservoirs to avoid surface and groundwater pollution.

7.5.3.1 Important Lessons Learnt

1. Water basins have a significant impact on the environmental well-being of their constituent ecologies, which in turn affects ecosystem services, necessitating a high significance in water resource management. Alternative income-generating options for basin societies have developed through the newly introduced eco-sustainable enterprises which have the ability to avert unfavorable expansions that have negative consequences for water resources.
2. For long-term sustainability, environmental quality, and the continuing flow of ecosystem products from a water resource, dealing with the many consequences that emerge from reservoir basin urbanization is a significant challenge. Sewage should be considered as a potential resource rather than a waste management problem, and harmful industrial effluents must be properly treated and disposed of.
3. Various stakeholders, non-governmental organizations (NGOs), and knowledge- and information-based networks all play an important role in integrated ecosystem conservation and management, including acting as whistle-blowers to ensure proper water and ecological governance.

7.6 Water Security for Sustainable Ecosystem Services: Way Forward

Understanding water ecosystem services necessitates an understanding of the interlinkages between hydrological, geographical, and ecological systems, as well as a conceptualization of how water affects human lives and welfare, as well as how human activities affect these ecosystems. Ecosystem service-based approaches attempt to comprehend these intricate linkages in order to aid in more efficient and long-term decision-making. Rather than just managing the resource, the society must understand the necessity to balance and manage the advantages gained from water resources. The management of catchments or watersheds necessitates an understanding of how land and water are intimately linked and that every land-use judgment is indeed a water use decision. It necessitates a perspective for management and planning which encompasses all parts of the ecosystem and their interconnections and acknowledges that water choices cannot be made in separation from the people who rely on those ecosystems.

7.6.1 Water Demand Mitigation

Rising population has increased the water demand which must be mitigated in order to control the resource use and for sustainable management of water resources. Water supply enhancement is one such approach that can be facilitated by present or approaching constructions, as well as those aimed at long-term sustainability. Water transfer actions between areas are used in short-term and urgent methods. Intercepting (rainwater harvesting), redirecting (water diversions), collecting (dams and reservoirs), and transporting water are the major techniques for increasing water supply by transferring water among basins or regions. The use of retrofitting water-saving fixtures in household, industrial production, and other public facilities where water consumption is required are examples of conservation efforts. The agricultural sector accounts for the majority of water loss, followed by the power generation industry. Irrigation methods that conserve water can make a major difference in this area. In power generation, between 30 and 75% of water withdrawals are wasted, particularly in cooling applications. The loss of water due to evaporation can be overcome by other approaches such as dry cooling, etc. Including the possibility of water reuse in the water supply and management scheme would enhance the understanding of the actual value of this approach's contribution. It can help with water policy creation and execution, which might lead to long-term resource sustainability (more particularly, availability of the resource).

7.6.2 Future Areas for Improving Water Conservation Policies

Numerous technological innovations have the potential to increase water resource management or monitoring, such as smart urban water management, which involves a multitude of detectors throughout all cities and near-real-time analysis and forecasts, drone technology improvements and cost reductions, and the use of media platforms to comprehend how people connect with wetlands (Taylor et al. 2021). Wetland management must place a greater emphasis on integrated water ecosystem-based evaluation and stakeholder involvement (Xu et al. 2018). Collaborations between government, scholars, business, and the community can help in the conservation of wetlands and their policies by restricting access to the majority of wetland areas and educating the public to combat the misconception that wetlands are wastelands. Exploring trade-offs among ES and linking them with stakeholders can help to determine the potential losers and winners of wetland management (Guida et al. 2016). In order to ensure the appropriate use of the nation's wetlands and meet international wetland conservation duties, the National Wetland Policy and any programs developed under it are linked to other land, soil, water, air, wildlife conservation, and economic development policies. Policy plans must address key areas that indicate the government's goals while also encouraging the required amount of assistance and participation from other stakeholders.

Conflict of Interest Authors declare no conflict of interest for this book chapter.

References

Ahluwalia I (2014) Transforming our cities: postcards of change. Harper Collins, New Delhi

Amano T, Székely T, Sandel B, Nagy S, Mundkur T, Langendoen T, Sutherland WJ (2018) Successful conservation of global waterbird populations depends on effective governance. Nature 553(7687):199–202

Blackstock KL, Martin-Ortega J, Spray CJ (2015) Implementation of the European water framework directive: what does taking an ecosystem services-based approach add? In: Launch of UNESCO CUP publication. Cambridge University Press, Cambridge, pp 57–64

Boretti A, Rosa L (2019a) Reassessing the projections of the world water development report. NPJ Clean Water 2:15. https://doi.org/10.1038/s41545-019-0039-9

Boretti A, Rosa L (2019b) Reassessing the projections of the world water development report. NPJ Clean Water 2(1):1–6

Burek P, Satoh Y, Fischer G, Kahil MT, Scherzer A, Tramberend S, Wiberg D (2016) Water futures and solution-fast track initiative. International Institute for Applied Systems Analysis, Laxenburg

Capon SJ, Chambers LE, Mac Nally R et al (2013) Riparian ecosystems in the 21st century: hotspots for climate change adaptation? Ecosystems 16:359–381

Carrasco P (2021) The knowledge status of coastal and marine ecosystem services—challenges, limitations and lessons learned from the application of the ecosystem services approach in management. Front Mar Sci 8:27. https://doi.org/10.3389/fmars.2021.684770

Craig LS et al (2017) Meeting the challenge of interacting threats in freshwater ecosystems: a call to scientists and managers. Elem Sci Anth 5:72. https://doi.org/10.1525/elementa.256

Das J, Jha S, Goyal MK (2020) On the relationship of climatic and monsoon teleconnections with monthly precipitation over meteorologically homogenous regions in India: Wavelet; global coherence approaches. Atmos Res 238:104889. https://doi.org/10.1016/j.atmosres.2020.104889

Díaz S, Pascual U, Stenseke M, Martín-López B, Watson RT, Molnár Z, Shirayama Y (2018) Assessing nature's contributions to people. Science 359(6373):270–272

Ellison D, Morris CE, Locatelli B, Sheil D, Cohen J, Murdiyarso D, Gutierrez V, van Noordwijk M, Creed IF, Pokorny J, Gaveau D, Spracklen DV, Bargués Tobella A, Ilstedt U, Teuling AJ, Gebrehiwot SG, Sands DC, Muys B, Verbist B, Springgay E, Sugandi Y, Sullivan CA (2017) Trees, forests and water: cool insights for a hot world. Glob Environ Chang 43:51–61, ISSN 0959-3780. https://doi.org/10.1016/j.gloenvcha.2017.01.002

FAO (2015) FAOSTAT. Food and Agriculture Organization of the United Nations, Rome, Italy. http://faostat.fao.org/default.aspx

Finlayson CM, Capon SJ, Rissik D, Pittock J, Fisk G, Davidson NC, Bino G (2017) Policy considerations for managing wetlands under a changing climate. Mar Freshw Res 68(10): 1803–1815

Goyal MK, Ojha CSP (2010a) Evaluation of various linear regression methods for downscaling of mean monthly precipitation in arid Pichola watershed. Nat Res 01(01):11–18. https://doi.org/10.4236/nr.2010.11002

Goyal MK, Ojha CSP (2010b) Evaluation of various linear regression methods for downscaling of mean monthly precipitation in arid Pichola watershed. Nat Resour 01(01):11–18. https://doi.org/10.4236/nr.2010.11002

Goyal MK, Ojha CSP (2012) Downscaling of precipitation on a lake basin: evaluation of rule and decision tree induction algorithms. Hydrol Res 43(3):215–230. https://doi.org/10.2166/nh.2012.040

Goyal MK, Panchariya VK, Sharma A, Singh V (2018) Comparative assessment of SWAT model performance in two distinct catchments under various DEM scenarios of varying resolution, sources and resampling methods. Water Resour Manag 32(2):805–825. https://doi.org/10.1007/s11269-017-1840-1

Goyal MK, Gupta AK, Jha S et al (2022) Climate change impact on precipitation extremes over Indian cities: non-stationary analysis. Technol Forecast Soc Change 180:121685. https://doi.org/10.1016/j.techfore.2022.121685

Grizzetti B, Lanzanova D, Liquete C, Reynaud A, Cardoso A (2016) Assessing water ecosystem services for water resource management. Environ Sci Pol 61:194

Gude VG (2017) Desalination and water reuse to address global water scarcity. Rev Environ Sci Biotechnol 16:591–609. https://doi.org/10.1007/s11157-017-9449-7

Guida RJ, Remo JW, Secchi S (2016) Tradeoffs of strategically reconnecting rivers to their floodplains: the case of the Lower Illinois River (USA). Sci Total Environ 572:43–55

Hackbart VC, de Lima GT, dos Santos RF (2017) Theory and practice of water ecosystem services valuation: where are we going? Ecosyst Serv 23:218–227

He C, Liu Z, Wu J et al (2021) Future global urban water scarcity and potential solutions. Nat Commun 12:4667. https://doi.org/10.1038/s41467-021-25026-3

Hoelscher K (2016) The evolution of the smart cities agenda in India. Int Area Stud Rev 19(1): 28–44

IPBES (2019) Summary for policymakers of the global assessment report on biodiversity and ecosystem services of the intergovernmental science-policy platform on biodiversity and ecosystem services. IPBES Secretariat, Bonn. https://www. ipbes. net/global-assessment-report-biodiversity-ecosystem-services

Kaika M (2017) 'Don't call me resilient again!': the new urban agenda as immunology... Or... What happens when communities refuse to be vaccinated with 'smart cities' and indicators. Environ Urban 29(1):89–102

Kandel P, Chettri N, Chaudhary S, Sharma P, Uddin K (2021) Ecosystem services research trends in the water tower of Asia: a bibliometric analysis from the Hindu Kush Himalaya. Ecol Indic 121: 107152. ISSN 1470-160X

Keeler BL, Polasky S, Brauman KA et al (2012) Linking water quality and Well-being for improved assessment and valuation of ecosystem services. Proc Natl Acad Sci 109(45):18619–18624

Lele S (2009) Watershed services of tropical forests: from hydrology to economic valuation to integrated analysis. Curr Opin Environ Sustain 1(2):148–155

Liu S, Yan B, Wang L (2011) The layer effect in nutrient removal by two indigenous plant species in horizontal flow constructed wetlands. Ecol Eng 37(12):2101–2104

Liu S, Crossman ND, Nolan M, Ghirmay H (2013) Bringing ecosystem services into integrated water resources management. J Environ Manag 129:92–102

Martin-Ortega J, Ferrier R, Gordon I (2015) Water ecosystem services: a global perspective. UNESCO, Paris

Mauerhofer V, Kim RE, Stevens C (2015) When implementation works: a comparison of Ramsar convention implementation in different continents. Environ Sci Pol 51:95–105

Mekonnen MM, Hoekstra AY (2016) Four billion people facing severe water scarcity. Sci Adv 2(2):e1500323
Netzer MS, Sidman G, Pearson TR, Walker SM, Srinivasan R (2019) Combining global remote sensing products with hydrological modeling to measure the impact of tropical forest loss on water-based ecosystem services. Forests 10(5):413
Paerl HW, Otten TG (2013) Harmful cyanobacterial blooms: causes, consequences, and controls. Microb Ecol 65(4):995–1010
Pascual U, Balvanera P, Díaz S, Pataki G, Roth E, Stenseke M, Yagi N (2017) Valuing nature's contributions to people: the IPBES approach. Curr Opin Environ Sustain 26:7–16
Pérez-Sánchez J, Senent-Aparicio J, Segura-Méndez F, Pulido-Velazquez D, Srinivasan R (2019) Evaluating hydrological models for deriving water resources in peninsular Spain. Sustainability 11(10):2872
Piesse M (2020) Global water supply and demand trends point towards rising water insecurity. Future Directions International, Nedlands
Poeta G, Battisti C, Acosta AT (2014) Marine litter in Mediterranean sandy littorals: spatial distribution patterns along Central Italy coastal dunes. Mar Pollut Bull 89(1–2):168–173
Poonia V, Goyal MK, Gupta BB, Gupta AK, Jha S, Das J (2021) Drought occurrence in different river basins of India and blockchain technology based framework for disaster management. J Clean Prod 312:127737. https://doi.org/10.1016/j.jclepro.2021.127737
Punjabi B, Johnson CA (2019) The politics of rural–urban water conflict in India: untapping the power of institutional reform. World Dev 120:182–192, ISSN 0305-750X,. https://doi.org/10.1016/j.worlddev.2018.03.021
Rabalais NN, Diaz RJ, Levin LA, Turner RE, Gilbert D, Zhang J (2010) Dynamics and distribution of natural and human-caused hypoxia. Biogeosciences 7(2):585–619
Renton DA, Mushet DM, DeKeyser ES (2015) Climate change and prairie pothole wetlands: mitigating water-level and hydroperiod effects through upland management. US Department of the Interior, US Geological Survey, Reston, VA
Richey AS, Thomas BF, Lo MH, Reager JT, Famiglietti JS, Voss K, Rodell M (2015) Quantifying renewable groundwater stress with GRACE. Water Resour Res 51(7):5217–5238
Sahu BK, Pati P, Panigrahy RC (2014) Environmental conditions of Chilika Lake during pre and post hydrological intervention: an overview. J Coast Conserv 18:285–297. https://doi.org/10.1007/s11852-014-0318-z
Shah M (2016) Urban water systems in India. A way forward. ICRIER. Working paper no 23. http://icrier.org/pdf/Working_Paper_323.pdf. Accessed 30 Sept 2021
Smith E, Gordon R, Madani A, Stratton G (2006) Year-round treatment of dairy wastewater by constructed wetlands in Atlantic Canada. Wetlands 26(2):349–357
Sordo-Ward Á, Granados I, Martín-Carrasco F, Garrote L (2016) Impact of hydrological uncertainty on water management decisions. Water Resour Manag 30(14):5535–5551
Subramoniam SR, Ravindranath S, Rakkasagi S, Ram H (2022) Water resource management studies at micro level using geospatial technologies. Springer, Cham, pp 49–74
Swyngedouw E (2004) Social power and the urbanization of water-flows of power. Oxford University Press, Oxford
Taylor NG, Al Grillas P, Hreisha H et al (2021) The future for Mediterranean wetlands: 50 key issues and 50 important conservation research questions. Reg Environ Change 21:33. https://doi.org/10.1007/s10113-020-01743-1
The Economist (2014). A planet of suburbs" the economist 6 December 2014, pp. 49–55. http://www.economist.com/suburbs
Tiwari Dubey J, Subramanian V, Kumar N (2020) Policy interventions in achieving water security in India. In: Shukla V, Kumar N (eds) Environmental concerns and sustainable development: volume 1: air, water and energy resources. Springer, Singapore, pp 275–291. https://doi.org/10.1007/978-981-13-5889-0_14
Tortajada C (2020) Contributions of recycled wastewater to clean water and sanitation sustainable development goals. NPJ Clean Water 3:22. https://doi.org/10.1038/s41545-020-0069-3

UNEP (2014) Valuing plastics: the business case for measuring, managing and disclosing plastic use in the consumer goods industry. U.N.E. Program, Nairobi

Wada Y, Flörke M, Hanasaki N, Eisner S, Fischer G, Tramberend S et al (2016) Modeling global water use for the 21st century: the water futures and solutions (WFaS) initiative and its approaches. Geosci Model Dev 9(1):175–222

Xu X, Jiang B, Tan Y, Costanza R, Yang G (2018) Lake-wetland ecosystem services modeling and valuation: progress, gaps and future directions. Ecosyst Serv 33:19–28

Zhang W, Jiang F, Ou J (2011) Global pesticide consumption and pollution: with China as a focus. Proc Int Acad Ecol Environ Sci 1(2):125

Chapter 8
Rejuvenation of Rivers in India: A Case Study on Efforts for Rejuvenation of River Ganga

Jitesh N. Vyas, Supriya Nath, R. B. Deogade, and Prabhat Chandra

Abstract Ganga, our national river, which is considered sacred by the Hindus, is facing the worst level of pollution. Its water which was once used for drinking has deteriorated to such an extent that it has become unfit even for bathing at most of the stretches of river. Its water quality deteriorates as it progresses from upstream to downstream. Anthropogenic activities like disposal of untreated sewage and industrial effluents, bathing, washing clothes, the bathing of animals, construction of dams, religious activities, waste generated due to tourism, disposal of half-burnt dead bodies, etc. are the main reason behind the pollution of River Ganga. Self-cleansing capacity of River Ganga is badly influenced by the pollution caused due to increase in population density, unplanned industrialization, and urbanization.

The government has been continuously working on various ways to improve the quality of River Ganga for more than three decades but didn't get the expected results. A number of initiatives have been undertaken to clean the River Ganga which are the Ganga Action Plan, Namami Gange Programme, etc. Namami Gange aims at Ganga rejuvenation by focusing on all the present efforts and preparing an effective action plan for the future. Academic institutes like the Indian Institute of Technology and non-governmental organizations like the India Water Portal, Centre for Science and Environment, Energy and Resources Institute (TERI), Tarun Bharat Sangh, Save Ganga Movement, and Ganga Mahasabha are also involved in efforts for rejuvenation of River Ganga. All the initiatives taken by the government can only be successful with active community involvement in mission mode for rejuvenating Ganga.

Keywords Pollution · Anthropogenic activities · Water quality · Self-cleansing capacity · Namami Gange programme · Ganga action plan

J. N. Vyas (✉) · S. Nath · R. B. Deogade · P. Chandra
Central Water and Power Research Station, Pune, India

A. K. Gupta et al. (eds.), *Ecosystem Restoration: Towards Sustainability and Resilient Development*, Disaster Resilience and Green Growth,
https://doi.org/10.1007/978-981-99-3687-8_8

8.1 Introduction

Our country, endowed with many river systems and its fertile alluvial flood plains, formed the basis for the origin of the Indus Valley civilization at a riverbank. Most of our daily requirements of water get fulfilled by surface water. These rivers have served us for generations as our economy, culture, and festivals have been centered on our rivers. Their overexploitation and pollution have led to their deterioration. Their deterioration has caused devastating impacts on wildlife, flora, and fauna as well as on our economy. Generation of massive wastewater from households, industries, and agricultural farms (Kumar et al. 2020) as well as surface runoff from surrounding catchment areas deposits organic and inorganic matter in the river. These deposits lead to further deterioration of water quality which causes the extinction of river biota (Kumar et al. 2020) and causes health impacts on humans. River rejuvenation is defined to be an effort that can restore the poor health of rivers that are polluted and overexploited (Shekhar 2016). Many rivers in India are perennial and peninsular. We are mainly dependent on rivers for irrigation and drinking purpose.

In India, the River Ganga is considered the most sacred and important river as our cultural ethos has been centered on it (Chaudhary and Walker 2019). It was declared to be the national river in 2008 by the Indian government (Ganga River Basin Environment Management Interim Report 2013). Forty-four percent of the Indian population living in approximately 30 cities, 70 towns, and several thousands of villages beside the river (Nandi et al. 2016; Kumar et al. 2020) depends on the River Ganga (Chaudhary et al. 2017; Chaudhary and Walker 2019). It covers a large catchment area (Xun et al. 2017) and crosses five states, viz., Uttarakhand, Uttar Pradesh (UP), Bihar, Jharkhand, and West Bengal (Chaudhary and Walker 2019). It is regarded as a living Goddess and worshipped as Ganga Maa by Hindus (Xun et al. 2017). The banks of River Ganga are used for cremation by Hindus as they believe that Ganga Maa provides "moksha" to the soul of dead bodies burnt on its bank (Naskar 2014; Chaudhary and Walker 2019). The river water is renowned for its use as antimicrobial and medicinal properties (Nautiyal 2009; Chaudhary and Walker 2019) due to bacteria like *Escherichia*, *Pseudomonas*, and *Enterobacter* and bacteriophages in the river water (Dwivedi et al. 2020; Kumar et al. 2020) which provides self-cleaning property to River Ganga (Kumar et al. 2020).

The Ganga covers approximately 8,61,404 km^2 (Chaudhary and Walker 2019; Das et al. 2020), which constitutes 26% of the geographic area of India (IAAD 2017). It has the largest river basin in India (Goyal and Ojha 2010; Goyal and Ojha 2012; Kumar et al. 2020), the 15th largest in Asia, and 29th in the world (Chaudhary and Walker 2019). Various major historic and sacred cities like Haridwar, Kanpur, Allahabad, Varanasi, and Kolkata are located along the bank of River Ganga (Joshi et al. 2009; Chaudhary and Walker 2019; Kumar et al. 2020).

8.1.1 Need of Rejuvenation

Rivers are dynamic and have an ecosystem that results from the interaction between biotic and abiotic constituents (Shekhar 2016; Shekhar and Prasad 2009). Hydrological variables like changes in environmental flow determine the ecological function of rivers (Shekhar 2016). This environmental flow when impeded by anthropogenic or natural factors can impact directly on river health. The natural flow gets disturbed during the construction of dams, and diverted water is used for various purposes like domestic, industrial, or agricultural use (Poonia et al. 2021; Goyal et al. 2018; Shekhar 2016). Activities like sand mining, deforestation in the catchment area, and construction activities around the river have also adversely impacted the river health (Shekhar 2016). Agricultural farming, overgrazing by animals, and arbitrary deforestation result in increased surface runoff leading to soil erosion (Sinha et al. 2013). The increased surface runoff leads to siltation of rivers which causes a reduction in flow and converts these perennial rivers into seasonal (Sinha et al. 2013). The situation has worsened due to the pollution of rivers.

8.1.2 Factors Responsible for Deterioration of River Ganga

Various problems that influence the water quality of rivers are indiscriminate urbanization, industrialization, excessive use of chemicals in agriculture, deforestation, excessive use of water for agricultural and industrial use (IAAD 2017; Kumar et al. 2020), inadequate and mostly inefficient sewage treatment plants (Xun et al. 2017; Reymond et al. 2020), insufficient funding for development of water quality infrastructure, and inefficient administrative execution (IAAD 2017; Kumar et al. 2020). Approximately one billion liters of untreated wastewater discharged every day into the River Ganga from numerous drainage discharges deteriorates its water quality (Chaudhary and Walker 2019). This deteriorated water quality causes a reduction in the flow of rivers and over-dependence on groundwater and thus disturbs the hydrological cycle (Khan et al. 2017; Kumar et al. 2020). It also causes floods in the river basin (Chaudhary and Walker 2019). The River Ganga is considered to be the most polluted river in the world despite being a lifeline for almost 50 million people (Chaudhary and Walker 2019). Sources of pollution in Ganga include both inorganic and organic pollutants consisting of urban sewage, industrial wastes, runoff from farmlands, and religious wastes (Chaudhary and Walker 2019). Among all the sources of pollution, urban sewage and industrial wastes are the most significant cause of degradation of the water quality of River Ganga (Dwivedi et al. 2018; Chaudhary and Walker 2019). Approximately three-fourths of pollution in the river is caused by the dumping of unprocessed municipal waste from expanding urban residential areas in the river basin (Das 2011; Chaudhary and Walker 2019). The high load of municipal waste in River Ganga is due to the location of 29 large cities which has populations greater than one million. Almost 8250 million liter per day (MLD) of municipal wastewater comes from cities in the Ganga basin; however, treatment plants have the capacity to treat 3500 MLD, and around 2550 MLD of the untreated waste is dumped into the Ganga (Chaudhary and Walker 2019).

8.1.3 Status of Pollution in River Ganga

As indicated in Fig. 8.1, the most polluted stretch of River Ganga is the middle stretch as most of the industrial cities like Kanpur and historic cities like Allahabad and Varanasi, which attract tourists for pilgrimage, are situated in this region. Also, the flow of the river reduces drastically here as the river reaches the plain surface. This sudden reduction in flow and inputs of industrial and municipal waste from densely populated cities located in this stretch, solid wastes in the form of plastics, wastes dumped by tourists, burning of dead bodies, atmospheric depositions through rainfall (Pandey et al. 2014; Yadav and Pandey 2017), etc. also make the River Ganga the most polluted in this part. Recently there have been lot of development activities for expanding Varanasi without following proper technical standards while construction of roads. There is requirement of more sewage treatment plants to treat

Fig. 8.1 River Ganga, its major tributaries, and its most polluted stretch

waste water, better garbage collection system, and drainage system (Pandey and Singh 2017).

8.1.4 Problems Due to Deterioration of Water

Various parameters indicating water quality, viz., pH, electrical conductivity, biological oxygen demand, chemical oxygen demand, and dissolved oxygen, show deterioration when untreated wastewater gets discharged in the river (Chaudhary and Walker 2019). The Central Pollution Control Board (CPCB) found that microbial counts were more than acceptable limits in drinking (50 MPN/100 mL) and bathing water (500 MPN/100 mL) (CPCB (Central Pollution Control Board) 2009; Chaudhary and Walker 2019). These have made water quality deteriorate to such an extent that it is unsuitable for drinking purposes, and at most of the stretches, it is even not worthy for bathing. The aesthetic appearance of the rivers degrades that discourages tourism.

8.2 Methodology for River Rejuvenation

The methodology adopted for river rejuvenation has the following steps (Das 2018):

1. The first step involves identification of the natural course of a river and understanding the factors which affect the flow like geology, geomorphology, geohydrology, pedology, and structure of the terrain.
2. Identification of factors responsible for river deterioration.
3. Preparation of proposal of rejuvenation of the river.
4. Preparation of action plan by using modern technologies like GIS, GPS, remote sensing, and mobile applications.
5. Implementation of the action plan and bringing community awareness among the local and government officials in a proper schedule.
6. For proper monitoring, young leaders from each village and relevant government officials get trained.
7. Assessment of impact by gathering the dynamic parameters like geohydrological, land use, and agricultural data periodically through mobile applications for better and sustainable management of water reserves.

The most popular river rejuvenation methodology used is Managed Aquifer Recharge (MAR) (Das 2018). There is potential to utilize the surplus water in rivers during the monsoon by artificially recharging so that base flows in the rivers can be sustained and reestablish environmental flows in rivers (Das 2018).

8.3 Rejuvenation Strategies for River Ganga

Excessive consumption of water in various productive activities leads to competitive demand reduced flow of rivers (CPCB 2017; Kumar et al. 2020). Various leaders in Indian governments in the last three decades took the initiative to clean the River Ganga (Xun et al. 2017). Then the Prime Minister, Mr. Rajiv Gandhi, introduced the Ganga Action Plan I (GAP-I) in 1985 for 5 years (budget of US$33 million) with help from various voluntary native and global organizations (Hamner et al. 2006; Kumar et al. 2020) to resolve water pollution of River Ganga. The GAP-I had the objective of expansion of sewage treatment plants (STPs) in urban as well as remote places and the creation of many electric crematoriums (Birol and Das 2010; Kumar et al. 2020). GAP-I helped in the decline of direct release of sewage waste in Ganga, although the water quality of the river was still unsuitable for bathing (Tare et al. 2003; Kumar et al. 2020). The government continued it till 2000 as the problems were not resolved. GAP Phase II started in the year 1993 to include tributaries of River Ganga and 25 towns that have population greater than 100,000 exempted under GAP Phase I (Xun et al. 2017; Kumar et al. 2020). GAP-II couldn't achieve the objectives mainly because of improper administrative preparation and less participation of the community (Kumar et al. 2020). Both GAP-I and GAP-II created better sewage treatment facility, but it did not result in visible improvement in water quality (Xun et al. 2017).

Prime Minister Narendra Modi merged the National Ganga River Basin Authority (NGRBA) into the Ministry of Water Resources (MoWR) and formed the Ministry of Water Resources, River Development and Ganga Rejuvenation, creating a platform named "Namami Gange" (National Mission for Clean Ganga) in 2014 to work in mission mode to rejuvenate Ganga (Xun et al. 2017; Vyas and Nath 2021). In 2015, the Union Cabinet permitted a budget of approximately US$3 billion for the first 5 years (Xun et al. 2017). The objective of Namami Gange was to integrate earlier and present efforts for the rejuvenation of Ganga and its tributaries (Kumar et al. 2020; Vyas and Nath 2021).

8.4 Efforts Made by NGOs and Community

Tarun Bharat Sangh (TBS) is an NGO that has gained popularity by constructing more than 10,000 rainwater harvesting structures (RWHS) on 7 rivers—Arvari, Sabi, Ruparel, Jahajwali, Sarsa, Bhagani, and Maheshwara rivers – in different geographical regions of Rajasthan, including Alwar, Jaipur, Jodhpur, Jaisalmer, Karauli, Pali, Sawai Madhopur, and Tonk districts (Sinha et al. 2013). These efforts have successfully resolved the water crisis for 8600 sq. km covering several thousands of villages. Since 2007, it has started working enthusiastically on *Nirmal Aviral Ganga*. TBS in tandem with communities made them self-reliant in water resource management by encouraging local people participation (Sinha et al. 2013).

Their initiatives like organizing several camps at different places generated awareness about the significance of the River Ganga and encouraged youth to actively volunteer in Ganga rejuvenation. This has gained momentum, and several people at different locations are working to restore the health of the Ganga River. TBS under the guidance of Dr. G. D. Agrawal launched a national movement for maintaining an unaltered flow of the river Bhagirathi in Uttarakhand (Sinha et al. 2013). This movement resulted in making River Ganga a national river by the Government of India, and NGBRA was formed in February 2009 as co-coordinating authority for the Ganga (Sinha et al. 2013).

Various programs like Ganga Utsav, Ganga Amantran, Great Ganga Run, Ganga Quest, and Cleanliness Drive are being organized by the government in which community participation and youth involvement are encouraged. Ganga Utsav was organized on 4 November 2019 to celebrate the declaration of Ganga as a national river. Multiple activities engaging students and youth like river cinemas, quizzes, storytelling, games on ecological learnings, group discussions, etc. were conducted to create awareness among students and youth. Social outreach programs through adventure sports like Ganga Amantran connected with lakhs of people in 34-day long river rafting expedition covering over 2500 km of River Ganga from Devprayag to Gangasagar. Marathon (Great Ganga Run) to create awareness about the River Ganga was organized in New Delhi on 15 September 2019. Almost 20,000 people participated in the marathon. Cleanliness Drive is organized regularly in association with the local organization, NGOs, and community volunteers at several places along Ganga. For the rejuvenation of River Ganga, people from local communities like Ganga Praharis and task force like Ganga Mitra work together to achieve the objective of restoring the "Nirmal and Aviral Dhara."

8.5 Significance of River Rejuvenation and Its Positive Impacts

There have been several socioeconomic, cultural, and environmental impacts of rejuvenating rivers in India. The augmentation of water in rivers led to increased water levels and renewal of flow in rivers (Sinha et al. 2013). These were deliberate changes associated with water availability. There were also unintended changes in quality of life as socioeconomic factors are associated with water abundance in the region (Sinha et al. 2013). It also resolved several environmental issues and created aesthetic green spaces with the return of flora and fauna when rivers became cleaner and abundant with water. It caused an increase in the biodiversity of land in the aquatic and terrestrial environment.

The revival of several rivers has solved the water crisis problem in several villages in Rajasthan, directly enriching the lives socially, culturally, and economically. It has saved several hours spent in search of water and to fetch it daily from far-off places. It has also increased the food security and well-being of people as the villagers had easy access to clean water nearby their villages.

Various indirect positive impacts of river rejuvenation on the social, economic, and environmental aspects have been identified, for example, their is drastic reduction in hard work for collection of drinking water by women and children. This resulted in increase in quality time for family, reduction in distress migration caused due to water crisis, and increase in the percentage of turnout in schools (Sinha et al. 2013). Significant societal changes were observed like changes in approach toward girl education and societal issues like dowry with better education (Sinha et al. 2013).

8.6 Conclusion and Recommendations

Overexploitation and anthropogenic activities have resulted in the deterioration of the water quality of rivers. There have been several attempts by the government and NGOs to rejuvenate River Ganga for the last few decades. Ganga rejuvenated itself during lockdown when industrial waste (Kumar et al. 2020; Dutta et al. 2020) and agricultural runoff were not discharged (Dutta et al. 2020). It indicates that industrial pollution needs to be controlled as industrial waste is the major source of pollution in River Ganga. The tendency of humans to disturb the natural ecosystem is dangerous and it needs to be curbed. All the initiatives can only be successful if community participation makes it a mass movement and citizens are involved in mission mode. Even after achieving the target, we should work together involving children to spread awareness so that they become torchbearers and ensure that River Ganga does not get recontaminated. It is only possible with collective effort and cultural shifts in our behavior.

To minimize the waste effluent which is discharged in rivers, we should strictly follow zero liquid discharge. We must treat and reuse the wastewater until it gets fully utilized. For achieving this goal, the following measures are recommended:

1. Chemicals from laboratories should not be discharged directly into the drain, and proper guidelines must be followed for its disposal.
2. Industries should strictly recycle and reuse their wastes until it gets consumed completely.
3. Effluent from sewage treatment plants should undergo tertiary treatment before discharging into water bodies.
4. Organic farming should be encouraged, and pesticides and insecticides should be avoided so that chemicals do not enter water bodies through runoff.
5. Desiltation of rivers should be done periodically at places where siltation is common.
6. The practice of burning dead bodies near the banks of the River Ganga should be completely banned.

7. Open defecation is a common practice on the river banks which should be restricted through community awareness.
8. Disposal of solid wastes should be restricted in River Ganga.

References

Birol E, Das S (2010) Estimating the value of improved wastewater treatment: the case of river Ganga, India. J Environ Manage 91(11):2163–2171. https://doi.org/10.1016/j.jenvman.2010.05.008

Chaudhary M, Walker TR (2019) River Ganga pollution: Causes and failed management plans (correspondence on Dwivedi et al. 2018. Ganga water pollution: a potential health threat to inhabitants of Ganga basin. Environ. Int. 117, 327–338). Environ Int 126:202–206. https://doi.org/10.1016/j.envint.2019.02.033

Chaudhary M, Mishra S, Kumar A (2017) Estimation of water pollution and probability of health risk due to imbalanced nutrients in river Ganga, India. Int J River Basin Manag 15(1):53–60. https://doi.org/10.1080/15715124.2016.1205078

CPCB (2017) Restoration of polluted river stretches. Central Pollution Control Board, Government of India, New Delhi. http://www.indiaenvironmentportal.org.in/files/file/CPCB%20Report%20on%20Restoration%20of%20Polluted%20Rivers%20-%20July%202017.pdf. Accessed 21 May 2021

CPCB (Central Pollution Control Board), 2009. Ganga water quality trends. In: Monitoring of Indian aquatic resources, CPCB, New Delhi. http://cpcb.nic.in/upload/NewItems/. Accessed 21 May 2021

Das S (2011) Cleaning of the Ganga. J Geol Soc India 78(2):124–130. https://doi.org/10.1007/s12594-011-0073-9

Das S (2018) River rejuvenation: innovative methodology and socioeconomic impact. J Geol Soc India 91:641–642. https://doi.org/10.1007/s12594-018-0916-8

Das J, Jha S, Goyal MK (2020) On the relationship of climatic and monsoon teleconnections with monthly precipitation over meteorologically homogenous regions in India: wavelet & amp; global coherence approaches. Atmos Res 238:104889. https://doi.org/10.1016/j.atmosres.2020.104889

Dutta V, Dubey D, Kumar S (2020) Cleaning the River Ganga: impact of lockdown on water quality and future implications on river rejuvenation strategies. Sci Total Environ 743:140756. https://doi.org/10.1016/j.scitotenv.2020.140756

Dwivedi S, Mishra S, Tripathi RD (2018) Ganga water pollution: a potential health threat to inhabitants of ganga basin. Environ Int 117:327–338. https://doi.org/10.1016/j.envint.2018.05.015

Dwivedi S, Chauhan PS, Mishra S, Kumar A, Singh PK, Kamthan M, Chauhan R, Awasthi S, Yadav S, Mishra A, Mallick S (2020) Self-cleansing properties of Ganga during mass ritualistic bathing on Maha-Kumbh. Environ Monit Assess 192(4):1–15. https://doi.org/10.1007/s10661-020-8152-2

Ganga River Basin Environment Management Interim Report (2013). https://jalshakti-dowr.gov.in/sites/default/files/GRBEMPInterimReport_0.pdf. Accessed 20 May 2021

Goyal MK, Ojha CSP (2010) Evaluation of various linear regression methods for downscaling of mean monthly precipitation in arid Pichola watershed. Nat Res 01(01):11–18. https://doi.org/10.4236/nr.2010.11002

Goyal MK, Ojha CSP (2012) Downscaling of precipitation on a lake basin: evaluation of rule and decision tree induction algorithms. Hydrol Res 43(3):215–230. https://doi.org/10.2166/nh.2012.040

Goyal MK, Panchariya VK, Sharma A, Singh V (2018) Comparative assessment of SWAT model performance in two distinct catchments under various DEM scenarios of varying resolution, sources and resampling methods. Water Resour Manag 32(2):805–825. https://doi.org/10.1007/s11269-017-1840-1

Hamner S, Tripathi A, Mishra RK, Bouskill N, Broadaway SC, Pyle BH, Ford TE (2006) The role of water use patterns and sewage pollution in incidence of water—borne/enteric diseases along the Ganges river in Varanasi, India. Int J Environ Health Res 16(2):113–132. https://doi.org/10.1080/09603120500538226

IAAD (2017) Report of the comptroller and auditor general of India on rejuvenation of river Ganga (Namami Gange). In: River development & Ganga rejuvenation report (39) of 2017 (performance audit). Ministry of Water Resources; Indian Audit and Account Department (IAAD), Government of India, New Delhi. http://www.indiaenvironmentportal.org.in/files/file/Performance_Audit_on_Ministry_of_Water_Resources,_River_Development_&_Ganga_Rejuvenation_Union_Government.pdf. Accessed 21 May 2021

Joshi DM, Kumar A, Agrawal N (2009) Assessment of the irrigation water quality of river Ganga in Haridwar District. Rasayan J Chem 2(2):285–292. http://rasayanjournal.co.in/vol-2/issue-2/7.pdf. Accessed 21 May 2021

Khan MY, Gani KY, Chakrapani GJ (2017) Spatial and temporal variations of physicochemical and heavy metal pollution in Ramganga river-a tributary of river Ganges, India. Environ Earth Sci 76(5):231. https://doi.org/10.1007/s12665-017-6547-3

Kumar A, Mishra S, Taxak AK, Pandey R, Yu ZG (2020) Nature rejuvenation: long-term (1989–2016) vs short-term memory approach based appraisal of water quality of the upper part of Ganga River, India. Environ Technol Innov 20:101164. https://doi.org/10.1016/j.eti.2020.101164

Nandi I, Tewari A, Shah K (2016) Evolving human dimensions and the need for continuous health assessment of Indian rivers. Curr Sci 111:263–271. https://www.jstor.org/stable/24908613

Naskar S (2014) The river where swimming lessons can be a health hazard. https://www.bbc.com/news/magazine-28112403. Accessed 21 May 2021

Nautiyal CS (2009) Self-purificatory Ganga water facilitates death of pathogenic Escherichia coli O157:H7. Curr Microbiol 58(1):2. https://doi.org/10.1007/s00284-008-9260-3

Pandey J, Singh R (2017) Heavy metals in sediments of Ganga river: up-and downstream urban influences. Appl Water Sci 7(4):1669–1678. https://doi.org/10.1007/s13201-015-0334-7

Pandey J, Pandey U, Singh AV (2014) Impact of changing atmospheric deposition chemistry on carbon and nutrient loading to Ganga river: integrating land–atmosphere–water components to uncover cross-domain carbon linkages. Biogeochemistry 119(1):179–198. https://doi.org/10.1007/s10533-014-9957-2

Poonia V, Goyal MK, Gupta BB, Gupta AK, Jha S, Das J (2021) Drought occurrence in different river basins of India and blockchain technology based framework for disaster management. J Clean Prod 312:127737. https://doi.org/10.1016/j.jclepro.2021.127737

Reymond P, Chandragiri R, Ulrich L (2020) Governance arrangements for the scaling up of small-scale wastewater treatment and reuse systems–lessons from India. Front Environ Sci 8:72. https://doi.org/10.3389/fenvs.2020.00072

Shekhar S (2016) Environmental flows and river rejuvenation. J Geol Soc India 88:673–824. http://www.geosocindia.org/index.php/jgsi/issue/view/7114,88, 813

Shekhar S, Prasad RK (2009) The groundwater in the Yamuna flood plain of Delhi (India) and the management options. Hydrogeol J 17(7):1557. https://doi.org/10.1007/s10040-008-0430-y

Sinha J, Sinha MK, Adapa UR (2013) Flow—river rejuvenation in India, impact of Tarun Bharat Sangh's work Sida Decentralised Evaluation. 28, Sida. https://cdn.sida.se/publications/files/sida61632en-flow-river-rejuvenation-in-india-impact-of-tarun-bharat-sanghs-work.pdf. Accessed 21 May 2021

Tare V, Bose P, Gupta SK (2003) Suggestions for a modified approach towards implementation and assessment of Ganga action plan and other similar river action plans in India. Water Qual Res J 38(4):607–626. https://doi.org/10.2166/wqrj.2003.039

Vyas JN, Nath S (2021) The role of government and the public in water resource management in India. In: Pollutants and water management: resources, strategies and scarcity. Wiley, Hoboken, NJ, pp 399–415. https://doi.org/10.1002/9781119693635.ch16

Xun W, James WR, Ora-orn P (2017) Ganga rejuvenation: governance challenges and policy options. World Scientific, London. https://www.perlego.com/book/854361/ganga-rejuvenation-pdf. Accessed 20 May 2021

Yadav A, Pandey J (2017) Water quality interaction with alkaline phosphatase in the Ganga River: implications for river health. Bull Environ Contam Toxicol 99(1):75–82. https://doi.org/10.1007/s00128-017-2108-4

Chapter 9
Rejuvenation of Kukrail Drain

L. Ar. Poonam Upadhyay

Abstract Water is the most important factor in the sustainability of urban areas. In the current year, because of the rise in urban settlement, advanced agricultural activities, and industrialization, environmental degradation has become a major concern. The rapid settlement of the natural landscape, first through agriculture and then through infrastructure and urban development, comes at the expense of destroying life by giving natural resources, which provide sustainability for human habitats and insurance for large-scale investments. Therefore, the study is done on the Kukrail drain which is a prominent tributary of the Gomati River, and in the current scenario, the stream faces many stresses which are basically caused by nearby settlements in terms of sewerage and stormwater drainage. There is also a destruction of riparian vegetation and degradation of the stream banks. This paper will focus on the current issues of drain and reviving the Kukrail drain by proposing some guidelines.

Keywords Agriculture · Current · Drain · Kukrail · Natural · Settlement · Stream · Sustainability

9.1 Introduction

Kukrail drain is in the midstream region of the Gomati-Ghaghara River in the central plain of the Ganges (Kumar et al. 2015). It emerged from the Kukrail Reserve Forest at a height of 122 m and, after a travel distance of 26 km, joined the Gomati River at a height of 108 m (Kumar et al. 2015) (Fig. 9.1)

The Kukrail drain is the only important tributary, confluence on the north bank of the Gomati (CGWB 2011). On the Trans-Gomati side, there used to be a small stream named Kukrail, which now carries the rainwater of this area and discharge into the river upstream of the dam (Feedback Ventures Pvt. Ltd 2006). The Kukrail

---

L. Ar. P. Upadhyay (✉)
AIT - School of Architecture and Planning, Greater Noida, Uttar Pradesh, India

A. K. Gupta et al. (eds.), *Ecosystem Restoration: Towards Sustainability and Resilient Development*, Disaster Resilience and Green Growth,
https://doi.org/10.1007/978-981-99-3687-8_9

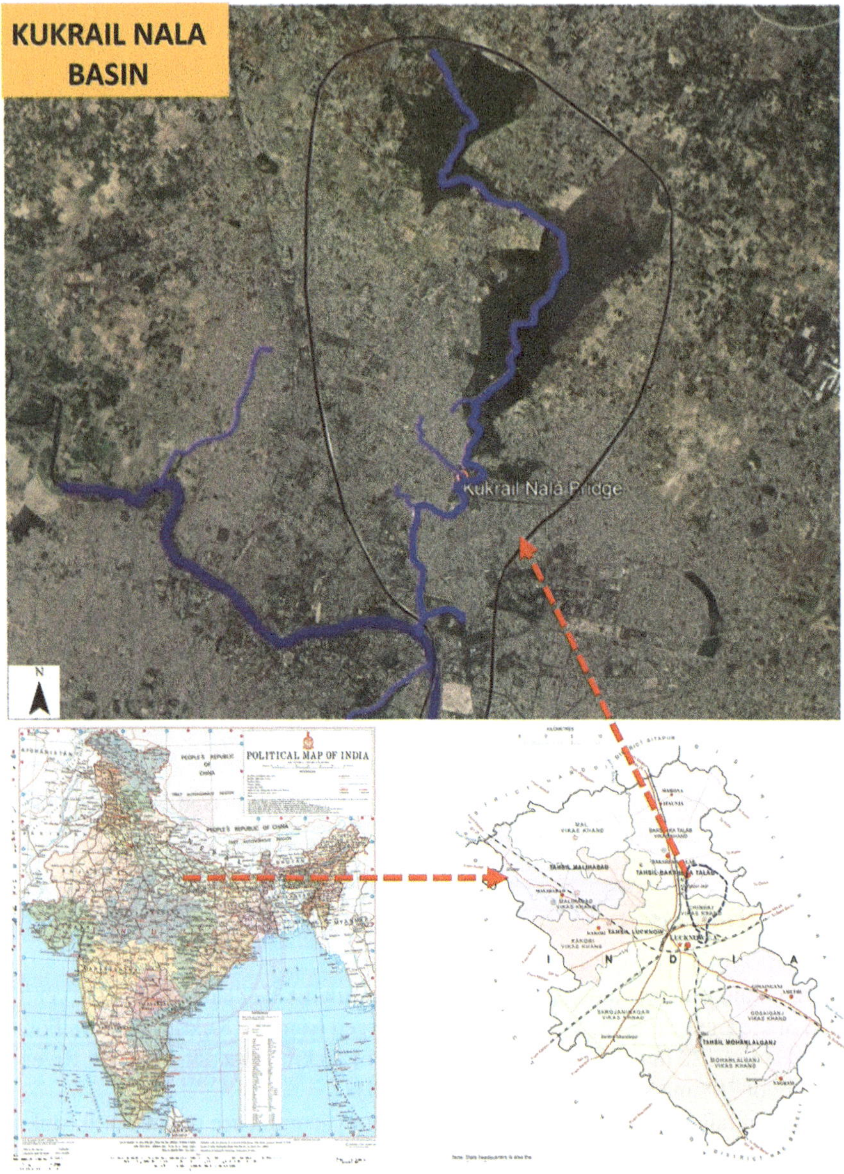

Fig. 9.1 Kukrail drain basin location in Lucknow. Source (Kumar et al. 2015; Survey of India 2020; Census of India 2011 Uttar Pradesh 2011; Google Earth 2021)

drain is a perennial stream (Singh et al. 2020a, b) and the largest drain on the Trans-Gomati side (Tokyo Engineering Consultants Co., Ltd, and CTI Engineering International Co., Ltd 2005). The river runs on a narrow bed surrounded by steep banks

(Singh et al. 2020a, b). Compared to Loni, its tributary drains are few in number (Sharma 1959). As it approaches the Gomati River, its banks retract into a narrow strip on either side of Faizabad Road and merge with Gomati near the Paper Mill Colony in Lucknow (Singh et al. 2020a, b).

The starting point coordinates of Kukrail drain are latitude 26°55′58.36″ N and longitude 80°59′58.47″ E (UPPCB 2019). It is a quaternary tributary of the Gomati River and is critical as it carries large amounts of water into the Gomati River while passing through the center of Lucknow (Bastia et al. 2021). The Kukrail catchment has an 86.75 km^2 area, and it is dominated by first-order streams (Kumar et al. 2015). A few decades back, the Kukrail drain was shaped like a perennial stream fed with underground water that comes from natural sources and extra water released through escape channels of the Irrigation Department (UPSBB 2015). But now, it's a seasonal stream fed only with drained water during heavy rainfall (UPSBB 2015).

Kukrail tributary is also the most prominent natural drainage system of the Lucknow river, the Gomati River (Dutta et al. 2010). Lucknow had the highest flooding level at 113.2 m in 1960, flooding most of the city (Tangri et al. 2018). The embankment was built on the banks of the Gomati River and also in Kukrail to the top of 114.4 m to protect the habitat (Tangri et al. 2018).

Sewer gates on embankments are installed to discharge drains in normal weather and prevent the flow back in floods (Feedback Ventures Pvt. Ltd 2006), and the embankment is approximately 25–30 years old (Feedback Ventures Pvt. Ltd 2006). But in 1960, due to heavy monsoon, water flowing backward by these drains into large areas of the city caused flooding (Feedback Ventures Pvt. Ltd 2006). When the Gomati River levels rise to the drain's invert level, these gates get closed, so there should be no backflow (Feedback Ventures Pvt. Ltd 2006) (Fig. 9.2).

The flood also occurred in the year 2008 (Feedback Ventures Pvt. Ltd 2006). In the current scenario, from the Google Earth image shown in Fig. 9.3, it is quite evident that the stream is seen from the forest, entering the urban area near Abrar Nagar. The stream winds its way from Khurram Nagar, Sarvodaya Nagar, and Kukrail Pul and finally flows into the river at Jugauli. Kukrail drain has a catchment area on the fastest-growing Kursi Road, Indiranagar, Mahanagar Colony, and Faizabad Road that contributed 80 MLD during the rainy season, which is set to rise at a much faster speed than shared by any other drain (Dutta et al. 2010).

9.2 Methods and Methodology

9.2.1 Aim

To determine the current condition of the Kukrail drain and its issues and rejuvenate its ecological significance in the city

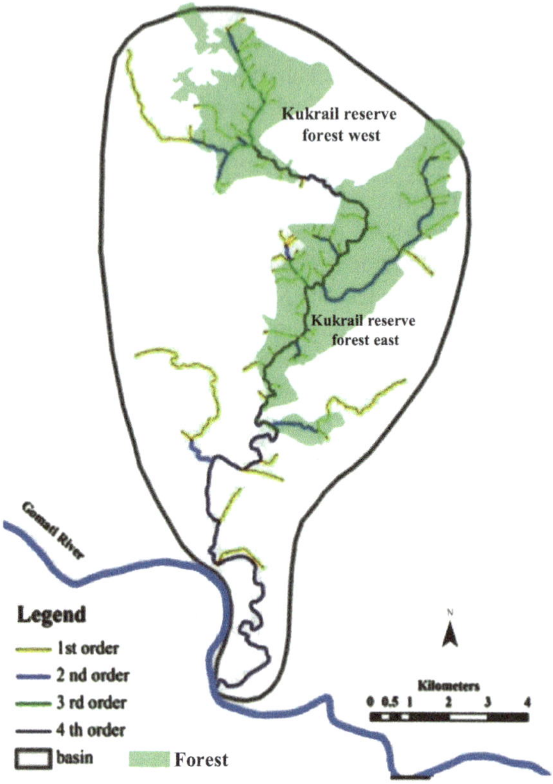

Fig. 9.2 Drainage map of Kukrail basin. Source (Kumar et al. 2015; Google Earth 2021)

9.3 Objective

9.3.1 Step 1

Comprehension of the importance of the Kukrail stream in the Gomati River

9.3.2 Step 2

Prospecting the Kukrail stream value in Lucknow city

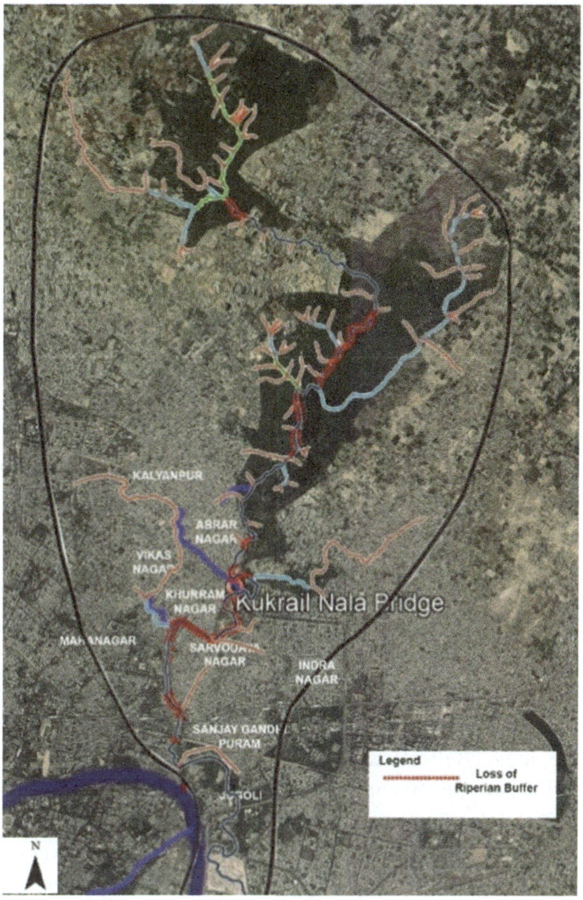

Fig. 9.3 Kukrail drain. Source (Kumar et al. 2015; Google Earth 2021)

9.3.3 Step 3

Understanding the various important aspects of Kukrail drain like topography, slope, geomorphology, hydrology, water quality, volume and course of the stream, vegetation, flora and fauna, and the urban fabric

9.3.4 Step 4

Understanding the causes which lead to the degradation of the stream through literature study and identifying the major issues which are significant for the lake ecology

9.3.5 Step 5

Formulating the guidelines for the landscape development of the Kukrail stream concerning the current issues of the stream and proposing the strategies and program development for the stream and implementation of the rational solution through the landscape

9.4 Importance of Kukrail Stream in Gomati River

The Kukrail drain is the biggest drain of the Trans-Gomati region (Tokyo Engineering Consultants Co., Ltd, and CTI Engineering International Co., Ltd 2005). One of the main drains entering the Gomati River, the Kukrail drain (Kumar et al. 2019), with a basin perimeter of 49.46 km (Kumar et al. 2015), is a major tributary that passes by Lucknow (Kumar et al. 2019). The number of the first-, second-, and third-order tributaries is 77, 14, and 3, respectively (Kumar et al. 2015). The first-, second-, third-, and fourth-order streams are 40.55 km long, 12 km, 4 km, and 23 km, respectively (Kumar et al. 2015). It is an important Lucknow district water system and has great value for the Gomati River (BallotBox India 2018).

Historically, the Kukrail drainage system has not only played a very important role in draining the rainwater from the Lucknow district into the Gomati River, which feeds the river, but it is also an important source of recharge of groundwater in the area (BallotBox India 2018). It saves the region from flood by absorbing rainwater runoff and drains in Gomati, which expands along with the floodplain and enriches the soil, and maintains the water system in the Ganga basin (BallotBox India 2018). Earlier, the high groundwater level contributed to the flow of this natural drainage and also carried excess rainwater into the Gomati River during the floods (Pandey 2020).

9.5 Value of the Kukrail Stream

There are many beliefs associated with the Kukrail Bridge, some people may be aware of it, and some may not because history has become blurred by the changing times (Raj 2018). Believe it now or superstition, it is claimed that a person bitten by a dog does not need any medication and gets treatment after bathing in this Kukrail drain (Raj 2018). The tradition of bathing on the Kukrail Bridge after dog bites in Lucknow is not the last few years or decades but generations old (Raj 2018). The villagers still come here because of this tradition (Raj 2018).

9.6 Understanding of Geomorphology

The main tributary of the Gomati River is the rivulet Kukrail (Kanaujia et al. 2015), and the Kukrail basin has the oldest alluvium which is represented by the mid to late Pleistocene Varanasi alluvium and comprises gray to brown clay, silt, and sand with or without Kankar (SENES Consultants India Pvt. Ltd 2015). Therefore, the Kukrail basin has high water permeability (BallotBox India 2018), and as an outcome of urban development (Google Earth 1987, 2004, 2021), it is quite evident that it was buried and made impenetrable (Kumar et al. 2019). The older alluviums occur in high-terrain areas and won't get flooded (CGWB 2011).

In the watershed region, the average bifurcation ratio is 4.36, indicating that the pattern of the drainage is natural and not substantially affected by structures of geology (Kumar et al. 2015). Because of its geographical features, the Kukrail drain basin is expected to comprise a higher volume of water downstream, but unplanned land use, increase in impermeable areas, and changes in rainfall patterns have led to permanent waterlogging in the area (Kumar et al. 2019).

9.7 Topography and Slope Facts of the Kukrail Stream

In general, the slope in the Kukrail area is gradual, which flows toward the stream (UPSBB 2015), and, for this catchment area, the frequency of the stream is 1.09 km^2, which basically depicts low relief (Kumar et al. 2015). Higher altitude is found in the range of forests and lower near the convergence of the stream into the Gomati River (Kumar et al. 2015). A part of that large volume of flooding is expected in the Kukrail drain basin due to the features of the basin (Bastia et al. 2021).

9.8 Hydrological Information

The peak discharge of the Kukrail drain is 425 m^3/s into the Gomati River, and the flow from the Kukrail drain joins the Gomati River by gravity and is not pumped like other drains (Ahmad 2013). Kukrail has a 200-m-wide floodplain (Pandey 2020). In recent years, rapid urbanization and concretization have been observed in the Kukrail catchment area, which lies on the Trans-Gomati side, resulting in more floods (Bastia et al. 2021). In the Kukrail River basin, built-up areas have expanded, making up 22% of the city (Kumar et al. 2019). Urbanization is a determining factor in the extent of flooding (Kumar et al. 2019). The effect of anticipated climate change in 2030 could increase the area which is exposed to urban floods by 20% over a period of 100 years (Kumar et al. 2019). Studying the results of the return periods of 50 and 100 years, it was found that at the mouth of the Kukrail River, the volume of peak discharge can increase by the range of 10–75%, which basically depends on the emission scenario (Kumar et al. 2019). The Kukrail stream catchment shows a tendency to increase the maximum discharge volume and runoff from the current scenarios to extreme cases (Kumar et al. 2019). Flood damage has been assessed in the Kukrail stream catchment in Lucknow, and in the future, flood damage will increase by 56% (Kumar et al. 2019).

9.9 Water Quality and Course of the Stream

This drain is a mixed drain that carries industrial wastewater and effluents from M/s HAL Limited and M/s C.P. Milk Private Limited (UPPCB 2019). Domestic wastewater is transported in the Kukrail drain from the Indira Nagar, Nishatganj, Kalyanpur, and Shakti Nagar watershed area of Lucknow and also effluent from Food Products, Pvt. Ltd. Kursi Road, Lucknow (UPPCB 2019). At SPS Kukrail as well, sewage discharge was observed in the Gomati River (Eastern U.P Rivers and Water Reservoirs Monitoring Committee 2019). There is a flow of water in it only from the sewer, which also comes from the drainage of Sarvodaya Nagar, Rahim Nagar, Manas Vihar, and Sanjay Gandhi Puram. Wastewater drains without any sort of proper and preventive treatment (Gupta 2020). The way of this drain, extending to the forest of Kukrail, is flowing dirty sewage (UPSBB 2015). The water came from several canals on the Kukrail River, but after illegally encroaching as well, it emerged as a drain (Kukrail Nala will be rehabilitated as a river, preparing to build a riverfront 2019). The Kukrail drain is a very old drain, and when the outlet of the Sharda Sahayak canal increases, it is also discharged into the Kukrail drain (Kukrail drain sewage flows into Gomati again 2018). The drain that already carries so much sewage cannot retain so much excess water by the canal and also releases a large amount of mixed sewage into the Gomati (Kukrail drain sewage flows into Gomati again 2018). But currently, the Kukrail drain is considered the main cause of creating pollution in the Gomati River (BallotBox India 2018).

According to the CPCB team, with the help of RO UPPCB Lucknow, Sri Ram Karan supervised eight drains, among which was also mentioned the Kukrail drain which falls into the Gomati River and discharges most of the uncleaned sewage (Eastern U.P Rivers and Water Reservoirs Monitoring Committee 2019) (Table 9.1).

Table 9.1 Kukrail drain discharging untreated sewage into Gomati River. Source (Eastern U.P Rivers and Water Reservoirs Monitoring Committee 2019)

Sr. no.	Drain discharging in Gomati River	Coordinates	Catchment area/near Gomati River
1	Kukrail drain at Kukrail barrage, Lucknow	26°51′10″ 80°57′25″	Kukrail picnic spot area, Lucknow
2	Kukrail drain at Abrar Nagar and Khurram Nagar, Lucknow	26°51′28″ 80°57′20″	Abrar Nagar and Khurram Nagar area
3	Kukrail drain near Kukrail SPS, Lucknow	26°51′32″ 80°57′40″	Paper Mill Colony

Kukrail drain's total discharge is 150 MLD, and around 90 million litres per day goes to the sewerage treatment plant of Bharwara for treatment, and about 60 million litres per day spill over into the Gomati River (UPPCB 2019). About 20 million liters per day is the mean dry weather flow of this drain (Parveen and Singh 2016). As per the team of CPCB, below mentioned are the following characteristics of the water sample collected at various locations during drain monitoring (Eastern U.P Rivers and Water Reservoirs Monitoring Committee 2019) (Table 9.2).

As per the CPCB water quality criteria, the BOD must be three mg/L or below for the class of water-B Outdoor bathing (Eastern U.P Rivers and Water Reservoirs Monitoring Committee 2019). The BOD results above show a higher BOD value for surface water compared to the established standards (Eastern U.P Rivers and Water Reservoirs Monitoring Committee 2019). Low BOD levels indicate quality water and high BOD levels depict polluted water (Eastern U.P Rivers and Water Reservoirs Monitoring Committee 2019). Therefore, in the Kukrail case, high levels indicate contaminated water.

According to the Google Earth image of the years 1987, 2004, and 2021, the form of the Kukrail stream basin is reduced as compared to its earlier size. Also, the huge land of healing water of this drain is reduced to a drop of dirty water (Jain et al. 2011). The Kukrail drain is being filled with garbage, and its depth has remained only 1–1.5 m (Gupta 2020). Three hundred to Four hundred meters in the middle of Kursi Road is dry (Gupta 2020).

As per the Google Earth image of the year 2004, it is evident that the rivulet is meeting the Gomati River, and the stream looks broader and healthy. It was still fluent and connected, but as the years progressed, the condition deteriorated (Google Earth 1987, 2004, 2021). In the year 2016, the Kukrail River is closed, and the channel capacity of Gomati is reduced (Prasad n.d.). The above Google Earth image exhibits the miserable condition of the Kukrail drain. The water rivulet is almost dry and is filled with earthly substances to block its flow (Prasad n.d.). Now it is a dead stream that is not naturally connected to the Gomati River because of the concrete riverfront (Pandey 2020). The effects of rivulet drought can be devastating, as rivers and aquatic ecosystems are affected (Bakker and Anderson 2005).

The water is also polluted due to the dumping of garbage and plastic waste in the Kukrail drain near Abrar Nagar and Khurram Nagar (Eastern U.P Rivers and Water Reservoirs Monitoring Committee 2019). The Kukrail drain appeared to be in a deplorable state, with solid, non-biodegradable waste floating and obstructing the flow of the drain beside Swarn Jayanti Park on the Ring Road (Srivastava n.d.). Dead animals and plastic waste are piled up on the sidelines. Pigs and dogs live there, and they are said to be breeding grounds for deadly insects and mosquitoes (Srivastava n.d.).

Table 9.2 During the drain monitoring, characteristics of the water sample were collected. Source (Eastern U.P Rivers and Water Reservoirs Monitoring Committee 2019)

Date	Drain discharging in Gomati River	pH	Color (Hazen)	SS (mg/L)	TDS (mg/L)	Cl^- (mg/L)	$SO4^-$ (mg/L)	Oil and grease (mg/L)	$NH3^-N$ (mg/L)	BOD (mg/L)	COD (mg/L)	T coliform (MPN/100 mL)	F coliform (MPN/100 mL)
5/2/2019	Kukrail drain at Kukrail barrage. Picnic spot, Kukrail, Lucknow	7.27	25	24.2	24.7	8.2	15.6	BDL	5.98	7.0	41.3	2.4×10^6	1.3×10^6
5/2/2019	Kukrail drain at Abrar Nagar, Khurram Nagar, Lucknow	7.14	35	16.5	32.9	17.6	25.2	BDL	10.2	18.9	81.9	7.9×10^6	1.7×10^6
5/2/2019	Kukrail drain near Kukrail SPS, Lucknow	7.07	25	51	36.2	27.9	32.4	BDL	13.3	22	60.8	3.5×10^7	1.3×10^7

9.10 Flora and Fauna of the Stream

Near to the Kukrail stream, the Kukrail Reserve Forest is situated in Indira Nagar/ Khurram Nagar, which is adjacent to the Colony of Shivpuri (Tripathi et al. 2015) on the northeastern (Kumari et al. 2016) picnic spot road, 9 km from the center of Lucknow (Tripathi et al. 2015), with an area of 5000 hectares and at coordinate N.-26054′399′, E-80059′046′ (Kumar et al. 2017). As colonization progressed, it is now very close to Indira Nagar (UPSBB 2015). In the past, about 30 years ago, it was a much-deteriorated forest (Kumari et al. 2016). High biological pressures, animal grazing, exploitation of vegetation, and nutrient deposition have disrupted the natural succession of forests (Kumari et al. 2016). As a result, over the last three decades when the state government acquired and protected the area as protected forests, species richness has declined significantly, and the predominance has been concentrated on a small number of species (Kumari et al. 2016). With the rapid decline in crocodile numbers in India at the time, there was also a need for crocodile protection (Tripathi et al. 2015). The reserve is world-renowned for its gharial breeding and rehabilitation program (Kumar et al. 2017). Kukrail picnic spots belong to natural forests and sanctuaries, with many types of deer (Tripathi et al. 2015). Kukrai Center was founded in 1978, and the Forest Department of Uttar Pradesh funded it in alliance with the Indian Ministry of Environment and Forests. This marked the starting of the efforts of state governments to protect alligators when there were only 300 remaining (Kumar et al. 2017). The Kukrail Forest Reserve is the largest recreational resource based on Lucknow users that, to a considerable extent, has compensated for the loss of parks, open spaces, and gardens in the landscape at the urban level (Singh 1992).

The Kukrail basin has a rich phyto-diversity (Gangwar and Lavania 2018). Vegetation in and surrounding Kukrail forest is tropical mixed deciduous forest along with vegetation of open land (Tripathi et al. 2015), and it consists of *Acacia catechu*, *Azadirachta indica*, *Bauhinia variegata*, *Butea monosperma*, *Cassia fistula*, *Dalbergia sissoo*, *Ficus religiosa*, *Holoptelea integrifolia*, *Mitragyna parviflora*, *Pongamia pinnata*, and *Syzygium cumini* (Tripathi et al. 2015). The shrubs in the region are *Clerodendrum viscosum*, *Jatropha glandulifera*, *Morinda pubescens*, and *Prosopis juliflora* that grow in a scattered manner (Tripathi et al. 2015). The herbs are *Ammannia baccifera*, *Cyperus rotundus*, *Indigofera linifolia*, *Justicia simplex*, and *Ocimum americanum* (Tripathi et al. 2015). The aquatic species are *Nymphoides* species, *Potamogeton indicus*, *Typha* species, and *Lemna polyrrhiza* (Tripathi et al. 2015). All the mentioned species often grow in annual and perennial water-logged areas (Tripathi et al. 2015). The tree species which are dominant in the area are mainly *Pongamia pinnata*, *Acacia nilotica*, and *Shorea robusta* (Kumari et al. 2016). As per the Awadh Forest Department, there are some exotic species in the region like *Leucaena leucocephala*, *Cassia auriculata*, *Khaya senegalensis*, *Kigelia pinnata*, *Acacia auriculiformis*, *Populus* species, *Peltophorum ferrugineum*, and *Parkinsonia aculeata*. The common parasites on the trees of the forest are *Dendrophthoe falcata* Ettings and *Cuscuta reflexa* Roxb. (Singh et al.

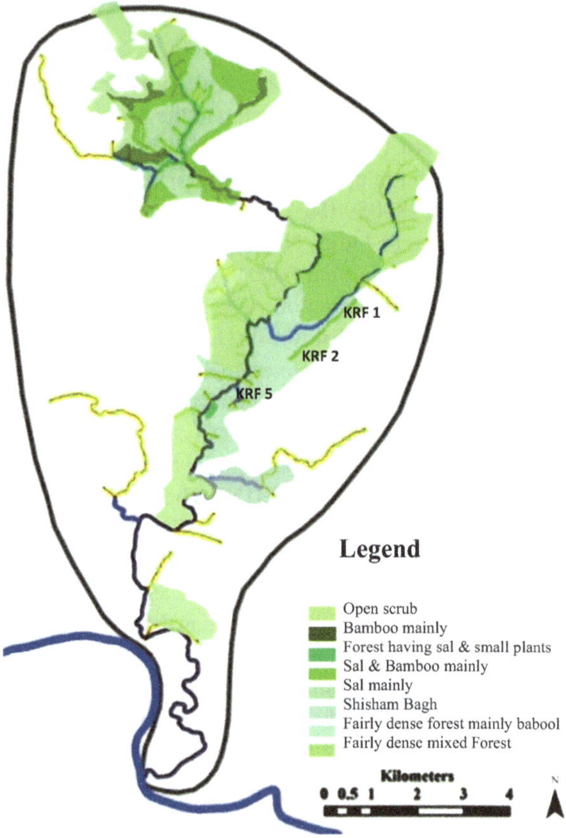

Fig. 9.4 Forest map of Kukrail drain. Source (Kumari et al. 2016; Kumar et al. 2015)

2020a, b). A particular species, i.e., *Pongamia pinnata*, with a ground coverage of 78%, is easily formed by the degraded alluvium. It drops a lot of shades, which does not like the growth of several other species (Mishra et al. 2014).

As per the Uttar Pradesh State Biodiversity Board, the region has the presence of great biodiversity of fauna which is mentioned below (Fig. 9.4):

Chital deer, rock python, shrew and porcupine, turtles, orb weaver, Indian marsh crocodile, fox, Indian cobra, jackal, five-striped squirrels, common house rat, Indian mole rat, house mouse, brown hare, blue bull, rhesus monkey, ambushers, common mongoose, small Indian mongoose, and various snakes like cobra, python, common krait, rat snake, and water snake (UPSBB 2015).

Many varieties of fish, i.e., Katla, Mangur, Rohu, Mirla, Chilwa, and Girai, are present in the nearby Kukrail drain (UPSBB 2015).

The spiders of this forest can be divided into seven feeding guilds based on feeding behavior (Kumar et al. 2017). They are blind, 17 species of orb weavers, 18 species of stalkers, 5 species of ground runners, 2 species of foliage runners, 4 species of sheet web builders, 4 species of space web builders, and 11 species of ambushers (Kumar et al. 2017). The 18 species of stalkers were the dominant guild

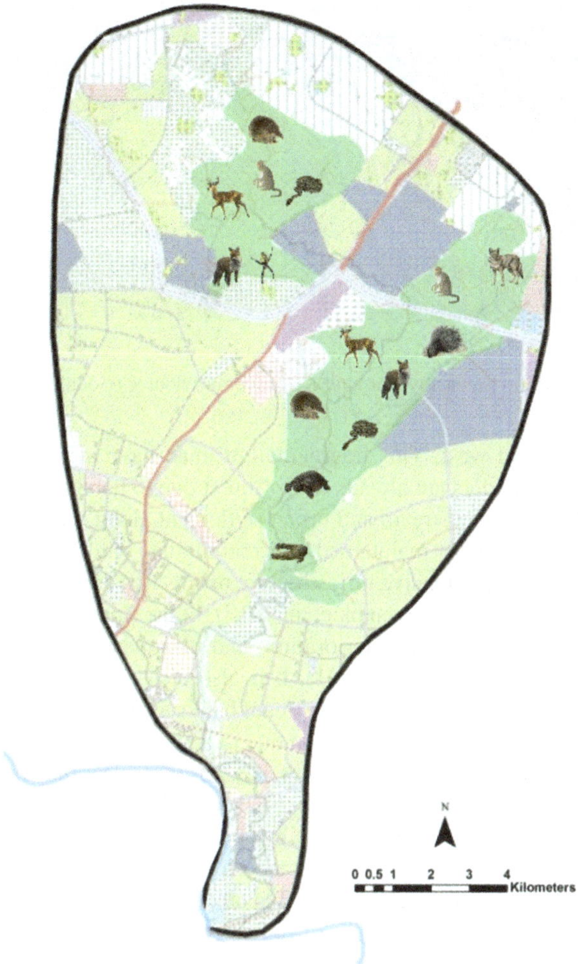

Fig. 9.5 Map showing fauna species of Kukrail region. Source (Kumar et al. 2015; UPSBB 2015; Lucknow Development Authority n.d.)

which was followed by 17 species of orb weavers, and guild leaf runners were scarce with 2 species (Kumar et al. 2017). Several threats to spiders have also been identified which contain destruction of habitat, water and terrestrial pollution, uncontrolled use of synthetic pesticides, encroachment on habitation, and so on (Kumar et al. 2017) (Fig. 9.5).

From the Google Earth image, it is evident that Kukrail Reserve Forest east has some areas fairly dense, and in the urban context, the Kukrail course is barely covered with vegetation. The anthropocentric development process has led to presumed changes in biodiversity. From the above map, it is also quite evident that the tree cover near the Kukrail region has been lost from the year 1987 to 2021, which has led to the decline of biodiversity in the zone (Fig.9.6).

From the year 2021 Google Earth image, it is clear that the Kursi Road that cut through the forest was one of the main causes of fragmentation and division in parts

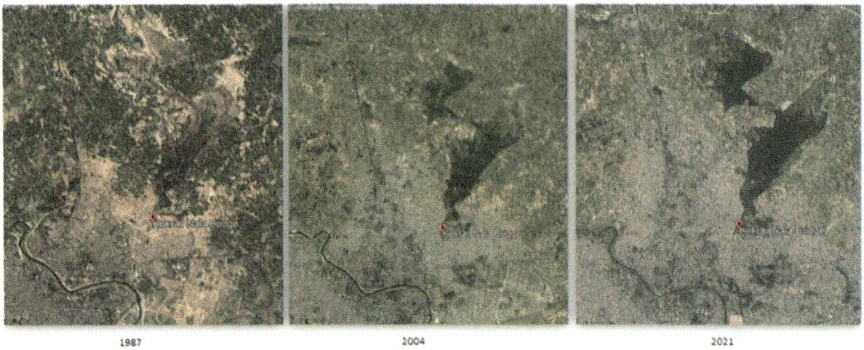

Fig. 9.6 Green cover around Kukrail drain. Source (Google Earth 1987, 2004, 2021)

east and west. This fragmentation must have affected the routes of many species since spiders are already under threat due to habitat destruction (Kumar et al. 2017). Between the stream and urban settlement along the entire stretch, there is no proper riparian buffer maintained to accommodate wildlife. The expansion of cities accompanied by extensive infrastructure results in a decrease and primary fragmentation of riparian forests (Asanok et al. 2017). In the riparian zone, a rise in urbanization leads to fragmentation, deterioration, and disappearance of habitat, thus reducing biodiversity (Singh et al. 2021). It is also clear from the above Google Earth map that there is the degradation of stream banks as well as the edge of the reserve forest. The forest also faces degradation due to an increase in human settlement, and they are at the stage of decline. From Fig. 9.3, it is very clear that there is a loss of riparian buffer along the stream (Das et al. 2020; Goyal et al. 2018; Poonia et al. 2021).

In addition, the pattern of behavior of modern pollen deposition has been made so that in turn it could help to assess the extent of pastoral practice and the depth of natural forest degradation (Tripathi et al. 2015). The influence of anthropogenic activity on the spectrum of pollen is seen in the presence of cereals high in values, which are 14–29%; Brassicaceae, which is 2–5%; and *Xanthium* pollen in the open ground (Tripathi et al. 2015). Tubuliflorae palynospecies have on average 7% of pollen which in the open ground represents severe pastoral activities as members of this family escape to pasture due to their inestimable nature for cattle and goats (Tripathi et al. 2015).

Cattle grazing and livestock breeding are practiced in large numbers as the embankment areas are vacant and are inadequately maintained and secured (Srivastava n.d.). One of the major devastating forces is unrestricted grazing for riparian ecosystems as it reduces the quality of flowing water, damages the soil, degrades vegetation, improves soil erosion, alters the morphology of the channels, and reduces biodiversity (Davis 1982).

Livestock does concrete damage to riparian vegetation by treading and grazing (Kauffman and Krueger 1984). Cattle grazing does compaction of soil due to runoff rise and, as a result, reduces plant access to water (Kauffman and Krueger 1984). A substantial increase in runoff leads to a decrease in water availability for aquifer recharge (Singh et al. 2021). Consequently, in the riparian zone, unrestricted and unregulated entry of cattle has eliminated the vegetation at large-scale in the Kukrail region (Srivastava n.d.), but it has also reduced the water quality, degraded the vegetation, damaged the soil, enhanced soil erosion, and altered the morphology of the channel. It has also caused soil compaction which has ultimately increased the runoff and, as a result, reduced the availability of water to the plant, and an increase in the runoff also resulted in a reduction in the inflow of water into aquifers (Goyal and Ojha 2010, 2012). The leaching from waste piles adds to water contamination (Srivastava n.d.). The loss of riparian vegetation can be attributed to two factors: intrusion and water and soil contamination. Soil erosion is also in the form of runoff from the fallow lands into the basin and floodplain, as well as from the barren edges of the stream (Srivastava n.d.). In addition, in December 2014, soil samples were collected from various locations from the nearby agricultural fields and forests (Kumari et al. 2016): KRF 1 A-1 26°55.361′ N and 81°00.227′ E agricultural field near the forest; KRF 2 A-1 26°55.021′ N and 80°59.968′ E agricultural field near the forest; and KRF 5 F-1 26°54.415′ N and 80°58.972′ E near Kukrail barrage (Kumari et al. 2016). The soil samples of this study consist of approximately 84% silt-sized particles, 8% clay-sized particles, and 8% sand (Kumari et al. 2016). The high silt content in the samples indicates that they are all part of a floodplain deposition (Kumari et al. 2016). Also, in the Kukrail Reserve Forest, the soil is good for the growth of plants (Kumari et al. 2016).

9.10.1 Urban Fabric

Urbanization in the current scenario is one of the main drivers transforming landscapes around the world (Ives et al. 2013). Urbanization affects ecosystems by replacement of vegetation with urban buildings such as infrastructure and roadways and indirectly modifying the configuration and formation of vegetation through deterioration and fragmentation (Pennington et al. 2010; McKinney 2002).

Urbanization poses an adverse impact, threatening the anatomy and functions of the ecosystem and affecting the increasing demands of ecology to facilitate sustainable ecosystem services (Song and Deng 2017; Peng et al. 2017). After the 1960s, only the Kukrail catchment was under development.

In 1960, this region was part of a floodplain, as evidenced by the flood events that led to the construction of embankments (Tangri et al. 2018). Unfortunately, the riparian zone was not taken into consideration before the construction of dams, which were built immediately after the widening of the stream (Pandey 2020).

The above factors also contributed greatly to the deteriorating health of the city's stream. The urban environment is strongly influenced by anthropogenic events. Significantly, more attention needs to be paid to monitoring land use in cities.

Fig. 9.7 shows the pattern of land use near the Kukrail region, i.e., mostly residential on both sides of the stream. Another important feature is the road, i.e., cutting across the Ring Road and providing a shorter route to the Gomati Riverside

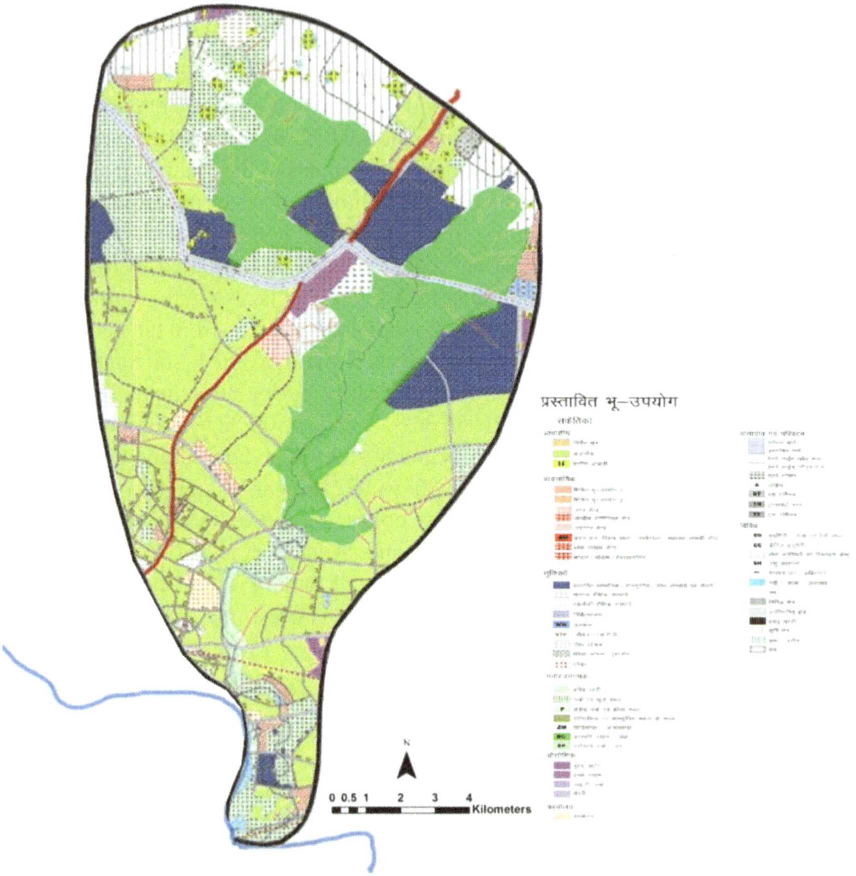

Fig. 9.7 Map showing land use around Kukrail region. Source (Kumar et al. 2015; Lucknow Development Authority n.d.)

Figure 9.3 shows the dense settlement structure, with little or no space for green open spaces in the settlements by the stream, and the dense urban development in the surrounding area has swallowed up any land for plant growth.

Apart from these, on both the banks of Kukrail, there are also encroachments (Bureau 2021) and illegal slums in Kukrail (Rai 2017). The stream is faced with the load from these settlements in the form of an urban surface. In many places, the drainage encroachment was carried out by the inhabitants of the land (Pandey 2020).

In the Kukrail stream, water comes from several canals, but after the occupation of illegal drains, it comes in the form of a drain from the stream (Kukrail Nala will be rehabilitated as a river, preparing to build a riverfront 2019).

As per the map of the urban sprawl of Lucknow city shown in Fig. 9.8, different layers of the built-up area growth patterns can be noticed near the Kukrail region, and it is expanding in the vastness of the large city.

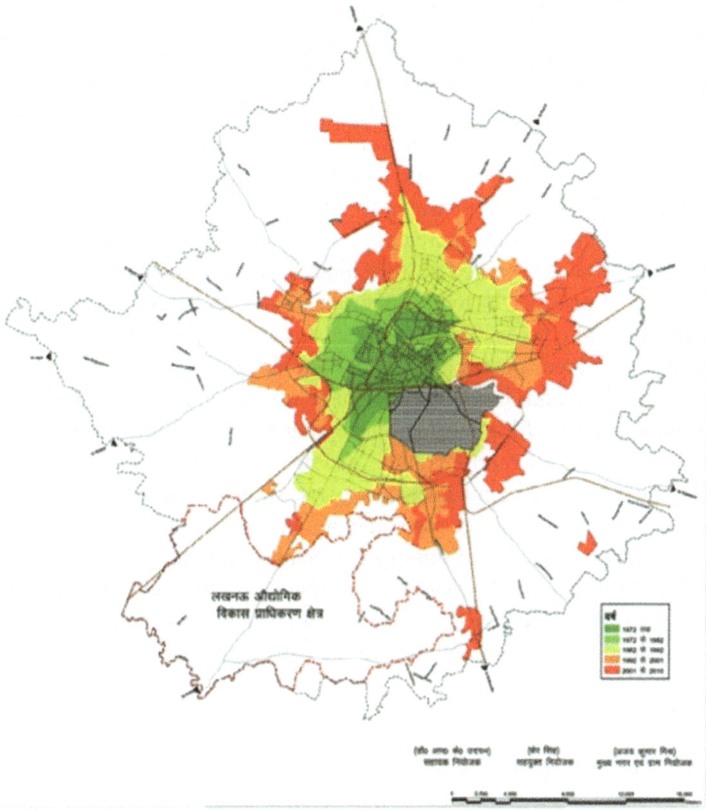

Fig. 9.8 Map showing urban sprawl of Lucknow. Source (Lucknow Development Authority n.d.)

9.10.2 Issues

Therefore, on the basis of all the abovementioned facts, the Kukrail basin is impenetrable due to an increase in urban settlement, and the form of the Kukrail stream basin is reduced as compared to its earlier size, and the water is contaminated. Today, solid waste dumping, urban stormwater, and its sewage are all subject to the burden of the Kukrail stream and tend to pollute it. In the urban area, encroachments and misuse of the stream regime are the main issues that need to be considered. The banks of streams in the urban section lack vegetation cover and are exposed to strong erosion. With the limited availability of vegetation in the stream's urban pockets, undesirable activities are increasing throughout the urban environment, which is also damaging the stream's health. The lack of open spaces degrades the urban experience. Riparian corridor fragmentation is a major issue for health and biodiversity, and local biodiversity is destroyed by the takeover of invasive species like *Pongamia pinnata*. Due to invasions, the riparian forest and the banks of the stream are endangered.

9.10.3 Approach and Rational Strategies

The entire watershed region needs to be conserved so as to rejuvenate the presence of the stream. In order to protect the stream banks, afforestation should be done and could be one of the possibilities to prevent soil erosion. It will also revive biodiversity and stream ecology. Afforestation should also be done in the floodplain regions. A combination of buffer and bioswales could also be a feasible solution to minimize the soil erosion in the catchment area as well as the streams feeding into the stream. Suitable provision of designated pasture land within the catchment area and along the stream would also act in terms of groundwater recharge and conservation of biodiversity and soil erosion. There is a need for a regulated delineation of community access extent. Tree species like *Acacia nilotica* and *Shorea robusta* need to be proposed in areas that are fairly dense or have no vegetation; they are native to the region. Implementing infiltration facilities using an environmental perspective can have different effects on runoff volume. Nature trails should be introduced in the region through a permeable pavement (as a stormwater facility) to explore the avian biodiversity in the region. There are some potential areas that can be developed for public open spaces such as the right bund of Sanjay Gandhi Puram, the left bund of Paper Mill Colony, the left bund of PAC Colony, and the right bund of Sarvodaya Nagar and also Abrar Nagar. These areas can be developed into successful green open spaces to allow people to indulge in the river landscape of the Kukrail stream.

9.10.4 Development of Program

Since urban growth is considered to be a major cause of flood levels, local governments should adopt appropriate policies for urban planning. In the future, the increase in flooding in the city depicts the need to control the same through implementing flood risk management strategies. In the watershed region, groundwater recharge is a naturally occurring process and should be planned for water resource management in order to maintain it. Effective land-use planning may safeguard the Varanasi alluvial soil of the region since the soil is also good for plant growth. Solid waste management should be adopted by source segregation of solid waste with organic matter dedicated to composting. Remove toxic and biodegradable waste from the basin for disposal.

A provision of a decentralized biological wastewater treatment system should be adopted to prevent wastewater and polluted solids from entering the stream. Treated water can further be used for the growth and maintenance of vegetated areas other than the agricultural area in the Kukrail region. A stormwater management plan must be implemented to curb the increased runoff, which will help rainwater penetration to deeper levels and increase the aquifer's water levels. Therefore, planners and policymakers in Lucknow should consider stormwater facilities in the master plan for the future. Another important component to the success of rainwater management systems is public awareness, as the community is often unaware of the benefits of this management system. Urban sprawl near the Kukrail region should only be addressed through proper planning of land use and analysis of land suitability and bearing capacity. The forest department must initiate the afforestation activities to be extended in the whole watershed region of the stream. Demarcate the entire floodplain, right from the origin to the confluence with the Gomati and the floodplains as an eco-zone. Riparian corridors should be used to move and network the outdoor recreation areas. The corridor will enhance the city's quality of experience, support biodiversity, and add value in terms of refreshing experiences.

9.10.5 Guidelines

An overhead forest pass should be imparted for fauna movement over the Kursi Road. The provision of a 50 m riparian buffer should be provided on both the stream banks and a 20 m buffer for all first-order streams which are draining into the stream. No construction should be allowed in the region from the 100 m to the right of the road on both sides of the stream, and any illegal structure within the region should be demolished. The collection of rainwater at the neighborhood level in open spaces should be mandatory in the basin and the RWH tanks to be encouraged in individual houses through programs. No direct wastewater should enter the stream or any of its tributaries. Polythene bags should be banned in the city to avoid choking of the drainage. The species *Pongamia pinnata* is to be eradicated and used for biodiesel production.

9.11 Conclusion

Water resources in urban areas are in poor condition and need to be maintained for the long-term sustainability of their cities. The study revealed that the water quality of the Kukrail stream is determined to be polluted by sewage. That is why wastewater treatment is needed for good water ecology. Regulatory controls and planning should be taken to prevent intrusion along the stream. The entire watershed area should be preserved. The stream area should be protected by the provision of a riparian buffer. It is also important to create awareness among people related to the management of the environment and resources of water and supports the level of conservation to the awareness at the house level.

References

Ahmad Z (2013) Hydrological study for Gomti River front development. Indian Institute of Technology Roorkee, Civil Engineering, Roorkee. https://ddmtn57ju2md7.cloudfront.net/media/bbifiles/1181924/38eeafb5-ca82-4c74-8c64-65170953eb36.pdf. Accessed Sept 2021.

Asanok L, Kamyo T, Norsaengsr M, Salinla-um P, Rodrungruang K, Karnasuta N, Kutintara U (2017) Vegetation community and factors that affect the woody species composition of riparian forests growing in an urbanizing landscape along the Chao Phraya River, Central Thailand. Urban For Urban Green, pp 138–149. https://doi.org/10.1016/j.ufug.2017.10.013

Bakker M, Anderson EI (2005) Steady flow to a well near a stream with a leaky bed. Ground Water 43:926. https://www.researchgate.net/publication/8981189_Steady_Flow_to_a_Well_Near_a_Stream_with_a_Leaky_Bed. Accessed Sept 2021

BallotBox India (2018). https://ballotboxindia.com/ap/kukrail-nala-lucknow/5182050549/. Accessed Sept 2021

Bastia J, Mishra BK, Kumar P (2021) Integrative assessment of Stormwater infiltration practices in rapidly urbanizing cities: a case of Lucknow City, India. Hydrology 8(2):93. https://www.mdpi.com/2306-5338/8/2/93

Bureau L (2021) 11 km long drain will be built with 50 crores to stop the dirty water of 17 drains from falling in Kukrail. Amar Ujala https://www.amarujala.com/lucknow/eleven-kilometer-lnog-interseptic-drain-is-proposed-to-stop-sevage-in-kukrail-lucknow-news-lko5899404147. Accessed Sept 2021

Census of India 2011 Uttar Pradesh (2011) District census handbook lucknow. https://censusindia.gov.in/2011census/dchb/0926_PART_B_DCHB_LUCKNOW.pdf. Accessed Sept 2021.

CGWB (2011) Groundwater scenario in major cities of India. Central ground water board, Ministry of Water Resources, New Delhi. https://www.indiawaterportal.org/articles/groundwater-scenario-major-cities-india-report-central-ground-water-board. Accessed Sept 2021

Das J, Jha S, Goyal MK (2020) On the relationship of climatic and monsoon teleconnections with monthly precipitation over meteorologically homogenous regions in India: Wavelet &; global coherence approaches. Atmos Res 238:104889. https://doi.org/10.1016/j.atmosres.2020.104889

Davis JW (1982) Livestock vs. riparian habitat management-there. Proceedings of the wildlife-livestock relationships symposium, 175–184. https://agris.fao.org/agris-search/search.do?recordID=US201302582049. Accessed Sept 2021

Dutta V, Singh A, Prasad N (2010) Urban sprawl and water stress with respect to changing landscape: study from Lucknow, India. J Geogr Reg Plann 2:84–105. https://www.researchgate.net/publication/228354247_Urban_sprawl_and_water_stress_with_respect_to_changing_landscape_Study_from_Lucknow_India. Accessed Sept 2021

Eastern U.P Rivers and Water Reservoirs Monitoring Committee (2019). River gomti interim report (I). https://swmc-up.in/Upload/Upload/ProjectDocument/1_636970650529169528%5ERiver%20Gomti,%20Interim%20Report_compressed.pdf. Accessed Sept 2021

Feedback Ventures Pvt. Ltd (2006) City development plan lucknow. https://lmc.up.nic.in/pdf/nnfinal.pdf. Accessed Sept 2021

Gangwar P, Lavania S (2018) Assessment of phytodiversity of Kukrail forest, in the perspective of urban development. Indian Forester 144(11):1040–1048. http://www.indianforester.co.in/index.php/indianforester/article/view/141115. Accessed Sept 2021

Google Earth (1987) Accessed 2021

Google Earth (2004) Accessed 2021

Google Earth (2021) Accessed 2021

Goyal MK, Ojha CSP (2010) Evaluation of various linear regression methods for downscaling of mean monthly precipitation in arid Pichola watershed. Nat Resour 1(1):11–18. https://doi.org/10.4236/nr.2010.11002

Goyal MK, Ojha CSP (2012) Downscaling of precipitation on a lake basin: evaluation of rule and decision tree induction algorithms. Hydrol Res 43(3):215–230. https://doi.org/10.2166/nh.2012.040

Goyal MK, Panchariya VK, Sharma A, Singh V (2018) Comparative assessment of SWAT model performance in two distinct catchments under various DEM scenarios of varying resolution, sources and resampling methods. Water Resour Manag 32(2):805–825. https://doi.org/10.1007/s11269-017-1840-1

Gupta A (2020) Kukrail will get the name of the river, the running water will clean the dirt: know how the rejuvenation will happen. Jagran. https://www.jagran.com/uttar-pradesh/lucknow-city-work-plan-to-prepare-kukrail-drain-in-lucknow-ready-jagran-special-21025224.html. Accessed Sept 2021

Ives C, Taylor M, Nipperess D, Hose G (2013) Effect of catchment urbanization on ant diversity in remnant riparian corridors. Landsc Urban Plan 110:155–163. https://doi.org/10.1016/j.landurbplan.2012.11.005

Jain A, Abdullah TA, Ashraf K, Nisar Z (2011) A conceptual approach for development of Kukrail Nala, Lucknow. Ins Town Plann India J 8:67–72. https://www.researchgate.net/publication/271122755_A_Conceptual_Approach_for_Development_of_Kukrail_Nala_Lucknow. Accessed Sept 2021

Kanaujia A, Kumar A, Kushwaha S, Kumar A (2015) Threats to wetlands in and around Lucknow, Uttar Pradesh, India. Nat Environ 20:45–52. https://www.researchgate.net/publication/327691782_Threats_to_Wetlands_in_and_around_Lucknow_Uttar_Pradesh_India. Accessed Sept 2021

Kauffman JB, Krueger WC (1984) Livestock impacts on riparian ecosystems and streamside management implications a review. J Range Manag 37:430–438. https://doi.org/10.2307/3899631

Kukrail drain sewage flows into Gomati again (2018) Times of India. https://timesofindia.indiatimes.com/city/lucknow/kukrail-drain-sewage-flows-into-gomti-again/articleshow/64442522.cms. Accessed Sept 2021

Kukrail Nala will be Rehabilitated as a River, Preparing to Build a Riverfront (2019) Navbharat Times. https://navbharattimes.indiatimes.com/metro/lucknow/development/kukrail-nala-will-be-rehabilitated-as-river-dpr-ordered/articleshow/71780828.cms. Accessed Sept 2021

Kumar D, Singh DS, Mishra M (2015) Implication of drainage basin parameters of Kukrail Nala, ganga plain, using remote sensing and GIS techniques. Int J Appl Remote Sens GIS 2(2):1–9. https://www.researchgate.net/publication/344586395_Implication_of_Drainage_Basin_Parameters_of_Kukrail_Nala_Ganga_Plain_using_Remote_Sensing_and_GIS_techniques. Accessed Sept 2021

Kumar A, Kanaujia A, Kumar A, Kumar V, Mishra H (2017) Diversity of spiders in Kukrail reserve Forest, Lucknow, Uttar Pradesh, India. J Environ Sci Technol 4:101–104. https://www.researchgate.net/publication/327861275_Diversity_of_Spiders_in_Kukrail_Reserve_Forest_Lucknow_Uttar_Pradesh_India. Accessed Sept 2021

Kumar P, Masago Y, Mishra BK, Jalilov S-M, Kefi M, Dilley M, Fukushi K (2019) Future outlook of Urban water environment in Asian cities. United Nations University, Tokya. https://www.iges.or.jp/en/pub/future-outlook-urban-water-environment-asian/en. Accessed Sept 2021

Kumari M, Azam M, Tripath JK (2016) A geochemical approach towards understanding the present environmental status of alluvial soil of Kukrail Reserve Forest, Lucknow. J Appl Geochem 18(4):430–446. https://www.researchgate.net/publication/309210024_A_GEOCHEMICAL_APPROACH_TOWARDS_UNDERSTANDING_THE_PRESENT_ENVIRONMENTAL_STATUS_OF_ALLUVIAL_SOIL_OF_KUKRAIL_RESERVE_FOREST_LUCKNOW. Accessed Sept 2021

Lucknow Development Authority (n.d.) Lucknow Master Plan 2031. https://ldaonline.co.in/. Accessed Sept 2021.

McKinney ML (2002) Urbanization, biodiversity, and conservation: the impacts of urbanization on native species are poorly studied, but educating a highly urbanized human population about these impacts can greatly improve species conservation in all ecosystems. Bioscience 52(10): 883–890. https://doi.org/10.1641/0006-3568(2002)052[0883:UBAC]2.0.CO;2

Mishra AK, Behera SK, Singh K, Sahu N, Bajpai O, Kumar A, Singh B (2014) Relation of forest structure and soil properties in natural rehabilitated and degraded forest. J Biodivers Manag For 02:1–8. https://doi.org/10.4172/2327-4417.1000117

Pandey D (2020) Will make Kukrail drain a tributary again, it will be rejuvenated. Hindustan. https://www.livehindustan.com/uttar-pradesh/story-kukrail-drain-will-be-made-a-tributary-again-this-will-be-rejuvenated-3258186.html. Accessed Sept 2021

Parveen N, Singh SK (2016) Physico-chemical characteristics of Gomti river at Lucknow stretch, India. Int J Adv Res 4(7):749–759. https://doi.org/10.21474/IJAR01

Peng J, Tian L, Liu Y, Zhao M, Hu Y, Wu J (2017) Ecosystem services response to urbanization in metropolitan areas: thresholds identification. Sci Total Environ 607-608:706–714. https://doi.org/10.1016/j.scitotenv.2017.06.218

Pennington DN, Hansel JR, Gorchov DL (2010) Urbanization and riparian forest woody communities: Diversity, composition. Biol Conserv 143(1):182–194. https://doi.org/10.1016/j.biocon.2009.10.002

Poonia V, Goyal MK, Gupta BB, Gupta AK, Jha S, Das J (2021) Drought occurrence in Different River basins of India and blockchain technology based framework for disaster management. J Clean Prod 312:127737. https://doi.org/10.1016/j.jclepro.2021.127737

Prasad R (n.d.) BALLOTBOX INDIA. https://ballotboxindia.com/dp/kukrain-Nala-detachment-from-Gomti-images/5182158549/.

Rai R (2017) India today. UP CM Yogi Adityanath directs police to launch drive against illegal immigrants in the state. https://www.indiatoday.in/mail-today/story/yogi-adityanath-directs-authorities-to-deport-illegal-immigrants-1064978-2017-10-15. Accessed Sept 2021

Raj A (2018) Youth Ki Awaaz. https://www.youthkiawaaz.com/2018/01/superstitious-practices-i-kukrail-bridge-lucknow/. Accessed Sept 2021

SENES Consultants India Pvt. Ltd (2015) Revised City Development Plan for Lucknow city-2040 I. https://lmc.up.nic.in/pdf/Final%20CDP%20-%20Volume%20I.pdf. Accessed Sept 2021

Sharma VC (1959) Uttar Pradesh District gazetteers, Lucknow (Vol. XXXVII). Government of Uttar Pradesh, revenue department. https://indianculture.gov.in/flipbook/3686. Accessed Sept 2021

Singh S (1992) Cultural tourism in the Awadh region with special reference to Lucknow. University of North Bengal, Department of Geography and Applied Geography, Salt Lake. http://14.139.211.59/bitstream/123456789/1193/19/109209.pdf. Accessed Sept 2021

Singh BP, Krishna A, Singh SC, Kumar S (2020a) Angiospermic biodiversity of Lucknow areas of Uttar Pradesh, India. Curr J Appl Sci Technol 39:101–109. https://doi.org/10.9734/CJAST/2020/v39i1330686

Singh BP, Krishna A, Singh SC, Kumar S (2020b) Species diversity amongst aquatic/wetland bodies of Lucknow District, U.P., India. Int J Curr Microbiol Appl Sci 10:359–367. https://www.researchgate.net/publication/342145124_Species_Diversity_amongst_AquaticWetland_Bodies_of_Lucknow_District_UP_India. Accessed Sept 2021

Singh R, Tiwari AK, Singh GS (2021) Managing riparian zones for river health improvement: an integrated approach. Landsc Ecol Eng 17:195–223. https://doi.org/10.1007/s11355-020-00436-5

Song W, Deng X (2017, January) Land-use/land-cover change and ecosystem service provision in China. Sci Total Environ 576:705–719. https://doi.org/10.1016/j.scitotenv.2016.07.078

Srivastava A (n.d.) BallotBox India. https://ballotboxindia.com/dp/Kukrail-Nala-SITE-VISITThis-action-group/5182271549/. Accessed Sept 2021

Survey of India (2020). https://surveyofindia.gov.in/documents/polmap-eng-11012021.jpg. Accessed Sept 2021

Tangri AK, Kumar D, Singh DS, Dubey CA (2018) The Gomati river: lifeline of central ganga plain. Springer, Cham. https://doi.org/10.1007/978-981-10-2984-4_11

Tokyo Engineering Consultants Co., Ltd., CTI Engineering International Co., Ltd (2005) Water quality management plan for Ganga river, sewerage master plan for Lucknow City. https://openjicareport.jica.go.jp/pdf/11797859_02.pdf. Accessed Sept 2021

Tripathi S, Arya AK, Basumatary SK, Bera SK (2015) Modern pollen and its ecological relationships with the tropical deciduous forests of Central Uttar Pradesh, India. Palynology 40:264–279. https://doi.org/10.1080/01916122.2015.1045049

UPPCB (2019) Action plan for restoration of polluted stretch of river Gomti from Sitapur to Kaithi (District Ghazipur). http://www.uppcb.com/pdf/PRIORITY-III/RIVER-GOMTI-5july19.pdf. Accessed Sept 2021

UPSBB (2015) Kukrail, Lucknow. https://upsbdb.org/pdf/2015/10/14-10-2015-Important-Circular-regarding-Intension-to-declare-GRC-Kukrail-Lucknow.pdf. Accessed Sept 2021

Chapter 10
Hydrological Modelling Using HEC-HMS and Estimation of the Flood Peak by Gumbel's Method

Masood Zafar Ansari, Ishtiyaq Ahmad, Kuldeep Singh Rautela, Manish Kumar Goyal, and Pushpendra Kumar Singh

Abstract This paper aims to develop a rainfall-runoff hydrological model for the Hasdeo River basin which is tributary of the Mahanadi River, Chhattisgarh, and Aghanashini River, Karnataka, to examine how rainfall affects surface runoff and peak discharges. In this study, the basin is delineated from DEM in HEC-GeoHMS as an import data for HEC-HMS model. Land use/land cover map were created using supervised classification of Sentinel 2 satellite data and merged with soil hydrological groups map to develop the curve number (CN) map. The SCS-CN, SCS unit hydrograph, recession, and Muskingum routing methods are used to calculate runoff volume, peak runoff rate, baseflow, and flood routing, respectively. The Nash-Sutcliffe efficiency (NSE), root mean square error (RMSE), coefficient of determination (R^2), and percentage bias (PBIAS) of observed and simulated flow are used for the accuracy assessment of the model. The comparison and reliability of gridded precipitation data over rain gauge station data is shown, and fluvial flooding was assessed by Gumbel's distribution method for 5-year, 10-year, 25-year, 50-year, and 100-year return periods. Finally, it can be concluded that the model may be employed in hydrological modelling in the Hasdeo watershed and Aghanashini basin with reasonable approximation. This study would help the local authorities for the development of sustainable flood forecasting and water conservation policies.

M. Z. Ansari (✉) · I. Ahmad (✉)
Department of Civil Engineering, National Institute of Technology (NIT), Raipur, India
e-mail: iahmad.ce@nitrr.ac.in

K. S. Rautela
Department of Civil Engineering, Indian Institute of Technology Indore, Indore, Madhya Pradesh, India

M. K. Goyal
Department of Civil Engineering, Indian Institute of Technology Indore, Indore, Madhya Pradesh, India
e-mail: mkgoyal@iiti.ac.in

P. K. Singh
Water Resources Systems Division, National Institute of Hydrology (NIH), Roorkee, India

A. K. Gupta et al. (eds.), *Ecosystem Restoration: Towards Sustainability and Resilient Development*, Disaster Resilience and Green Growth,
https://doi.org/10.1007/978-981-99-3687-8_10

Keywords Hydrological modelling · HEC-HMS · HEC-GeoHMS · Nash-Sutcliffe efficiency · Gumbel's method · Gridded precipitation

10.1 Introduction

The land and water resource development for a river basin is one of the most critical issues nowadays (Poonia et al. 2021a, b). The river basins are facing a huge stress due to the anthropogenic and climatic factors (Rautela et al. 2022a). The computation of runoff from an ungauged or poorly gauged basin is a key concern in the developing countries due to the higher operational and maintenance cost of the associated instruments (Rautela et al. 2020). However, the hydrological processes within the river basins are not static; their properties, as well as the inputs that drive hydrology, change with time (Milly et al. 2008). The associated future risk on the river basins will be addressed with the help of hydrological models (Rautela et al. 2022b). The computer-based hydrological models are either lumped, semi-distributed or distributed in nature and developed based on the conceptual representation of the physical waterflow process (Lohani 2012). The applicability of these models is based on the availability of the datasets. Although the total amount of rainfall is anticipated to decline because of the simulation of basin hydrological processes, more catastrophic events are anticipated to occur, potentially endangering infrastructure, agriculture, and human life with prolonged droughts and frequent floods (Sharma and Goyal 2020; Shakarneh et al. 2022).

In the basin hydrological process, there are two forms of future risk that must be addressed, and rainfall-runoff models can help. The first is the difficulty of predicting whether a significant flood discharge with the potential to endanger life or property will occur in the immediate term. The second difficulty is predicting whether changes in basin characteristics or climate constitute a hazard to water supplies or flood and drought frequency over the longer term (seasonal or decadal). Both issues are reliant on inputs from weather and climate prediction models, which have large uncertainty in their forecasts (Goyal et al. 2022). The limits of hydrological measurement techniques are one of the key reasons to simulate the rainfall-runoff processes in hydrology (NHP report 2018). We can't measure everything about hydrological systems that we'd like to know. In fact, we only have a limited number of measurement techniques and a limited number of spatial and temporal measures. We need a methodology to extrapolate from available data in both spatial and temporal, especially to ungauged catchment (where data are not accessible) and into the future, to assess the potential impact of future hydrological changes (where measurements are not possible). However, the goal of employing models to predict hydrological problems must be to improve decision-making in areas such as water resources planning, flood protection, contaminant mitigation, abstraction licencing, and other areas (Goyal and Ojha 2010, 2012; Goyal et al. 2018). With increasing demands on water resources around the world, better decision-making in the face of changing weather patterns from year to year necessitates more accurate models (Das et al. 2020; Beven 2012). The Hydrologic Engineering Centre-Hydrological Model

System (HEC-HMS) is one of the crucial and decisive tools to estimate hydrological processes and the water resources availability (Fleming et al. 2013).

In this study, Hasdeo river basin and Aghanashini river basin flowing in Chhattisgarh and Karnataka state were selected, and the rainfall-runoff modelling of these river basins are simulated using the SCS curve number, SCS unit hydrograph, recession, and Muskingum method that have been used as the loss model, the transform model, the baseflow model, and the routing model, respectively. This paper compares the results of daily rainfall data at rain gauge stations with IMD gridded precipitation data for the basin to show how rainfall change has a distinct impact on determining discharge to peak and runoff depth with the river discharge at outlet, as well as the reliability of gridded precipitation data over rain gauge station data, which will be useful for ungauged catchments or missing precipitation data. The findings of this study will aid in improving watershed planning, basin management, hydraulic structure development, and forecasting by understanding climate change.

10.2 Study Area

10.2.1 Hasdeo Basin

The Mahanadi middle sub-basin (400–750 m above mean sea level) includes the Hasdeo River basin, which is the Mahanadi's second largest branch after the Sheonath River (Fig. 10.1). The Hasdeo River is the left bank tributary of the Mahanadi River. It flows through the districts of Korba and Janjgir-Champa in Chhattisgarh before entering the Mahanadi near Seorinarayan, a popular pilgrimage site (Water Year Book 2013). It begins in the mountainous region of Deogarh in Sonhat Taluka in the Koriya district of Chhattisgarh, at an elevation of 1052 m above sea level (Prasad 2015). The Hasdeo River stretches for 333 km. This river has a catchment area of 10,535.96 km^2. Upstream of the Minimata Bango Dam, the rivers Gej, Bamni, and Atem meet Hasdeo, whereas Tan and Ahiran meet it downstream. It flows from north to south and has seven watersheds between 21°45′ N and 23°37′ N latitude and 82°00′ E to 83°04′ E longitude. The seven separate watersheds include Upper Hasdeo, Gej Nala, Bamni Nadi, Tan Nadi, Chornai, Ahiran Nadi, and Lower Hasdeo (Mishra 2016).

10.2.2 Aghanashini Basin

The catchment of Aghanashini River in Uttara Kannada, Karnataka, is considered as the study area up to Santeguli station (Fig 10.2). It originates from Shankara Honda in the Gadihalli (Sirsi) at an altitude of 676 m above mean sea level (AMSL). It is

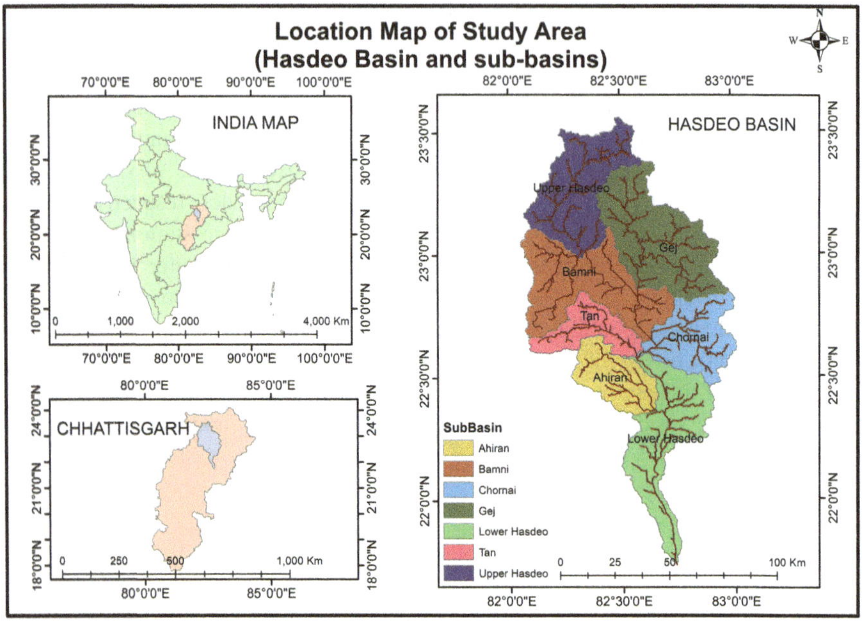

Fig. 10.1 Location map of study area, Hasdeo River basin

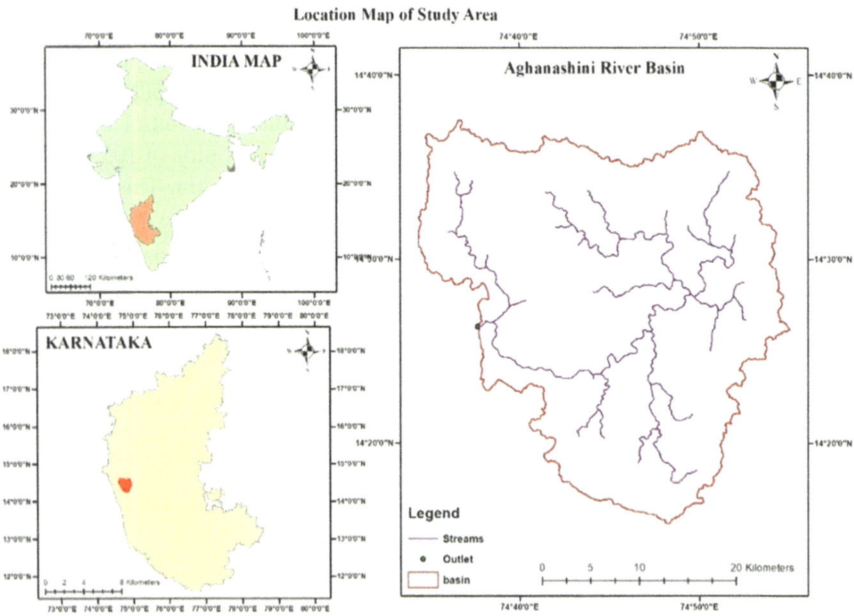

Fig. 10.2 Location map of study area, Aghanashini River basin

one of the virgin rivers of the world because of unobstructed flow through Western Ghats range and travels 117 km to join the Arabian Sea near Kumta and draining an area of 998.02 km^2 up to Santeguli. The study region is located between the longitudes 74° 34′ 21″ E to 74° 55′ 7″ E and the latitudes 14° 15′ 36″ N to 14° 37′ 33″ N. The marshes that are produced by the evergreen trees in the forest contribute to the river's continual flow. The Myristica wetlands function like a sponge, soaking up water during the monsoon and letting it out during the dry season. Since the land already has water in the soil pores, this also lessens the impact of the floods.

10.3 Materials and Methods

10.3.1 Data Collection

The data required and sources of acquisition of data for hydrological modelling have been shown in Table 10.1.

Table 10.1 Data sources

Data type and description	Source/provider of data
(a) Hasdeo basin	
Topographical map-digital elevation model	https://bhuvan.nrsc.gov.in/bhuvan_links.php/
Satellite imagery-Sentinel 2-LULC map	https://scihub.copernicus.eu/
Soil map-hydrological soil group map	https://www.nbsslup.in/
Daily rainfall data	Data Centre, WRD, Raipur
Discharge data	
Rain gauge station data	http://hydrologyproject.cg.gov.in/
Gridded precipitation data	Indian Meteorological Department
(b) Aghanashini basin	
Topographical map-digital elevation model	SRTM (https://earthexplorer.usgs.gov/)
Satellite imagery-Sentinel 2-LULC map (ESRI)	https://www.arcgis.com/apps/instant/media/index.html
Soil map-hydrological soil group map	https://www.fao.org/land-water/land/land-governance/land-resources-planning-toolbox/category/details/en/c/1026564/
Daily rainfall data	Indian Meteorological Department
Discharge data	WRIS (https://indiawris.gov.in/wris/)

10.4 Methodology

10.4.1 HEC-HMS Model Development

The methodology adopted for this study is systematically explained in Fig. 10.3. By pre-processing of DEM, the sub-basins and drainage network have been developed (Beighly and Moglen 2003). To create a project area of interest, place a project point near the desired outlet of the basin. Basin model created with ArcGIS and GeoHMS to show physical qualities of hydrological parameters including river length and slope, basin slope, longest flowpath to the basin, basin centroid, centroid elevation, and centroidal longest flowpath (Khadka and Bhaukajee 2018). There are different methods for loss flow and direct runoff calculation within the HEC-HMS (USACE-HEC 2006). In the study, the SCS curve number, SCS unit hydrograph, recession, and Muskingum method have been used as the loss model, the transform model, the baseflow model, and the routing model, respectively. The generated HEC-HMS model showing the sub-basin, reaches, and junctions is shown in Fig. 10.4a, b.

The SCS curve number method is used for actual infiltration calculations. Initial abstraction, curve number, and impervious (%) are inputs for this method

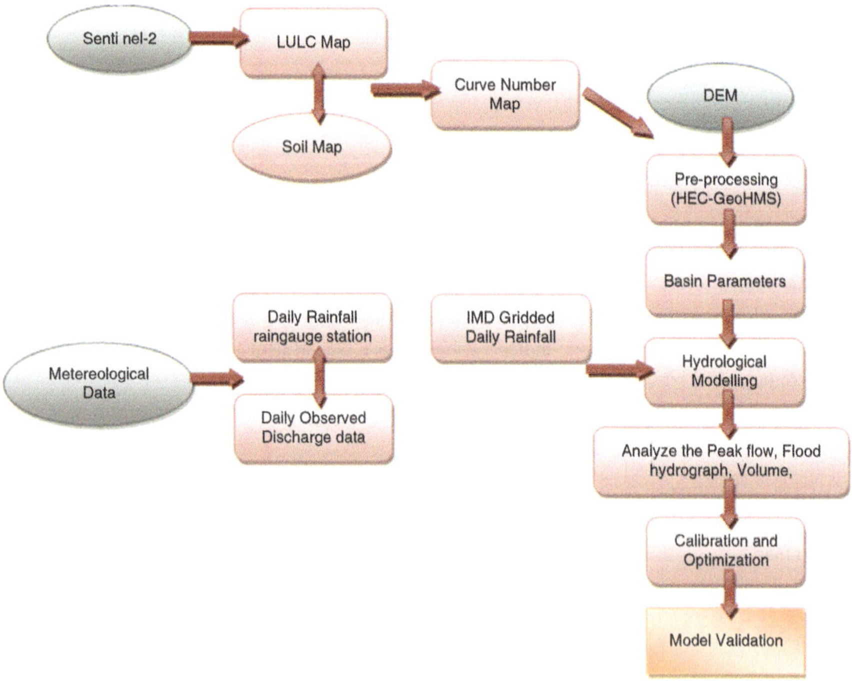

Fig. 10.3 Methodology

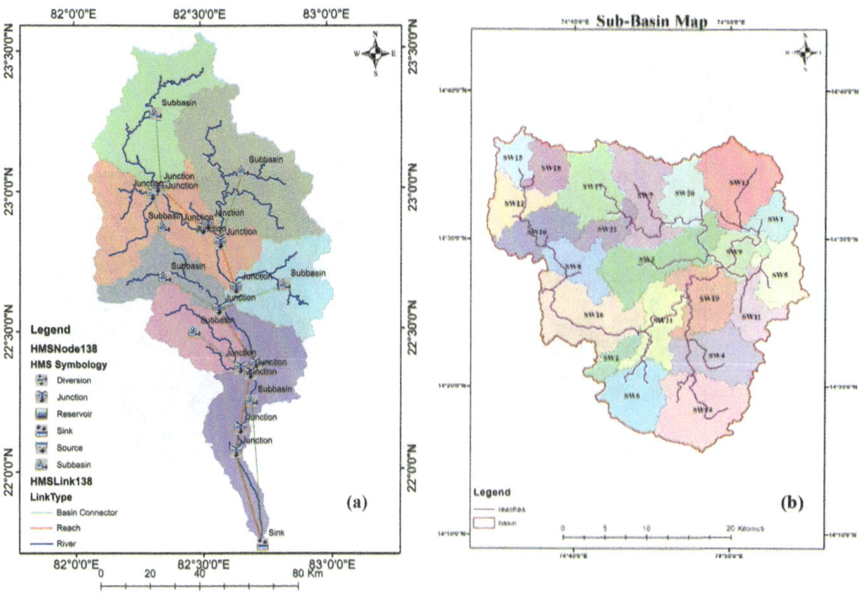

Fig. 10.4 Sub-basin maps for (a) Hasdeo basin and (b) Aghanashini basin

(Subramoniam et al. 2022). The equation for the SCS curve number technique is as follows:

$$Q = \frac{(P - I_a)^2}{(P - I_a + S)} \tag{10.1}$$

$$S = \frac{25400}{\text{CN}} - 254 \tag{10.2}$$

$$\text{I}_\text{a} = 0.2 \times \text{S} \tag{10.3}$$

where Q is runoff (mm); P is rainfall (mm); S is potential maximum retention after runoff begins; I_a is initial abstraction, i.e. the amount of rain that falls before runoff begins (mm); and CN is the curve number.

CN is basically a coefficient that reduces the rainfall to runoff. It depends upon two parameter land use and hydrologic soil group (HSG) (Ahmad and Verma 2015). The curve number in (Fig. 10.5a, b) is the union layer of soil cover and land use/land cover map. Land use/land cover is created by super-classification of Sentinel 2 data using maximum likelihood classification method. And soil map is classified in HSG based on soil taxonomy.

Impervious (%) is taken as per cent of built-up, as the built-up has minimum infiltration.

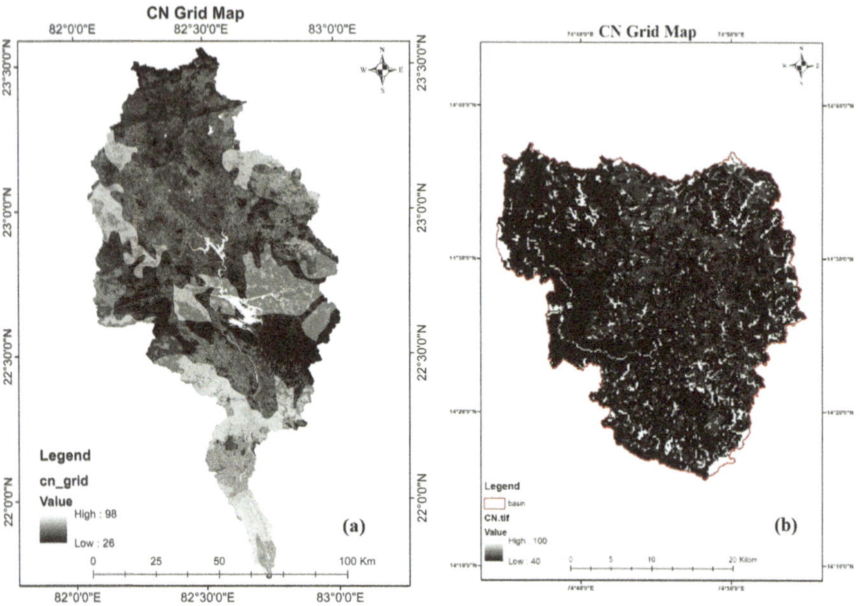

Fig. 10.5 Curve number map for (**a**) Hasdeo basin and (**b**) Aghanashini basin

The SCS unit hydrograph method is used for actual surface runoff calculations. Lag time is input for this method. It is the delay in time taken by water reaching at the outlet. Lag time (T_{lag}) is estimated as the equations given below:

$$T_{lag} = T_c \times 0.6 \tag{10.4}$$

$$T_c = \frac{L^{0.8} + (S + 1)^{0.7}}{1900 + Y^{0.5}} \tag{10.5}$$

where T_c represents time of concentration, i.e. the time it takes for runoff to travel from the hydraulically most distant point of the watershed to the outlet, which is estimated using the KIRPICH equation, L is the hydraulic length of the watershed, and Y represents slope at each sub-basin and calculated using HEC-GeoHMS tool.

The recession method is used for actual subsurface calculations. The recession constant describes the rate at which baseflow recedes between storm occurrences.

$$\text{Recession constant} = \frac{\text{baseflow at the current time}}{\text{baseflow one day earlier}}$$

Flood routing is a method for determining the timing and magnitude of a flood wave at a given place on a stream using data from one or more upstream points (Chow et al. 1988). Muskingum K is the travel duration across the reach, whereas with values ranging from 0.0 to 0.5, Muskingum X is the weighting between inflow and outflow effects.

10.4.2 Model Calibration and Validation

The Hydrologic Engineering Centre-Hydrologic Modeling System (HEC-HMS) (Fleming and Brauer 2016) is used in this study to provide a prediction of the hydrological response of a watershed.The developed HEC-HMS model after the trial runs was calibrated and validated against observed discharge data at the outlet. The calibration was performed in order to adjust model parameters so that the simulated flows have had better agreement with the observed flows. After calibration, the model was validated using the same input hydrological parameters obtained via model calibration and observed discharge data.

The Nash-Sutcliffe coefficient of efficiency (NSE) and coefficient of determination (R^2) parameters are used to assess the effectiveness of the created model. NSE (Eq. 10.6) values range from $-\infty$ to 1 (Nash and Sutcliffe 1970) and R^2 (Eq. 10.7) from 0 to 1.

$$NSE = 1 - \left[\frac{\sum_{i=1}^{n} (O_i - P_i)^2}{\sum_{i=1}^{n} (O_i - O_{avg})^2} \right] \tag{10.6}$$

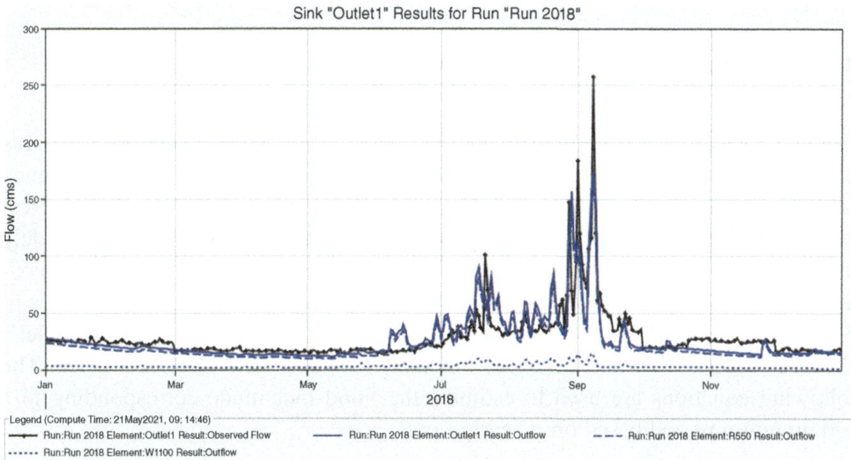

Fig. 10.6 Observed and simulated flow for calibrated year 2018 of rainfall station data

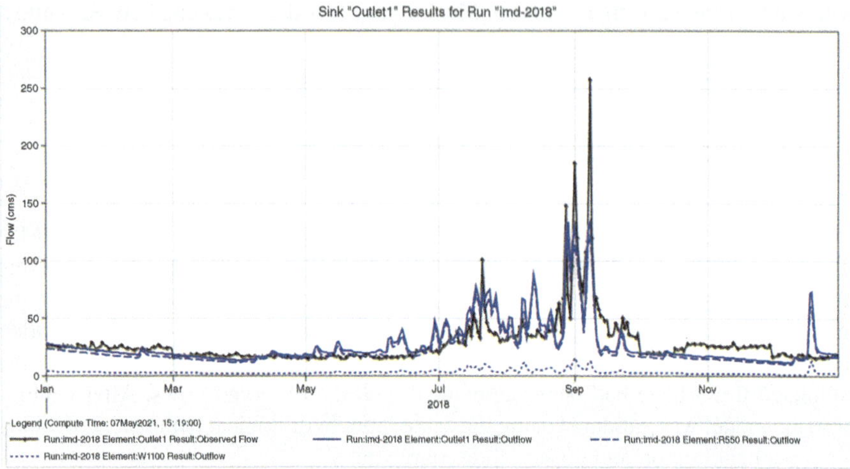

Fig. 10.7 Observed and simulated flow for calibrated year 2018 of satellite rainfall data

$$R^2 = \left[\frac{\sum_{i=1}^{n} \left(O_i - O_{avg}\right)\left(P_i - P_{avg}\right)}{\sum_{i=1}^{n} \left(O_i - O_{avg}\right)^2 \sum_{i=1}^{n} \left(P_i - P_{avg}\right)^2} \right]^2 \tag{10.7}$$

where O_i is the ith observed value, O_{avg} is the average observed value for the entire study period, P_i is the ith predicted (simulated) value, and P_{avg} is the average of the predicted value over the entire study period.

10.4.3 Flood Frequency Analysis

Frequency analysis is used to estimate the likelihood of certain values of a variable phenomenon occurring and to assess the accuracy of event prediction. The frequency analysis is commonly used to predict the return period (T) of a hydrological event (Subramanya 2008). Gumbel distribution method is one of the most widely used extreme value distribution functions for the prediction of flood peaks, maximum rainfalls, and so on. In the present study, the peak floods for return periods of 5, 10, 25, 50, and 100 years for the Hasdeo River are estimated using Gumbel's extreme value distribution function. For that, 18 years' data have been used. The following equations are used to estimate the flood magnitude corresponding to a certain return period based on a yearly flood series:

$$X_T = \overline{X} + K\sigma_{n-1} \tag{10.8}$$

$$K = \frac{y_T - \overline{y}_n}{S_n} \tag{10.9}$$

$$y_T = -\left[ln.\,ln\,\frac{T}{T-1}\right] \tag{10.10}$$

where X_T is the value of variate X of a random hydrologic series with return period T, $\overline{X}$ is the mean, σ_{n-1} is the std. deviation, K is the frequency factor, y_T is the reduced variate, $\overline{y}_n$ is the reduced mean, and S_n is the reduced std. deviation.

10.5 Results and Discussion

10.5.1 Hasdeo Basin

The time-series data such as daily rainfall station data and satellite rainfall data, the model has been run simultaneously (Figs. 10.6 and 10.7). To identify the best match between the simulation and observation, the HEC-HMS model was calibrated for various flood events. The calibration of hydrological models by employing local observed or remotely sensed data is in fact useful to improve accuracy of predictions (Halwatura and Najim 2013). Calibration has been done both manually and automatically (by optimization trial) of the simulated flow with observed flow (Figs. 10.6 and 10.7). After the calibration from 2015 to 2018, the calibrated final parameters were used to validate the model from 2013 and 2014 shown in Fig. 10.8 and Table 10.2. The results obtained for rainfall station data and gridded precipitation data for calibrated year 2018 have been shown in Fig. 10.7, and the summary

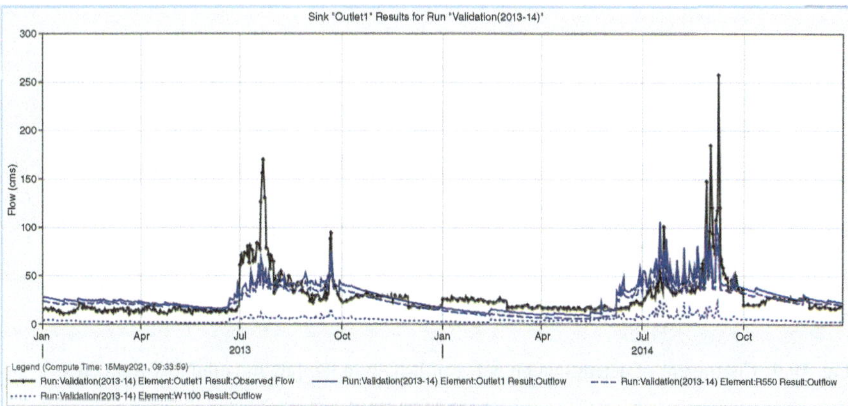

Fig. 10.8 Observed and simulated flow for validated years 2013–2014

Table 10.2 Summary of results for validated model 2013–2014

Observed volume (mm)	171.88
Calibrated volume (mm)	185.15
Nash-Sutcliffe efficiency(NSE)	0.43
RMSE	0.8
% bias	7.76

Table 10.3 Summary of results for calibrated model 2018

Observed volume (mm)	88.47
Calibrated volume (mm)	85.89
Nash-Sutcliffe efficiency (NSE)	0.61
RMSE	0.6
% bias	−2.91

Table 10.4 Summary of results for calibrated model 2018

Observed volume(mm)	88.47
Calibrated volume (mm)	88.76
Nash-Sutcliffe efficiency (NSE)	0.57
RMSE	0.7
% bias	0.35

of the results has been shown in Tables 10.3 and 10.4, respectively. These results are the close and shows good correlation between the observed and simulated flow.

Figure 10.9 shows the value of R^2 as 0.8, which is a good correlation of IMD satellite data over rain gauge station data. The result gives reliability of gridded precipitation data over rain gauge station data which will be useful for ungauged catchment or missing precipitation data.

Using Gumbel's method, peak flood discharges of the Hasdeo River are predicted in Table 10.5.

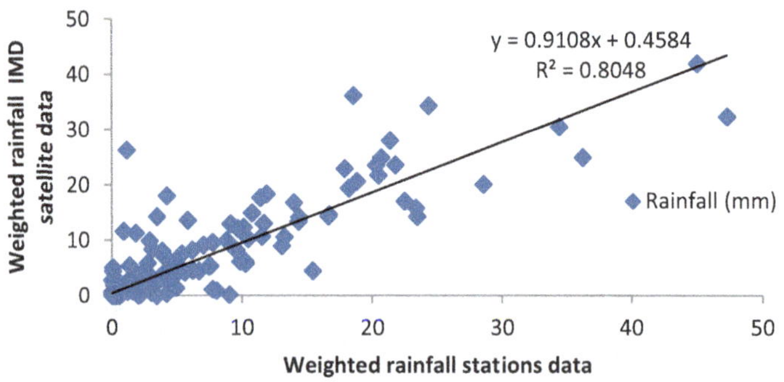

Fig. 10.9 Coefficient of determination between IMD grid rainfall and rainfall station data

Table 10.5 Predicted peak flood discharge for corresponding return periods

Return period (T)	Reduced variate (Y_t)	Frequency factor (K)	Expected flood (X_t) in cumec
5	1.50	0.97	1980.46
10	2.25	1.70	2855.71
25	3.20	2.63	3961.59
50	3.90	3.32	4782.00
100	4.60	4.00	5596.35

10.5.2 Aghanashini Basin

The hydrological soil classification is based on the soil's texture, and the soil is separated into the hydrological soil Groups B, C, and D. The catchment region is made up of Group B soil, which comprises 27.22 km^2; Group C soil, which includes 681.53 km^2; and Group D soil, which covers 289.26 km^2. The hydrological soil type Group C, which has a relatively high capacity for creating runoff, covers most of the region. The research area's unique soil features have a significant impact on Aghanashini River's baseflow. Due to the soil's ability to store moisture, the baseflow was greater. In general, Group B and Group C type soil store more moisture and release it as baseflow during dry periods, but during rainy seasons, the soil is fully saturated and produces more baseflow, which increases river discharge. The baseflow index is calculated to be 0.71, giving a total baseflow and runoff of 37053.84 mm and 15386.78 mm, respectively. The curve number for the basin varies from 40 to 96. The model simulation of the streamflow discharge was simulated using the HEC-HMS model for the years 2014–2018. The calibration is performed for the years 2014 to 2017, while validation is carried out for the year 2018. The SCS-CN-based model shows a good relation in between the observed and simulated discharge for calibration and validation period. Since the initial abstraction ratio is 0.3, the simulated discharge is significantly less than the observed discharge. A significant difference of -16.17 m^3/s is observed between average discharges, while the deviation of -336.89 m^3/s is observed between the peak discharges. The significant deviations between observed and simulated values are noticed in this method (Figs. 10.10 and 10.11). The coefficient of determination (R^2) of SCS-CN method was found to be 0.80 and 0.63 for the calibration and validation period, respectively (Fig. 10.12 and 10.13).

The performance of these models is evaluated through goodness-of-fit statistics such as NSE, RMSE, and PBIAS shown in Tables 10.6 and 10.7. As discussed by Moriasi et al. (2007), the model shows satisfactory performance as their NSE values are 0.79 and 0.64, RMSE values 1.68 and 1.78 cm, and PBIAS values 1.93 and 2.13% for the calibration and validation period, respectively.

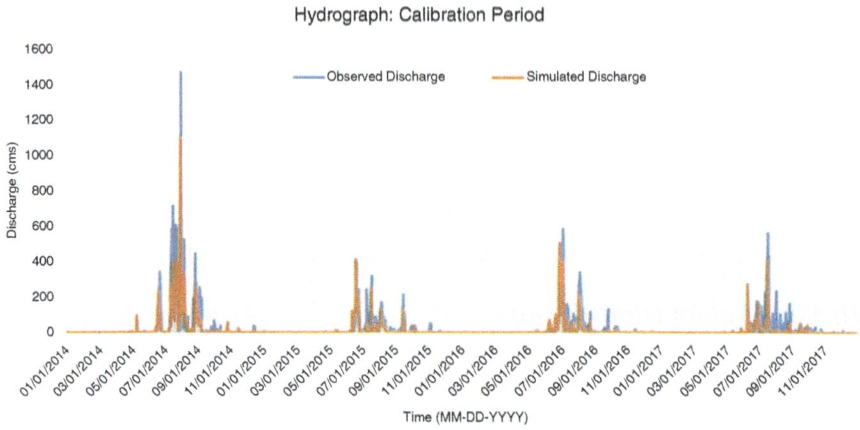

Fig. 10.10 Observed and simulated flow for calibrated years of rainfall satellite data

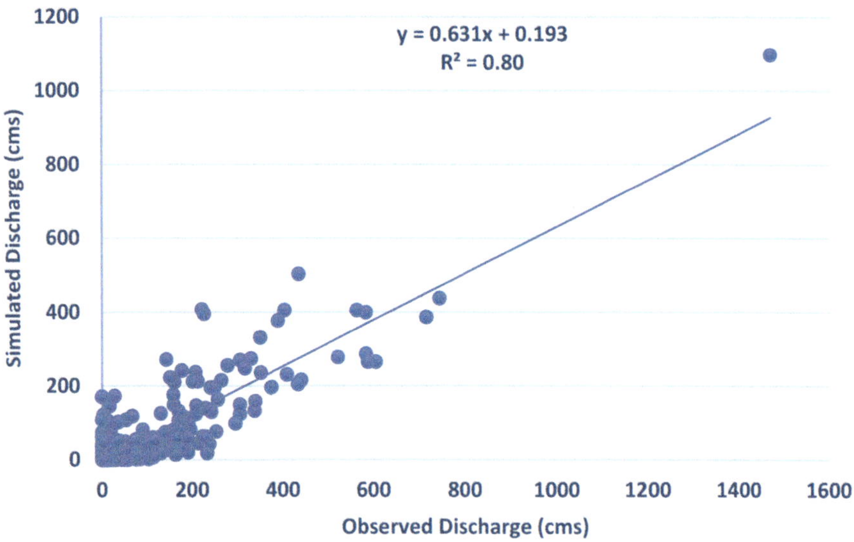

Fig. 10.11 Observed and simulated flow for validation year (2018)

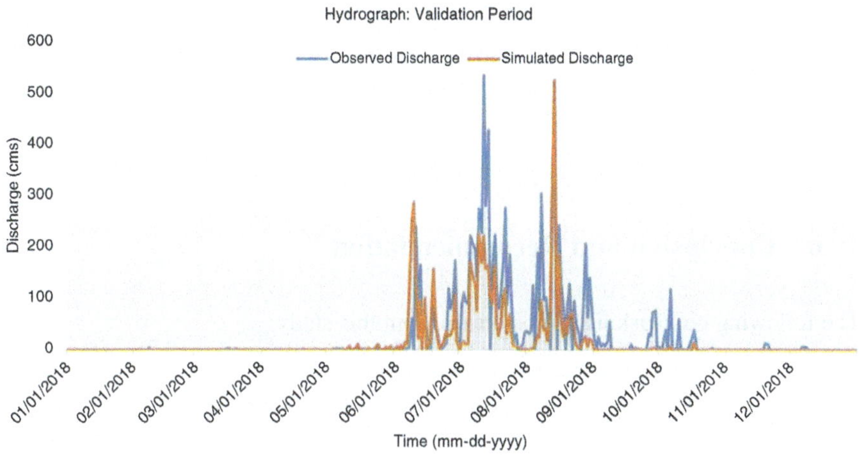

Fig. 10.12 Correlation in between observed and simulated flow for calibrated years (2014–2017)

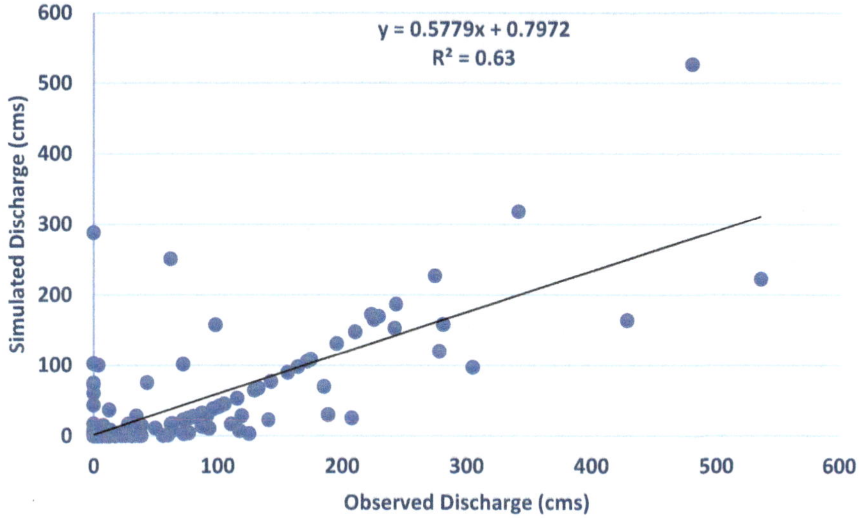

Fig. 10.13 Correlation in between observed and simulated flow for validation year (2018)

Table 10.6 Summary of results for calibrated model (2014–2017)

Observed discharge (cm)	22.13
Calibrated discharge (cm)	19.16
Nash-Sutcliffe efficiency (NSE)	0.79
RMSE	1.68
% bias	+1.93

Table 10.7 Summary of results for calibrated model (2014–2017)

Observed discharge (cm)	20.13
Calibrated discharge (cm)	18.32
Nash-Sutcliffe efficiency (NSE)	0.64
RMSE	1.78
% bias	+2.13

10.6 Conclusion and Recommendation

The following conclusions can be drawn from this study:

1. The model simulate well continuous flow of a 1-year period for the Hasdeo River catchment and 4 years for the Aghanashini basin. The models perform well for a single year and continuously for many years. Sometimes, continuous modelling is challenging to obtain NSE within a satisfied range.
2. In this study, the rainfall-runoff hydrological model of basin with different climatic and geographical characteristics was developed. SCS curve number method used for actual infiltration calculations are performed by a loss model, SCS unit hydrograph method to transform excess precipitation into the direct runoff, recession method for actual subsurface calculations is performed by a baseflow method and routing Muskingum model for routing the reach were found to be suitable methods for the Hasdeo River basin and Aghanashini basin.
3. The model shows the Nash-Sutcliffe efficiency criterion in the range of 0.43–0.64 and 0.64–0.79 for the Hasdeo River basin and Aghanashini basin, respectively. The percentage bias of the simulation model is in the range of 0.30–8.00% and 1.93–2.13% for the Hasdeo River basin and Aghanashini basin, respectively. These results of the rainfall-runoff simulation model indicate the close and the good correlation between simulated and observed flow in this study area.
4. IMD gridded precipitation shows good correlation with rainfall station data which shows reliability of data to be used in ungauged basins. Since the station data is missing in the Aghanashini basin, based on the results, the satellite-based precipitation products (SBPPs) could provide a reliable dataset. The SBPPs can be used further for the basins which have a similar hydrological characteristic.
5. This study is helpful for the concerned policy makers and stakeholders in runoff simulation for flood forecasting and water conservation based on the rainfall data.

Acknowledgement The work was supported by the Data Collection Centre, Water Resources Department, Raipur, for providing essential meteorological data for the Hasdeo River basin, such as the daily rainfall and discharge data and WRIS India for providing the discharge and rainfall data for Aghanashini basin, Karnataka. We are also grateful to the Director of the National Institute of Hydrology in Roorkee, as well as the Head of the Water Resources Division, for providing an institutional facility, especially the GIS lab.

References

Ahmad I, Verma MK (2015) Application of analytic hierarchy process in water resources planning: a GIS based approach in the identification of suitable site for water storage. Water Resour Manag 2018(32):5093–5114. https://doi.org/10.1007/s11269-018-2135-x

Beighly RE, Moglen GE (2003) Adjusting measured peak discharges from an urbanizing watershed to reflect a stationary land use signal. Water Resour Res 39(4):1–11

Beven KJ (2012) Rainfall-runoff modelling: the primer, 2nd edn. Wiley, Hoboken, NJ. ISBN 978-0-470-71459-1

Chow VT, Maidment DR, Mays LW (1988) Applied hydrology, 0–07–039732-5. McGraw-hill, New York

Das J, Jha S, Goyal MK (2020) On the relationship of climatic and monsoon teleconnections with monthly precipitation over meteorologically homogenous regions in India: Wavelet &; global coherence approaches. Atmos Res 238:104889. https://doi.org/10.1016/j.atmosres.2020.104889

Fleming M, Brauer T (2016) Hydrologic modelling system HEC-HMS quick start guide. Version 4.2. US Army Corps of Engineers, Institute of water Resources, Alexandria, VA

Fleming M et al (2013) HEC-GeoHMS geospatial hydrologic modelling extension. Version 10.2. US Army Corps of Engineers, Institute of water Resources, Alexandria, VA

Goyal MK, Ojha CSP (2010) Evaluation of various linear regression methods for downscaling of mean monthly precipitation in arid Pichola watershed. Nat Resour 01(01):11–18. https://doi.org/10.4236/nr.2010.11002

Goyal MK, Ojha CSP (2012) Downscaling of precipitation on a lake basin: evaluation of rule and decision tree induction algorithms. Hydrol Res 43(3):215–230. https://doi.org/10.2166/nh.2012.040

Goyal MK, Panchariya VK, Sharma A, Singh V (2018) Comparative assessment of SWAT model performance in two distinct catchments under various DEM scenarios of varying resolution, sources and resampling methods. Water Resour Manag 32(2):805–825. https://doi.org/10.1007/s11269-017-1840-1

Goyal MK, Gupta AK, Jha S et al (2022) Climate change impact on precipitation extremes over Indian cities: Non-stationary analysis. Technol Forecast Soc Change 180:121685. https://doi.org/10.1016/j.techfore.2022.121685

Halwatura D and Najim MMM (2013) Application of the HEC-HMS model for runoff simulation in a tropical catchment. Environ Model Softw 46:155–162. https://doi.org/10.1016/j.envsoft.2013.03.006

Khadka J, Bhaukajee J (2018) Rainfall-runoff simulation and modelling using HEC-HMS and HEC-RAS models: case studies from Nepal and Sweden. Thesis database (TVVR-18/5009)

Lohani AK (2012) Lecture notes—rainfall-runoff analysis and modeling

Milly PC, Betancourt J, Falkenmark M, Hirsch RM, Kundzewicz ZW, Lettenmaier DP, Stouffer RJ (2008) Stationarity is dead: whither water management?. Science 319(5863):573–574

Mishra M (2016) India River Week—Chhattisgarh

Moriasi DN, Arnold JG, Van Liew MW, Bingner RL, Harmel RD, Veith TL (2007) Model evaluation guidelines for systematic quantification of accuracy in watershed simulations. Transactions of the ASABE 50(3):885–900

Nash JE, Sutcliffe JV (1970) River flow forecasting through conceptual models part I—a discussion of principles. J Hydrol 10(3):282–290

NHP (2018) Hydrological modeling—current status and future directions. Report, vol 1. National Institute of Hydrology, Roorkee

Parhi PK (2018) Flood Management in Mahanadi Basin using HEC-RAS and Gumbel's extreme value distribution. J Inst Eng India Ser A 99:751

Poonia V, Jha S, Goyal MK (2021a) Copula based analysis of meteorological, hydrological and agricultural drought characteristics across Indian river basins. Int J Climatol 41(9):4637–4652

Poonia V, Goyal MK, Gupta BB, Gupta AK, Jha S, Das J (2021b) Drought occurrence in different river basins of India and blockchain technology based framework for disaster management. J Clean Prod 312:127737

Prasad P (2015) Report-hasdeo sub basin. Theme: Environmental flow

Rautela KS, Kuniyal JC, Kanwar N, Bhoj AS (2020) Estimation of stream hydraulic parameters and suspended sediment load of river Neola in the foothills of the Panchachuli Glacier during the ablation period. JHESD 15:1–16

Rautela KS, Kuniyal JC, Alam MA, Bhoj AS, Kanwar N (2022a) Assessment of daily streamflow, sediment fluxes, and erosion rate of a pro-glacial stream basin, central Himalaya, Uttarakhand. Water Air Soil Pollut 233(4):1–16

Rautela KS, Kumar M, Sofi MS, Kuniyal JC, Bhat SU (2022b) Modelling of streamflow and water balance in the Kuttiyadi River basin using SWAT and remote sensing/GIS tools. Int J Environ Res 16(4):1–14

Shakarneh MOA, Khan AJ, Mahmood Q et al (2022) Modeling of rainfall–runoff events using HEC-HMS model in southern catchments of Jerusalem Desert-Palestine. Arab J Geosci 15:127. https://doi.org/10.1007/s12517-021-09406-z

Sharma A, Goyal MK (2020) Assessment of drought trend and variability in India using wavelet transform. Hydrol Sci J 65(9):1539–1554

Subramanya K (2008) Engineering hydrology, 3rd edn. Tata McGraw-Hill Publishing Company, New Delhi

Subramoniam SR, Ravindranath S, Rakkasagi S, Ram H (2022) Water resource management studies at micro level using geospatial technologies. Springer, Cham, pp 49–74

USACE-HEC (2006) Hydrologic modeling system HEC-HMS v3.2 user's manual. US Army Corps of Engineers, Hydrologic Engineering Center (HEC), Davis, CA

Water Year Book (2013) Water resource department. Government of Chhattisgarh, Raipur

Chapter 11
Comparing Runoff of the NRCS-CN Method and Observed Runoff Data: A Case Study

Sinchana M. Sakalesh and Shivakumar J. Nyamathi

Abstract In the present study, morphometric characterization, runoff estimation, and comparison of observed and estimated runoff for the Tunga River catchment, tributary to the Tungabhadra River of the Krishna Basin, are carried out. Runoff estimation is carried out using the NRCS-CN method. The study is performed in a GIS environment by considering land use/land cover and soil texture. The area of the Tunga River catchment is 2950 Km^2. The catchment is delineated using Survey of India (SOI) topo maps of 1:50000 scale using GIS software. The average annual rainfall observed in the area is 2247.35 mm. The runoff data is collected from CWC, Bangalore. The study is carried out for soil infiltration rates of 0.2S and 0.3S. The estimated and observed runoff are being compared, and the reasons for the differences in runoff are discussed.

Keywords Runoff estimation · NRCS-CN (Natural Resource Conservation Service) · GIS (Geographical Information System)

11.1 Introduction

Estimating runoff from a watershed is a critical component in flood prediction and mitigation, water quality management, hydropower production, and a number of other water resource operations. Seepage from tanks, canals, streams, and functional irrigation are all significant sources of recharge. Hydrometeorological and hydrological data play a critical role in determining the accessibility of water sources for the planning and design of artificial recharge structures (Goyal and Ojha 2010; Das et al. 2020; Poonia et al. 2021). Rainfall runoff modeling can be done in a variety of ways (Goyal et al. 2022). One of the most basic and simple methods for rainfall runoff modeling is the Soil Conservation Services and Curve Number (SCS-CN) technique. Runoff data is essential for watershed management in order to conserve

S. M. Sakalesh (✉) · S. J. Nyamathi
Water Resources Engineering, Department of Civil Engineering, UVCE, Bangalore University, Bangalore, India

A. K. Gupta et al. (eds.), *Ecosystem Restoration: Towards Sustainability and Resilient Development*, Disaster Resilience and Green Growth,
https://doi.org/10.1007/978-981-99-3687-8_11

and develop natural resources (Goyal and Ojha 2012; Goyal et al. 2018). The morphometric analysis of the drainage basin and channel network is critical for understanding the drainage basin's geohydrological behaviors and reflecting the catchment's climate, geology, and geomorphology. Morphometric analysis gives information about the physical characteristics of the watershed and contributes in the calculation of catchment parameters. Morphometric analysis provides the physical characteristics of the watershed and also helps in the derivation of parameters within the catchment. The runoff estimation carried out by using the NRCS-CN method will help in proper planning and management of catchment yield for better planning of river basin (Subramoniam et al. 2022). Rainfall information is essential for estimations and calculations of design flows. Generally, this information is summarized in tables or depicted as graphs indicating the rainfall depth in a specific area for a specific duration. The main aim of the study is to estimate runoff by the NRCS-CN method for the Tunga catchment, compare observed runoff with the estimated runoff, and analyze the percentage difference in estimated runoff with respect to observed runoff.

11.2 Objectives of the Study

The main aim of the study is to estimate runoff by the NRCS-CN method for the Tunga catchment, compare observed runoff with the estimated runoff, and analyze the percentage difference in estimated runoff with respect to observed runoff.

11.3 Catchment Area and Data Products Used

11.3.1 Catchment Area

In this work, the hydrological process of the catchment was intended to be studied by constructing thematic maps utilizing remotely sensed data and a database using ArcGIS software.

The Tunga River runs through Shivamogga and Chikkamagaluru districts and is located at 75° 40′ 0″ longitude and 14° 0″ latitude. It has a total area of 2950 km^2. The location map of the research area is shown in Fig. 11.1. Using 1:50000 SOI toposheets, the catchment is delineated/digitized in ArcGIS software.

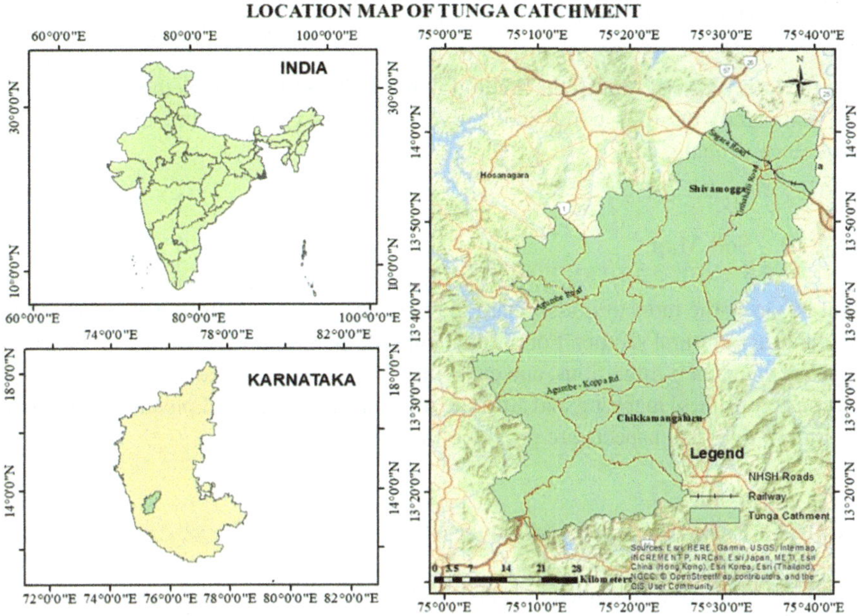

Fig. 11.1 Location map of the Tunga catchment

11.3.2 Data Sets

- Survey of India (SOI) topographical maps (1987–1988) at 1:50,000
- Soil map of the year 2005 from the National Bureau of Soil Survey and Land Use Planning, Hebbal, Bangalore
- Land use/land cover map of the year 2016 from the USGS Earth Explorer
- DEM SRTM (30 m resolution) from Bhuvan
- Daily rainfall data (1998–2018) from Director, Statistics Department, MS Building, Bengaluru
- Daily runoff data (1998–2018) from CWC, Bengaluru

11.4 Methodology

11.4.1 Morphometry

The drainage map was prepared using SOI topo maps on a 1:50000 scale. A stream order map is developed, and the morphometric parameters are discussed.

11.4.2 Runoff Estimation

For the estimation of runoff, the methodology of the NRCS-CN method is followed, the estimated runoff is compared with the observed runoff, and the percentage difference is analyzed.

11.4.2.1 Soil Map

Soil is one of the most important natural resources in the growth of both agricultural and non-agricultural sectors. For optimal land use planning, information on different types of soils, their limitations, and capacities is required to assess their suitability for various crops. Soil mapping and attribute information in terms of physical, chemical, and morphological aspects are provided through soil surveys.

11.4.2.2 Land Use and Land Cover Map

Land use/land cover knowledge is crucial for any planning and management operations, and it is also regarded as an essential component for modeling and comprehending the planet as a system. Land use refers to the human activities linked to a piece of land, such as agriculture, urbanization, and so on. The characteristics are referred to as land cover (Fig. 11.2)

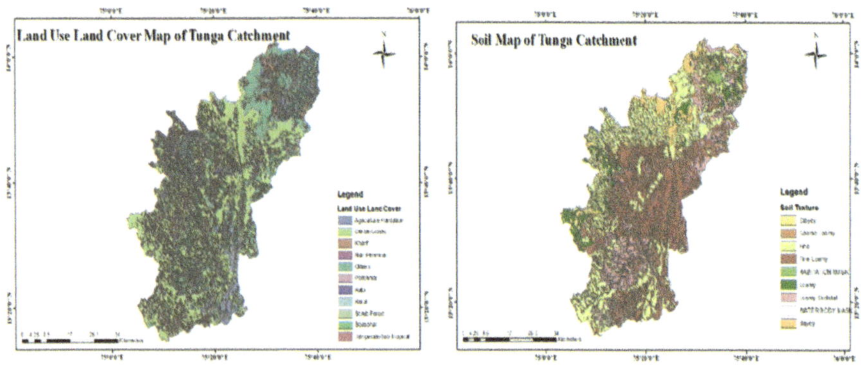

Fig 11.2 Land use/land cover map and soil map of the Tunga catchment

11.5 Results and Discussion

11.5.1 Morphometric Analysis

Using SOI toposheets, the catchment area and streams are delineated/digitized, and the stream ordering is determined. Basic morphometric properties such as stream numbers and length, as well as other data, are extracted and tabulated.

Quantitative morphometric analysis is used to investigate the various characteristics of the watershed. The methodology suggested by Strahler (1964) was used to rank streams in the present study. The highest stream order determined using Strahler's method of stream ordering is seventh order, and hence, the catchment is designated as seventh order. The overall length of stream segments is often highest in first-order streams and decreases as stream order increases. The drainage density reflects land use, infiltration effects, and the duration between precipitation and discharge in the catchment (Figs. 11.3 and 11.4, Table 11.1).

Tunga watershed has a stream order ranging from 1 to 6; stream number (Nu) 3535; a stream length (Lu) of 3747.53 m; a mean stream length ratio of 2.24; a mean bifurcation ratio (R_b) of 5.07, which is >5, hence to say that it is structurally uncontrolled; and a main channel length (Cl) of 96.35 m. Basin geometries are basin area of 2950 km^2; basin perimeter of 377.28 km; shape factor of 3.148, which is greater than 0.9, hence circular in shape; form ratio (Ff) of 0.317; elongation ratio

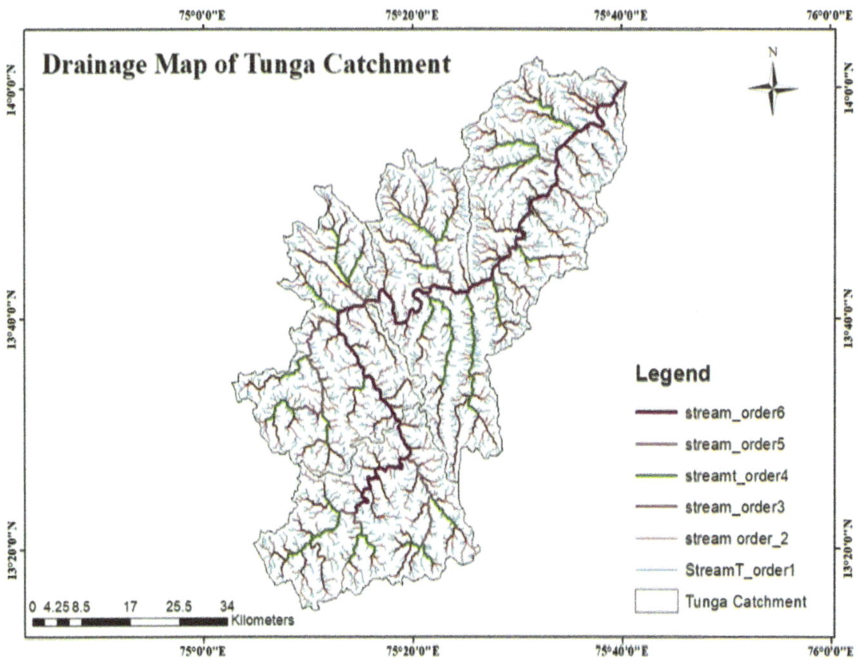

Fig 11.3 Drainage map of the Tunga catchment

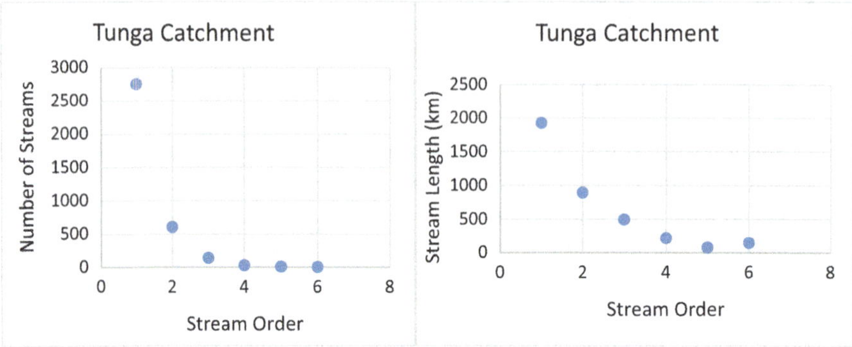

Fig. 11.4 Stream order vs. No. of streams and stream order vs. stream length for the Tunga catchment

Table 11.1 Morphometric characteristics of the Tunga catchment

Stream order	No. of streams	Stream length	Mean stream length	Mean stream length ratio	Bifurcation ratio
Su	*Nu*	*(Lu) (km)*	*(Lsm) (km)*	*(Lur)*	*(Rb)*
1	2752	1925.494	0.699		
2	608	892.162	1.467	0.476	4.526
3	138	493.250	3.574	0.410	4.405
4	28	217.0236	7.7508	0.461	4.928
5	8	74.952	9.369	0.827	3.5
6	1	144.635	144.635	0.0647	8

(Re) of 0.6362, hence circular and less elongated; and circularity ratio (Rc) of 0.2603, which implies infiltration is uniform. Drainage texture analysis is stream frequency (Fs) of 1.1983, which ranges from 0 to 5, thus low stream frequency. Drainage density (Dd) of 1.27 Km/Km^2 range is between 1.24 and 2.49, hence coarse texture. The relief characteristics are that the height of the basin mouth is 556 m, the maximum height of the basin is 1006 m, the total basin relief is 450 m, the relief ratio is 4.607, the relative relief ratio is 0.001, and the ruggedness number is 0.571.

11.5.2 Runoff Estimation

The primary parameters used to estimate runoff are land use/land cover, soil groups, rainfall, and curve numbers. The analysis is performed with Ia = 0.2 S as well as Ia = 0.3 S. For both scenarios, graphs are created, and the results are compared to the observed runoff. The depth of runoff in cumecs is translated to mm. The Thiessen polygon map and curve number maps are constructed using the soil and land use/land cover maps, and runoff is estimated (Figs. 11.5, 11.6 and 11.7).

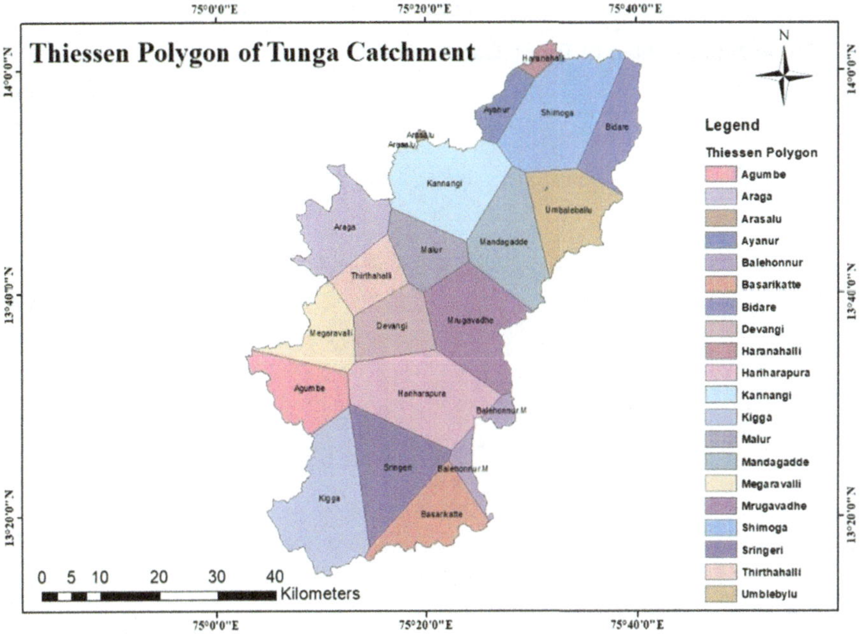

Fig. 11.5 Thiessen polygon of the Tunga catchment

The Tunga catchment receives an average annual rainfall of 2247.35 mm, and the observed average annual runoff is 278.08 mm. The maximum rainfall is in the year 2007, which is 3011.65 mm, and the minimum rainfall is in the year o2003, which is 1579.71 mm. The maximum observed runoff is 398.91 mm in 2013, and the minimum observed runoff is 180.1 mm in 2003.

11.6 Conclusions

11.6.1 Morphometric Analysis

- Tunga's catchment area is 2950 km2, with a boundary of 377.28 km2. The highest order assigned is sixth, making it a sixth-order catchment.
- The low drainage density of 1.27 km/km^2 shows that the catchment has a coarse drainage texture; the circularity ratio of 0.26 indicates that the catchment is circular in character, with little elongation.
- The term bifurcation ratio (Rb) refers to the relationship between the number of streams in one order and the number of streams in the next higher order. The bifurcation ratio of more than 5 (3–4.7) from first to second order shows stronger infrastructural erosion, implying that the catchment is not structurally controlled or that structural disturbance is minimal.

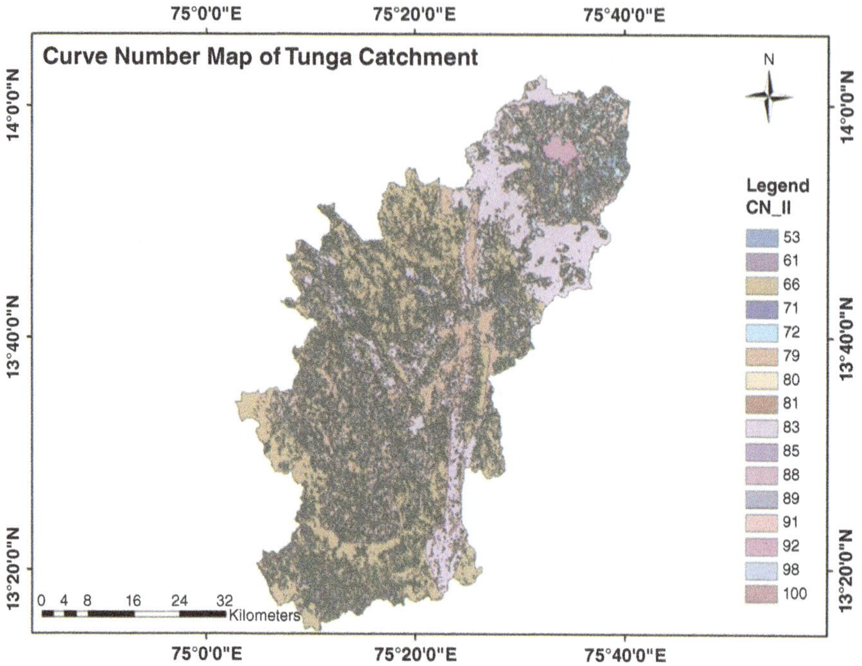

Fig. 11.6 Curve number map of the Tunga catchment

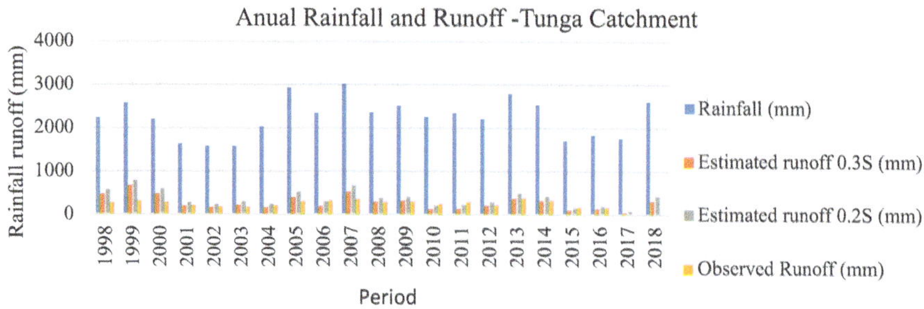

Fig. 11.7 Rainfall and runoff of the Tunga catchment

11.6.2 *Runoff Estimation*

The SCS model was used to simulate the surface runoff for the Tunga watershed. Daily rainfall data was obtained for a period of 20 years, from 1998 to 2018, with the data related to the southwest monsoon period of June to October, which accounts for 95% of the yearly rainfall in the area.

The soil being coarse-loamy in nature and the region having low runoff potential and maximum infiltration are also due to the thick vegetation in the region.

It is recorded that the estimated runoff is higher than the observed runoff. The difference in estimated runoff might be because of not considering any anthropogenic activities in the region.

The Tunga River receives an average annual rainfall of 2247.35 mm, with a maximum rainfall of 3011.35 mm and a minimum rainfall of 1579.50 mm.

The maximum observed runoff ratio is 14.29 mm, and the estimated runoff ratio is 17.83 mm.

The range of deviation in estimated runoff for 0.2S is varying between 18% and 0.5%, and for 0.3S, it is 13% to 0.2% with respect to observed runoff.

11.6.3 Limitations

The current study underlines the need for more research into the factors that influence and control runoff processes in the region, as well as the creation of design approaches that are appropriate for the region's hydrological characteristics.

A method for calculating the revised CN values required to simulate daily direct runoff is being developed.

References

Das J, Jha S, Goyal MK (2020) On the relationship of climatic and monsoon teleconnections with monthly precipitation over meteorologically homogenous regions in India: Wavelet &; global coherence approaches. Atmos Res 238:104889. https://doi.org/10.1016/j.atmosres.2020.104889

Goyal MK, Ojha CSP (2010) Evaluation of various linear regression methods for downscaling of mean monthly precipitation in arid Pichola watershed. Nat Resou 01(01):11–18. https://doi.org/10.4236/nr.2010.11002

Goyal MK, Ojha CSP (2012) Downscaling of precipitation on a lake basin: evaluation of rule and decision tree induction algorithms. Hydrol Res 43(3):215–230. https://doi.org/10.2166/nh.2012.040

Goyal MK, Panchariya VK, Sharma A, Singh V (2018) Comparative assessment of SWAT model performance in two distinct catchments under various DEM scenarios of varying resolution, sources and resampling methods. Water Resour Manag 32(2):805–825. https://doi.org/10.1007/s11269-017-1840-1

Goyal MK, Gupta AK, Jha S, Rakkasagi S, Jain V (2022) Climate change impact on precipitation extremes over Indian cities: non-stationary analysis. Technol Forecast Soc Change 180:121685. https://doi.org/10.1016/j.techfore.2022.121685

Poonia V, Goyal MK, Gupta BB, Gupta AK, Jha S, Das J (2021) Drought occurrence in Different River basins of India and blockchain technology based framework for disaster management. J Clean Prod 312:127737. https://doi.org/10.1016/j.jclepro.2021.127737

Strahler AN (1964) Quantitative geomorphology of drainage basins and channel networks. In: Handbook of applied hydrology. Mc Graw Hill Book Company, New York, pp 4–11

Subramoniam SR, Ravindranath S, Rakkasagi S, Ram H (2022) Water resource management studies at micro level using geospatial technologies. Springer, Cham, pp 49–74. https://doi.org/10.1007/978-3-030-98981-1_2

Chapter 12
Artificial Neural Network Models for Rainfall-Runoff Modeling in India: Studies From the Kolar and Kuttiyadi River Watersheds

Deepak Kumar Tiwari, Kuldeep Singh Rautela, H. L. Tiwari, and Manish Kumar Goyal

Abstract The last two decades have observed an increased usage of artificial neural networks (ANNs) for assessing and forecasting water resource parameters, as acknowledged in current studies. In the early phases of development, there is a widespread propensity for users to embrace ANNs as a "black box" model in the hopes of finding a suitable mapping between input and output variables. ANNs are defined as nonlinear mathematical models that can develop a network like that of human biological neurons and can train themselves to give better input and output relations and predicting capability. Several studies have done relevant research in the field of ANN. ANN-based hydrological models were effectively applied to hydrological studies for rainfall-runoff forecasting, streamflow modeling, basin inflow projection, rainfall foretelling, sediment deposition modeling, hydraulic energy estimation, etc. Various recent research sightsees various methods that can provide an understanding of the interconnections and interactions existing within the system. Several methodologies that appreciate the variable input involvement are evaluated in feature, and rule extraction methods for well-trained artificial neural networks are focused. This paper systematically studied, in brief, the gradual evolution of models in the field of rainfall-runoff analysis. Two different algorithms in two different river

D. K. Tiwari (✉)
Department of Civil Engineering, GLA University, Mathura, Uttar Pradesh, India
e-mail: deepak.tiwari@gla.ac.in

K. S. Rautela
Department of Civil Engineering, Indian Institute of Technology Indore, Indore, Madhya Pradesh, India

H. L. Tiwari
Department of Civil Engineering, Maulana Azad National Institute of Technology, Bhopal, Madhya Pradesh, India

M. K. Goyal
Department of Civil Engineering, Indian Institute of Technology Indore, Indore, Madhya Pradesh, India

A. K. Gupta et al. (eds.), *Ecosystem Restoration: Towards Sustainability and Resilient Development*, Disaster Resilience and Green Growth,
https://doi.org/10.1007/978-981-99-3687-8_12

watersheds (NARX and Levenberg-Marquart) are used to train models, and the various combinations of the input parameters related to rainfall-runoff modeling are analyzed based on their network details, functions, usage, timescale, and versatility. First, the Kolar River basin in Madhya Pradesh's Raisen district, which rises in the Vindhya Range of Sehore district and flows southwest to reach the Narmada at Nasrullaganj, has been examined for runoff modeling using ANNs. After that, the Kuttiyadi River basin, a west-flowing river in the southern India, was studied. It has been concluded that different models are required for different timescales and watershed characteristics, places, and circumstances, like extreme events, water storage, irrigation purposes, general water supply, flood management, reservoir recharge, groundwater recharge, etc.

Keywords Rainfall · Runoff modeling · ANN · Climate change · Neural network

12.1 Introduction

Rainfall-runoff (R-R) modeling and forecasting processes are essential for the planning and development of land and water management (Poonia et al. 2021a, b). Chiefly, it is characterized by two main time-based classes. Primarily hourly and daily forecast which are short-term forecasting and decisive for the dependable operation of flood and its mitigation structures (Tiwari et al. 2021a). Furthermore, weekly, monthly, and annual, which are known as long-term forecasting, are significant in the operations and planning of dam storages or reservoirs, hydroelectricity generation, sediment deposition and transport, management decisions in agricultural practices, water release scheduling, and other purposes (Das et al. 2020; Sharma and Goyal 2020; Gunathilake et al. 2021; Poonia et al. 2021a, b; Rautela et al. 2022a). Hydrologists in the past decades have been highly motivated to use ANN for modeling the complex natural phenomenon of the R-R process (Asadi et al. 2019; Rautela et al. 2022a). The climate change and anthropogenic impacts over the river basins significantly impact the hydrological process (Goyal and Ojha 2010). Runoff assessment processes have a wide range of entirely black-box model networks too much more detailed conceptual and physically based model's approach. Normally, models that have been used are of two types: the first one is deterministic/conceptual models, and the other type is theoretical systems-based or data-driven models (Goyal and Ojha 2012; Goswami et al. 2018). The technique may be troublesome since deterministic or conceptual R-R models frequently require a large number of discrete parameters, and the emphasis is likely to shift to theoretic approach-based or data-intensive models (Alizadeh et al. 2017). Various ANN models involved in the modeling, which are used by the researchers, include but are not limited to the following: back propagation neural network (BPNN) (Nacar et al. 2017), multilayer perceptron (Goyal et al. 2018; Kumar et al. 2005), genetic programming (Nourani et al. 2012), geomorphology-based ANN (Zhang and Govindaraju 2003), fuzzy logic (Lohani et al. 2011), evolutionary computation (EC) (Aytek et al. 2008), generalized regression neural network (GRNN) (Kisi 2011), radial basis neural networks (Shamseldin 2009), feed-forward back propagation (FFBP), NARX modeling (Tiwari et al. 2021b), etc.

This paper is divided into five parts. The first part involves an overview and general introduction to the topic. In Sect. 12.2, ANN-based models for rainfall-runoff modeling, which have different types of models, are discussed in detail in their working and network specifications. In Sect. 12.3, the evaluation and assessment of the models reviewed in Sect. 12.2 are presented with a general comparative analysis of the models. In Sect. 12.4, various recommendations for improvements in the next models are given, and lastly, the conclusion of the study is given.

12.2 ANN-Based Models for Rainfall-Runoff Modeling

Empirical models are data-driven models that give a relationship between rainfall and runoff without considering the physical explanation of the processes involved within a drainage basin. Initially, from 1850 to 1970, the model was based on some empirical formula (Todini 2007). With the introduction of transfer functions and, later on, the addition of various input-output combinations and data processing techniques, a number of data-driven hydrological models, such as ANN, fuzzy, ARIMA, NARX, etc., were acquainted with the research community. Tokar and Markus (2000) introduced ANN for the R-R with the back propagation training algorithm (BP) and compared it with a simple conceptual model (SCRR). This ANN model outperformed the SCRR model for various climatic conditions, geomorphology, and timescales.

Later on, more complex and advanced networks like multilayer perceptron (MLP) and radial basis functions (RBF) were developed (Dawson et al. 2002). RBF is more efficient than MLP, as it takes less time for training, works with less parameters, and often gives a more accurate result than MLP. RBF uses a Gaussian transfer function. But these models do not include the geological feature of the drainage basin, which was observed by Zhang and Govindaraju (2003), and close similarities between ANN architecture and basin geometry were found. So, a geomorphological ANN (GANN) was developed to combine the strengths of GIUH and ANN and help elevate the ANN model from an empirical to a geomorphology-based one (Singh et al. 2023).

In this same year, Jain and Prasad Indurthy (2003) compared ANN with the regression method and IUH. They suggest the ANN models were discovered to perform better than the other two techniques. Also, multiple hidden layer ANN was performing relatively better than a single hidden layer. Till now, most of the networks were static, which means that the weights associated with the networks remain the same after training and have no dynamic learning from past experiences. However, static networks are simple but cannot cope with major changes, which are not incorporated during training. This shortcoming was addressed by Chiang et al. (2004), where a static network was compared with dynamic real-time recurrent learning networks. This network helped in the dynamic updating of the network for unusual R-R variation characteristics. A typical recurrent network is one where the output of the network is used as the input of the hidden layer, which dynamically adjusts the weights associated with nodes. These models were performing well for the high magnitude of flow, but for the low magnitude, Jain and Srinivasulu (2004) introduced a new training algorithm, i.e., a genetic algorithm (GA), for training the network. This model is capable of addressing the low runoff condition, where the R-R relationship is complex and dynamic with no saturation level and climatic variations. Wu et al. (2005) established a distinct approach for streamflow prediction and river basin runoff modeling and can be functional for petite developed urban watersheds for the preliminary flooding warning system.

Although the ANN models are performing well, they were labeled as black-box model, i.e., their functioning is not easily interpretable. This problem was somewhat solved by Jacquin and Shamseldin (2006) with the introduction of fuzzy systems in R-R. However, it is also data-driven but still provides meaningful insight into their workings. In fuzzy system, subjective knowledge can also be incorporated with an introduction to if-then rules (Tiwari et al. 2021c).

Similarly, the Bayesian neural network (BNN) was introduced by Khan and Coulibaly (2006), and it was compared with regular ANN and a conceptual model. The BNN determines the relative importance of various inputs by allocating separate regularizing coefficients to each input. They suggest that separate uncertainty analysis is also not required as uncertainty is already incorporated into the parameters, but Beven (2015) suggested that this can't be true when epistemic uncertainties are involved in prediction. Later, combining the two systems of ANN and fuzzy, a hybrid intelligent system (HIS) was developed with a fuzzy interface system in a functionally equivalent ANN architecture. HIS is advantageous over ANFIS, ANN, and fuzzy for incorporating all of their benefits and with a lesser number of parameters, transparent rules, a lesser time requirement, etc. (Nayak et al. 2007). All these methods previously described are single objective models, which highlight a specific aspect of the hydrograph, but with multiple objective models, more information can be gathered, showing greater model consistency and reliability (De Vos and Rientjes 2008). Moreover, ANN used for data ranging from hours to decades (high-scale variations) leads to a high degree of non-stationarity and signal fluctuation. To address the issue, wavelet transformation came into the picture, which decomposes a time series into different scales and extracts useful information. A novel WANN for R-R prediction was given by Nourani et al. (2009), where runoff and precipitation data are first normalized between 0 and 1 and then disintegrated at various frequencies. It was a great model with a huge set of long-term input data. The mathematical modeling in ANN, which reveals the underlying link between input and output parameters, beats statistical approaches when comparing empirical methods with statistical or graphical methods for flood forecasting (Kar et al. 2010).

Time-lagged inputs of precipitation and discharge are also important for the prediction of discharges along with the antecedent moisture content (AMC) and can enhance the fuzzy-based neural network's performance (Lohani et al. 2011). The effectiveness of WNN can further be increased by genetic programming involvement (Nourani et al. 2012) and shows a schematic diagram of typical WGPNN, where first data preprocessing with wavelet transform is achieved, and before using it as input to the ANN unit, genetic programming is applied to determine the effective and dominant variables and seasonal variability.

With the advancement of satellite technologies, real-time R-R modeling for data-scarce regions is made possible using satellite rainfall products like TRMM. Nanda et al. (2016) developed a WNARX (wavelet-based nonlinear autoregressive with exogenous inputs) using this rainfall product and assessed it with further regression network models, like ARMA, ANN, and WNN. The WNARX network model is much more enhanced than other models with TRMM real-time rainfall products and provides a new prospect for real-time flood forecasting. Firstly, inputs are normalized between the values of 0 and 1 and then decomposed into various signals of different timescales or frequencies. After that, the frequencies are correlated with runoff and used as input for the network. However, these models are efficient but lack a physical explanation for the process and result. Rezaie-Balf et al. (2017) compared the performance of ANN with regression models such as model tree (MT) and multivariate adaptive regression splines (MARS). It was concluded that MT performed better than ANN and MARS, as MT requires fewer data inputs in volume but with increased parameters. Also, MT provides a better explanation for the interrelationship between input and output. Alizadeh et al. (2017) used projected rainfall for forecasting runoff, which is 2 months in advance by means of WANN. But the prediction for 1 month ahead was more accurate than for 2 months ahead, which is within acceptable limits. Kratzert et al. (2018) suggested a novel data-driven method long short-term memory (LSTM), which is a recurrent neural network. This network is used for snow-affected areas, where long-term dependency learning is required between inputs and outputs of the network. While comparing it with SAC-SMA, the SNOW-17 model (conceptual model) gives comparable results. This pretrained model can also be applied to ungauged sites or sites where a small number of observations are recorded.

12.3 Case Study of Kolar River Basin

The main purposes of the Kolar Dam are to provide water for the city of Bhopal, irrigation for the Raisen district's Jholiapur region, fishing, etc. The dam, which lies in the adjacent city of Bhopal, is also a popular tourist destination. The Kolar catchment is in the Sehore district of the Indian state of MP. The current study makes use of monthly rainfall data in millimeters for three locations, namely, Birpur, Brijesh Nagar, and Ichhawar, during a period of 30 years between 1988 and 2018. Additionally, average monthly temperature data in degrees Celsius is utilized for the same time period. Kolar Dam Authority, the government body in charge of managing the dam, provided discharge data for 19 years and 3 months, starting in October 1999 and ending in October 2018 (Fig. 12.1).

In this portion, the stepwise procedure used for the analysis and its significance are discussed. Three rainfall stations, Birpur, Brijesh Nagar, and Ichhawar

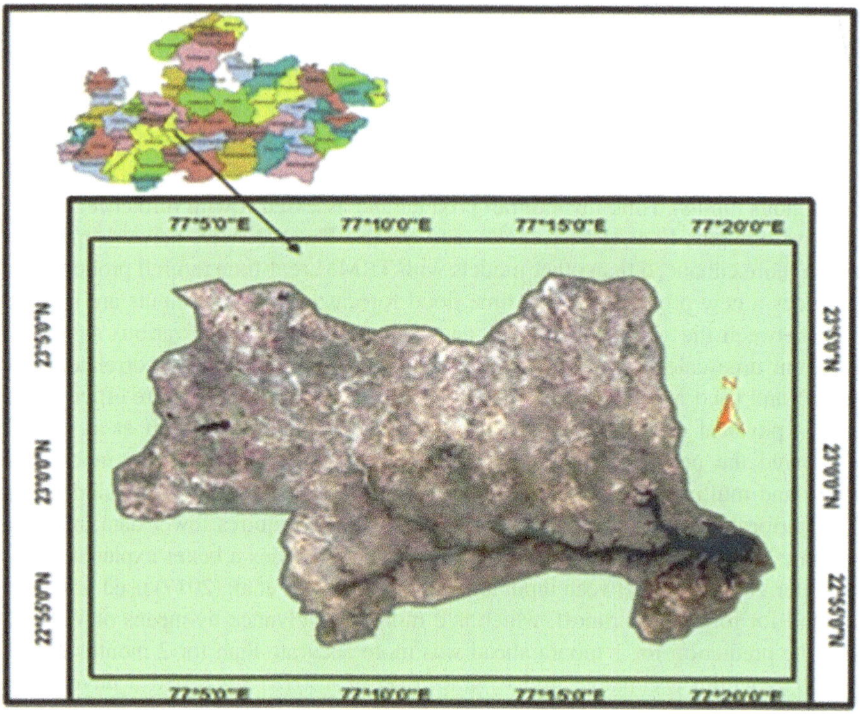

Fig. 12.1 Kolar River watershed map

(Fig. 12.2), as well as temperature, are utilized as inputs for the neural network's prediction, with the input's normalization determined. In the third stage, ArcGIS 10.5 is used to determine the geomorphological characteristics of the sub-watersheds, and ERDAS Imagine is used to preprocess the photos. Additionally, wavelet decomposition of the inputs is used as a data preprocessing approach following morphometric treatment. The suggested ANN model's multiple models are assessed based on morphology-based parameters and data preprocessing procedures in the fourth stage.

12.4 Case Study of Kuttiyadi River Catchment

The Kuttiyadi River (Fig. 12.3) originates from hilly regions of the Narikota Ranges in the Western Ghats, Kerala state of India, at a height of 1220 meters above mean sea level (AMSL) (Rautela et al. 2022b). The Kuttiyadi River is flowing in the west direction and merges into the Arabian Sea near the Kottakkal. The watershed of the Kuttiyadi River is bounded by latitudes N11°30′ to N11°45′ and longitudes E75°42′ to E75°59′ and covers a geographic area of 450 km^2. The watershed of the river is

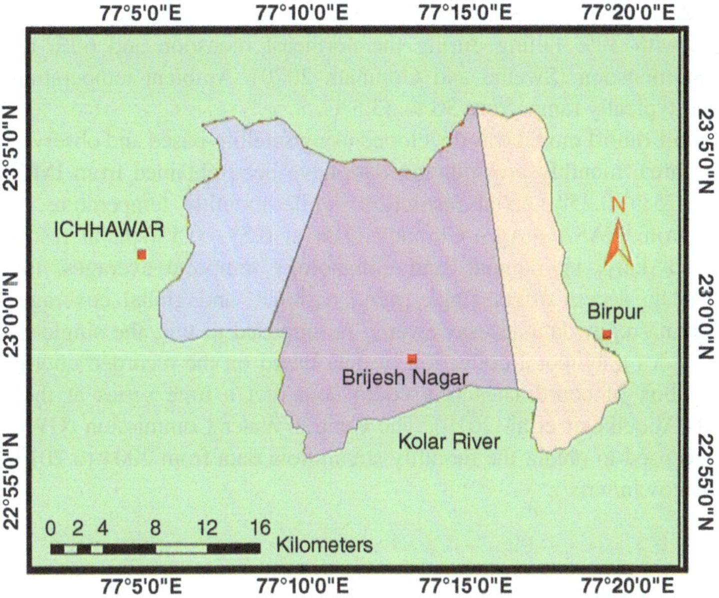

Fig. 12.2 Rain gauge stations at Kolar River watershed

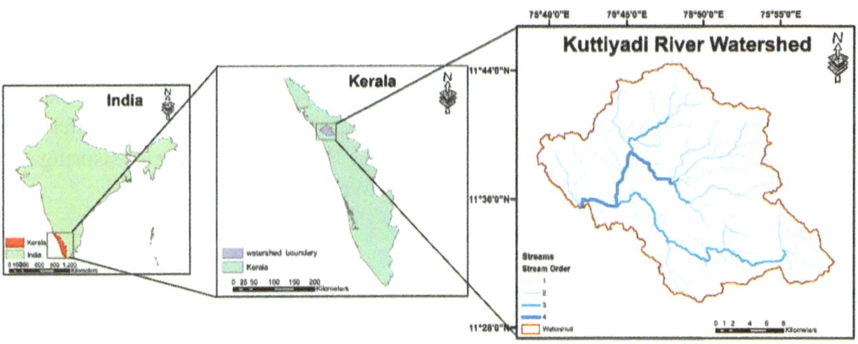

Fig. 12.3 Kuttiyadi River watershed map

categorized into three main zones: highland zone on the east, midland zone on the center, and lowland zone on the west, with varying elevations of 90–1220 amsl, 45–90 amsl, and 0–45 amsl, respectively. On the other hand, the watershed area of the river is classified into three parallel zones, such as the eastern, central, and western zones, consisting of Archean group, residual laterites, and recent sand and silt, respectively (Rautela et al. 2022c). The watershed receives a very erratic rainfall pattern. Two coastal towns, Bagdogra and Quilandy, receive an average annual rainfall of 3070 mm and 3660 mm, respectively. According to historical data, Kuttiyadi town's highest regions received 4522 mm of rainfall every year (Rautela et al. 2022b). The watershed receives an average of 5170 mm of precipitation annually, with 30% falling during the northeast monsoon and 60% during the southwest monsoon (Swetha and Gopinath 2020). Ambient temperatures in the watershed typically range from 30 to 33.5°C.

A rainfall-runoff model was developed using satellite-based and observed data. In the study area, monthly precipitation data have been obtained from IMD gridded data at 0.25°× 0.25° spatial resolution, while monthly temperature data were acquired from NASA's Agro-Climatic Data at 0.5°× 0.5° spatial resolution. In addition to daily, interannual, and climatology temporal averages, the NASA POWER data access offers single-point, regional, and global coverage (https://power.larc.nasa.gov/data-accessviewer/). As opposed to this, the single-point end-point provides a temporal collection of data based on the recorded coordinates. A bounding box of coordinates is used to construct a time series at the regional endpoint (Aboelkhair et al. 2019). The Central Water Commission (CWC) WRIS portal was used to obtain the monthly streamflow data from 2004 to 2013 (https://indiawris.gov.in/wris/).

12.5 Model Evaluation Criteria

In the current study, the three indices, such as root mean square error (RMSE), coefficient of determination (R^2), and Nash-Sutcliffe efficiency (E), have been used to evaluate the efficiency of the models prepared by the various combinations of the input datasets. However, the Nash-Sutcliffe efficiency (NSE) and root mean square error (RMSE) are used as parameters of efficiency for model validation and network testing, whereas the coefficient of determination (R^2) was used for measuring the dispersion between the observed and simulated values from the model.

12.5.1 Nash-Sutcliffe Efficiency (NSE)

$$\mathrm{NSE} = 1 - \frac{\sum_{t=1}^{T} \left(Q_m^t - Q_o^t\right)^2}{\sum_{t=1}^{T} \left(Q_o^t - Q_{mo}^t\right)^2} \tag{12.1}$$

where

Q_m^t = discharge modeled at time interval t
Q_o^t = discharge observed at time interval t
Q_{mo}^t = mean value of discharges observed.

12.5.2 Root Mean Square Error (RMSE)

RMSE is used for accuracy measurements and relates forecasting inaccuracies of several models for specific data and not among datasets, as it is scale-dependent. It is given by:

$$\text{RMSE} = \sqrt{\frac{\sum_{n=1}^{N} (P_i - A_i)^2}{N}} \tag{12.2}$$

where Pi = observed value, Ai = predicted value, and N = number of input data.

12.5.3 Coefficient of Determination (R^2)

$$R^2 = \frac{n(\sum Q_{obs} \times Q_{sim}) - (\sum Q_{obs}) \times (\sum Q_{sim})}{\frac{1}{2}\left[n(\sum Q_{obs}{}^2) - (\sum Q_{obs})^2\right] \times \left[n(\sum Q_{sim}{}^2) - (\sum Q_{sim})^2\right]\sqrt{}} \tag{12.3}$$

where Q_{obs} = discharge observed, Q_{sim} = simulated discharge, and n = total observations. The extent of R^2 lies sandwiched between 0 and 1, which stand for no correlation and absolute correlation between observed and simulated values.

12.6 ANN Models Performance Evaluation

In this study, NARX- and Levenberg-Marquardt (LM)-based ANN models are used for rainfall-runoff (R-R) modeling and prediction. The nonlinear autoregressive model with exogenous inputs (NARX) is appropriate to the class of RNNs, and they are well suited for predicting the nonlinear relationship of hydrological parameters, whereas the LM algorithm offers an iterative method for determining a multivariable function's base that is expressed as the sum of squares of nonlinear functions with real values (Levenberg 1944). It has several feedback connections, enfolding several network layers in the neural pathway. The outlining equations for the NARX- and LM-based ANN models are as follows (see Eq. 12.4 and 12.5):

$$\begin{aligned} Y = f\big(&y(t-1), y(t-2), y(t-3), \ldots, y(t-n_y), \\ &u(t-1), u(t-2), u(t-3), \ldots, u(t-n_u)\big) \end{aligned} \tag{12.4}$$

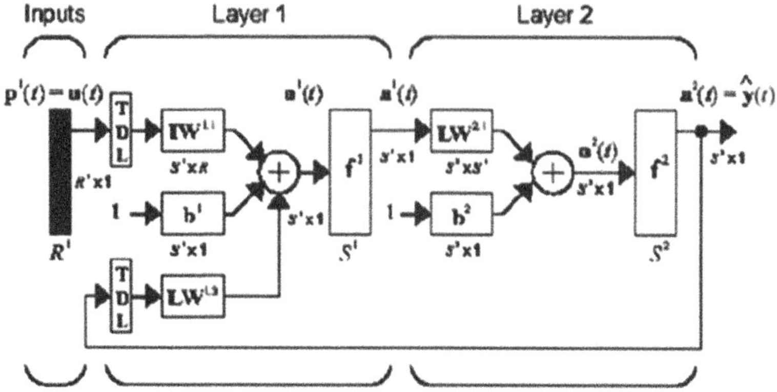

(a)

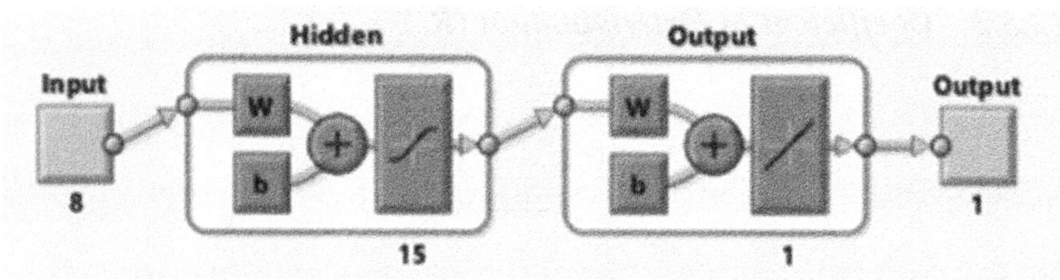

(b)

Fig. 12.4 (**a**) A two-layer feedforward NARX model network. (**b**) A three-layer back propagation feedforward (BPFF) model network

$$Y = f(x_t, y_t, z_t) \tag{12.5}$$

where the subsequent value of the dependent output signal Y(t) is regressed on previous values of the output signal and preceding values of an independent (exogenous) input signal. One can contrivance the NARX model by means of a feedforward neural network to optimize the function f. Similarly, in Eq. 12.5, the output is the function of various combinations of the input parameters, such as rainfall, temperature, previous day discharges, etc., with a varying number of neurons, hidden layers. The resulting network having a general diagram is presented below, where a feedforward network having two layers is used for the approximation. As shown in Fig. 12.4a and 12.4b, the resultant of output layer is passed through as feedback to input in hidden layer 1.

Table 12.1 shows various input combinations for the models to be incorporated while modeling. Here, Q(t) is predicted discharge for the current month, R is rainfall, T is for temperature, and t is the time period in months. These ten and nine models were used for the simulation of runoff by NARX- and LM-based models, respectively (Table 12.1).

Table 12.1 Input structure of neural network for rainfall-runoff modeling

Model No	Input combination for prediction Q (t) (NARX model)	Input combination for prediction Q (t) (LM-based model)
1	Q(t−1), R(t−1)	R(t), T(t)
2	Q(t−1), Q(t−2)	R(t), R(t−1), T(t)
3	Q(t−1), Q(t−2), Q(t−3)	R(t), R(t−1), T(t), T(t−1)
4	Q(t−1), Q(t−2), Q(t−3), Q(t−4)	R(t), R(t−1), R(t−2), T(t), T(t−1)
5	Q(t−1), Q(t−2), R(t−1)	R(t), R(t−1), R(t−2), T(t), T(t−1), R (t−2)
6	Q(t−1), Q(t−2), Q(t−3), R(t−1)	R(t), R(t−1), T(t), Q(t−1)
7	Q(t−1), R(t−1), T(t−1)	R(t), T(t−1), T(t), Q(t−1)
8	Q(t−1), Q(t−2), R(t−1), T(t−1)	R(t), R(t−1), T(t), T(t−1), Q(t−1)
9	Q(t−1), Q(t−2), Q(t−3) R(t−1), T(t−1)	R(t), R(t−1), R(t−2), T(t), T(t−1), R (t−2), Q(t−1), Q(t−2)
10	Q(t−1), Q(t−2), Q(t−3) R(t−1), R(t−2) T (t−1)	

12.7 Results and Discussion

12.7.1 Case Study of Kolar Dam Catchment

In the previous section, a new model was proposed and evaluated against NARX models and with ten model networks. The efficacy, competence, and field applicability of the proposed hydrological rainfall-runoff model are evaluated based on hydrology data for rainfall, runoff, and temperature at the Kolar Dam catchment. All the dataset is divided into three parts: 70% for training and testing, and the remaining 30% is used for cross-validation to avoid poor performance during forecasting and overfitting problems. The performance criteria used for the networks are root mean square error (RMSE), NSE, and R^2. The best-performing model for these networks is model no. 4, with a WGNARX network with NSE, RMSE, and R^2 values of 0.68, 5.21, and 0.88, respectively. The scatter plot in Fig. 12.5 shows the accuracy of regression analysis of observed and predicted data. Runoff data are in good alignment with the predicted data. This shows the predicting capability of ANNs (Table 12.2 and 12.3).

12.7.2 Case Study of Kuttiyadi River Catchment

A LM-based BPFF ANN model is used with 15 numbers of neurons and a single hidden layer to model the rainfall-runoff in the Kuttiyadi River watershed. Since the watershed has received very high precipitation due to the strong circulation of NE and SW monsoonal activities, the LM-based model has shown high efficiency in past studies (Gunathilake et al. 2021; Rautela et al. 2022a). The input datasets are first

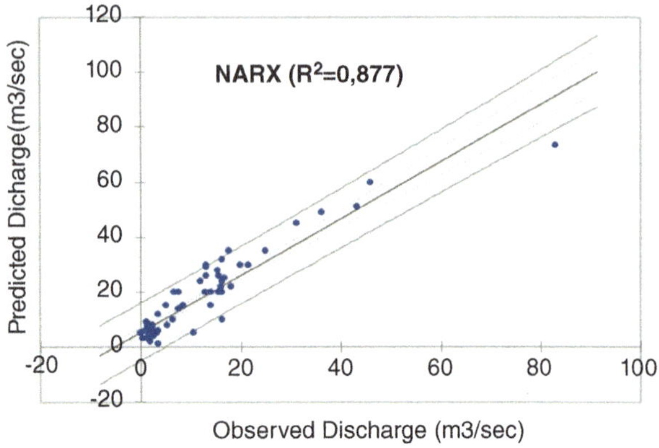

Fig. 12.5 The runoff prediction for 1 month ahead for the best-performing network NARX model

Table 12.2 Results for runoff modeling for NARX model with ten output combinations

Model no.	Input combination for prediction Q(t)	NARX model NSE	RMSE	R^2
1	Q(t−1), R(t−1)	0.58	6.20	0.59
2	Q(t−1), Q(t−2)	0.47	8.23	0.49
3	Q(t−1), Q(t−2), Q(t−3)	0.51	7.39	0.52
4	Q(t−1), Q(t−2), Q(t−3), Q(t−4)	**0.68**	**5.21**	**0.88**
5	Q(t−1), Q(t−2), R(t−1)	0.63	9.07	0.65
6	Q(t−1), Q(t−2), Q(t−3), R(t−1)	0.61	6.39	0.65
7	Q(t−1), R(t−1), T(t−1)	0.38	8.56	0.41
8	Q(t−1), Q(t−2), R(t−1), T(t−1)	0.61	8.94	0.62
9	Q(t−1), Q(t−2), Q(t−3) R(t−1), T(t−1)	0.64	6.24	0.64
10	Q(t−1), Q(t−2), Q(t−3) R(t−1), R(t−2) T(t−1)	0.36	3.96	0.39

Table 12.3 Results for runoff modeling for NARX model with ten output combinations

Model No.	Input combination for prediction Q(t)	LM-based ANN model NSE	RMSE	R^2
1	R(t), T(t)	0.43	35.98	0.45
2	R(t), R(t−1), T(t)	0.41	36.75	0.45
3	R(t), R(t−1), T(t), T(t−1)	0.56	31.87	0.57
4	R(t), R(t−1), R(t−2), T(t), T(t−1)	0.46	35.31	0.50
5	R(t), R(t−1), R(t−2), T(t), T(t−1), R(t−2)	0.62	29.37	0.64
6	R(t), R(t−1), T(t), Q(t−1)	0.67	27.47	0.70
7	R(t), T(t−1), T(t), Q(t−1)	0.73	24.91	0.76
8	R(t), R(t−1), T(t), T(t−1), Q(t−1)	0.62	29.27	0.68
9	R(t), R(t−1), R(t−2), T(t), T(t−1), R(t−2), Q(t−1), Q(t−2)	**0.79**	**21.93**	**0.80**

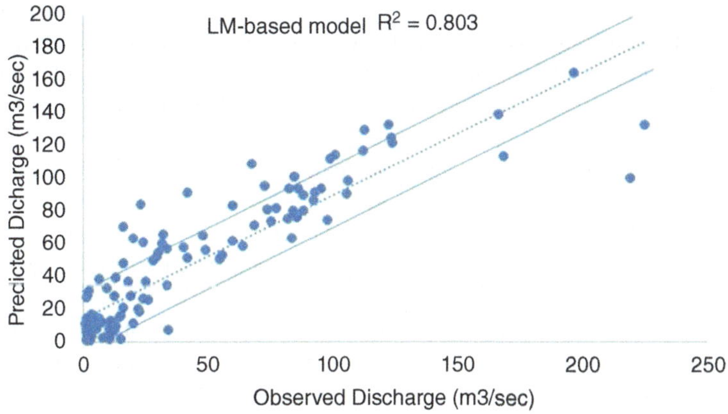

Fig. 12.6 The runoff prediction for 1 month ahead for the best-performing network LM-based ANN model

divided into a ratio of 70:15:15 for training, testing, and validation, respectively. After several iterations, the model will be stopped after getting a higher value for the model evaluation parameters. The best-performing model for these networks is model no. 9, with NSE, RMSE, and R^2 value of 0.68, 21.93, and 0.80, respectively. However, the scatter plot for the best model in Fig. 12.6 shows the accuracy of regression analysis between the observed and predicted data. The runoff data are in good alignment with the predicted data. This shows the predictive capability of ANNs.

Since the efficiency of the model can be enhanced further by adopting data processing techniques, such as wavelets, particle swarm optimization, bootstrapping, fuzzy logic, adaptive neuro-fuzzy interface system (ANFIS), etc. These case studies show that an artificial neural network (ANN) is one of the simplest but also the most vigorous model for drawing the nonlinear relation of rainfall-runoff, albeit it cannot signify the physical processes of the watersheds. Also, sometimes results obtained by an ANN-based model do not give accurate results, as the involving physical processes have been overlooked in the model for a complex watershed. Again, runoff variance cannot be explained by rainfall alone in an efficient way, but it can be endorsed to rainfall intensity, antecedent moisture content (AMC), and physical features of the catchment, such as geology, soil, slope, and land use/land cover (LULC) conditions (Hinge et al. 2018).

12.8 Evaluation and Assessment

This paper has systematically investigated the literature on the functions of data-driven models in streamflow model construction and forecasting. Various papers were scrutinized to relate the use of data-driven models with significant modeling and forecasting capabilities to the purpose of the model streamflow process. During the literature review, it was found that data processing techniques, such as model tree, fuzzy logic, wavelet transform, etc., improve the accuracy of intelligent methods to some extent through input time series data handling and preprocessing. The choice of variables used for input data for various models affects the performance in diverse conditions. The main advantage of this method is the requirement of single variables, whereas other models require several variables to produce accurate predictions. These case studies of Kolar and Kuttiyadi watersheds for runoff modeling are performed to access the predictive capability of the ANN model, which has given very promising results.

The model evaluation parameters, such as root mean square errors (RMSE), Nash-Sutcliffe efficiency (E), and coefficient of determination, are used as comparison criteria. R^2 measures statistically how close the data are to the fitted on the regression line. Basically, the correlation analysis between parameter inputs and the desired output would perform better if the calibration of the modeling depends on all the correlated antecedent values and abandons any lag time that was under the confident limitation (Yaseen et al. 2015).

12.9 Recommendation for Further Research Work

All the modeling work is taken into the study area for the general temporal variation and not considered for the extreme weather conditions. Models are also highly region-specific, which may not be the best-performing models for different catchments even with similar climatic conditions. The following are the recommendations based on the observation during the literature survey:

1. Dynamic ANN models are better than static neural networks, as they have better learning capability with the introduction of new data and are capable of self-regression. Dynamic models, like NARX, BPNN, etc., are better neural networks.
2. Input models may also consider temperature as one of the input parameters for the rainfall-runoff process.
3. Many data processing techniques, such as wavelet transform, particle swarm optimization, bootstrap sampling, and clustering, can be used, along with neural networks to increase model efficiency.
4. Sensitivity analysis should also be done for the models, and correlation analysis between input and output parameters must be performed to get better results with data; it will help in eliminating unnecessary input parameters, which have less correlation with output.

12.10 Conclusion

The continuously evolving research findings in the area of hydrological modeling have opened diverse pathways for the management of water resources, and the credibility of ANN models has increased. This review article critically analyses the various research works in hydrological modeling. This article also explored as honestly as possible the various hydrological models by taking a requisite number of literatures under review ranging from different timescales. A more thorough literature review is required in this field. Preprocessing of data in time series, selecting the most essential input variables through correlation between input and output parameters, and choosing the most fitting timescale were the key principles for accurate modeling. Lastly, this review also recommends the use of dynamic models instead of static ones, data preprocessing techniques such as wavelets and others, the use of climatic data like temperature and evapotranspiration in the input networks, and sensitivity analysis for the models. ANNs have the benefit of being simpler than other, more complex models. Therefore, the ANN technique offers the most practical alternative for flood forecasting in circumstances where information is scarce or challenging to get. By lowering the time spent studying the data, neural networks (ANNs) provide a way to lower the analytical expenses of topographical and hydrological information. The simple yet innovative technique adopted in this study for runoff modeling of the Kolar River with various combinations of input factors, along with rainfall data, gives better predictability. This method can be applied to data-scarce regions where datasets are limited and available for a shorter duration, which sometimes invites human errors.

References

Aboelkhair H, Morsy ME, Afandi G (2019) Assessment of agroclimatology NASA POWER reanalysis datasets for temperature types and relative humidity at 2 m against ground observations over Egypt. Adv Space Res 64(1):129–142

Alizadeh MJ, Kavianpour MR, Kisi O, Nourani V (2017) A new approach for simulating and forecasting the rainfall-runoff process within the next two months. J Hydrol 548:588–597

Asadi H, Shahedi K, Jarihani B, Sidle RC (2019) Rainfall-runoff modelling using hydrological connectivity index and artificial neural network approach. Water (Switzerland) 11(2):212

Aytek A, Asce M, Alp M (2008) An application of artificial intelligence for rainfall-runoff modeling. J Earth Syst Sci 117(2):145–155

Beven K (2015) Facets of uncertainty: epistemic uncertainty, non- stationarity, likelihood, hypothesis testing, and communication. Hydrol Sci J 69(9):1652–1665

Chiang YM, Chang LC, Chang FJ (2004) Comparison of static-feedforward and dynamic-feedback neural networks for rainfall-runoff modeling. J Hydrol 290(3–4):297–311

Das J, Jha S, Goyal MK (2020) On the relationship of climatic and monsoon teleconnections with monthly precipitation over meteorologically homogenous regions in India: wavelet & global coherence approaches. Atmos Res 238:104889

Dawson CW, Harpham C, Wilby RL, Chen Y (2002) Evaluation of artificial neural network techniques for flow forecasting in the river Yangtze, China. Hydrol Earth Syst Sci 6(4):619

De Vos NJ, Rientjes THM (2008) Multiobjective training of artificial neural networks for rainfall-runoff modeling. Water Resour Res 44(8):1–15

Goswami UP, Hazra B, Goyal MK (2018) Copula-based probabilistic characterization of precipitation extremes over North Sikkim Himalaya. Atmos Res 212:273–284

Goyal MK, Ojha CSP (2010) Evaluation of various linear regression methods for downscaling of mean monthly precipitation in arid Pichola watershed. Nat Resour 1(01):11

Goyal MK, Ojha CSP (2012) Downscaling of precipitation on a lake basin: evaluation of rule and decision tree induction algorithms. Hydrol Res 43(3):215–230. https://doi.org/10.2166/nh.2012.040

Goyal MK, Panchariya VK, Sharma A, Singh V (2018) Comparative assessment of SWAT model performance in two distinct catchments under various DEM scenarios of varying resolution, sources and resampling methods. Water Resour Manag 32(2):805–825. https://doi.org/10.1007/s11269-017-1840-1

Gunathilake MB, Karunanayake C, Gunathilake AS, Marasingha N, Samarasinghe JT, Bandara IM, Rathnayake U (2021) Hydrological models and artificial neural networks (ANNs) to simulate streamflow in a tropical catchment of Sri Lanka. Appl Comput Intell Soft Comput 2021: 6683389

Hinge G, Surampalli RY, Goyal MK (2018) Prediction of soil organic carbon stock using digital mapping approach in humid India. Environ Earth Sci 77(5):1–10

Jacquin AP, Shamseldin AY (2006) Development of rainfall-runoff models using Takagi-Sugeno fuzzy inference systems. J Hydrol 329(1–2):154–173

Jain A, Prasad Indurthy SKV (2003) Comparative analysis of event-based rainfall-runoff modeling techniques—deterministic, statistical, and artificial neural networks. J Hydrol Eng 8(2):93–98

Jain A, Srinivasulu S (2004) Development of effective and efficient rainfall-runoff models using integration of deterministic, real-coded genetic algorithms and artificial neural network techniques. Water Resour Res 40(4):1–12

Kar AK, Lohani AK, Goel NK, Roy GP (2010) Development of flood forecasting system using statistical and ANN techniques in the downstream catchment of Mahanadi Basin, India. J Water Resou Prot 02(10):880–887

Khan MS, Coulibaly P (2006) Bayesian neural network for rainfall-runoff modeling. Water Resour Res 42(7):1–18

Kisi Ö (2011) A combined generalized regression neural network wavelet model for monthly streamflow prediction. KSCE J Civ Eng 15(8):1469–1479

Kratzert F, Klotz D, Brenner C, Schulz K, Herrnegger M (2018) Rainfall—runoff modelling using long short-term memory (LSTM) networks. Hydrol Earth Syst Sci 22:6005–6022

Kumar ARS, Sudheer KP, Jain SK, Agarwal PK (2005) Rainfall-runoff modelling using artificial neural networks. Hydrol Process 19:1277–1291

Levenberg K (1944) A method for the solution of certain non—linear problems in least squares. Q Appl Math 2:164–168

Lohani AK, Goel NK, Bhatia KKS (2011) Comparative study of neural network, fuzzy logic and linear transfer function techniques in daily rainfall-runoff modelling under different input domains. Hydrol Process 25(2):175–193

Nacar S, Hınıs MA, Kankal M (2017) Forecasting daily streamflow discharges using various neural network models and training algorithms. KSCE J Civ Eng 00(0000):1–10

Nanda T, Sahoo B, Beria H, Chatterjee C (2016) A wavelet-based non-linear autoregressive with exogenous inputs (WNARX) dynamic neural network model for real-time flood forecasting using satellite-based rainfall products. J Hydrol 539:57–73

Nayak PC, Sudheer KP, Jain SK (2007) Rainfall-runoff modeling through hybrid intelligent system. Water Resour Res 43(7):1–17

Nourani V, Komasi M, Mano A (2009) A multivariate ANN-wavelet approach for rainfall-runoff modeling. Water Resour Manag 23(14):2877–2894

Nourani V, Komasi M, Alami MT (2012) Hybrid wavelet-genetic programming approach to optimize ANN modeling of rainfall-runoff process. J Hydrol Eng 17(6):724–741

Poonia V, Goyal MK, Gupta BB, Gupta AK, Jha S, Das J (2021a) Drought occurrence in different river basins of India and blockchain technology based framework for disaster management. J Clean Prod 312:127737

Poonia V, Jha S, Goyal MK (2021b) Copula based analysis of meteorological, hydrological and agricultural drought characteristics across Indian river basins. Int J Climatol 41(9):4637–4652

Rautela KS, Kumar D, Gandhi BGR, Kumar A, Dubey AK (2022a) Application of ANNs for the modeling of streamflow, sediment transport, and erosion rate of a high-altitude river system in Western Himalaya, Uttarakhand. RBRH 27:1

Rautela KS, Kumar M, Khajuria V, Alam MA (2022b) Comparative geomorphometric approach to understand the hydrological behaviour and identification of the erosion prone areas of a coastal watershed using RS and GIS tools. Discov Water 2(1):1–16

Rautela KS, Kumar M, Sofi MS, Kuniyal JC, Bhat SU (2022c) Modelling of streamflow and water balance in the Kuttiyadi River basin using SWAT and remote sensing/GIS tools. Int J Environ Res 16(4):1–14

Rezaie-Balf M, Zahmatkesh Z, Kim S (2017) Soft computing techniques for rainfall-runoff simulation: local non–parametric paradigm vs. model classification methods. Water Resour Manag 31(12):3843–3865

Shamseldin AY (2009) Investigation of internal functioning of the radial-basis-function neural network river flow. J Hydrol Eng 14(3):286–292

Sharma A, Goyal MK (2020) Assessment of drought trend and variability in India using wavelet transform. Hydrol Sci J 65(9):1539–1554

Singh S, Goyal MK, Jha S (2023) Role of large-scale climate oscillations in precipitation extremes associated with atmospheric rivers: nonstationary framework. Hydrol Sci J 68:395. https://doi.org/10.1080/02626667.2022.2159412

Swetha TV, Gopinath G (2020) Landslides susceptibility assessment by analytical network process: a case study for Kuttiyadi river basin (Western Ghats, southern India). SN Appl Sci 2(11):1–12

Tiwari MK, Deo RC, Adamowski JF (2021a) Short-term flood forecasting using artificial neural networks, extreme learning machines, and M5 model tree. In: Advances in streamflow forecasting. Elsevier, Amsterdam, pp 263–279

Tiwari DK, Tiwari HL, Nateriya R (2021b) Geomorphology-wavelet based approach to rainfall runoff modeling for data scarce semi-arid regions, Kolar River catchment, India. J Eng Res 6:1–16

Tiwari DK, Tiwari HL, Nateriya R (2021c) Meteorological spatiotemporal drought impact assessment on water storage of Kolar reservoir, India. J Eng Res 10:1–16

Todini E (2007) Hydrological catchment modelling: past, present and future. Hydrol Earth Syst Sci 11(1):468–482

Tokar AS, Markus M (2000) Precipitation-runoff modeling using artificial neural networks and conceptual models. J Hydrol Eng 5(2):156–161

Wu JS, Han J, Annambhotla S, Bryant S (2005) Artificial neural networks for forecasting watershed runoff and stream flows. J Hydrol Eng 10(3):216–222

Yaseen ZM, El-shafie A, Jaafar O, Afan HA, Sayl KN (2015) Artificial intelligence based models for stream-flow forecasting: 2000-2015. J Hydrol 530:829–844

Zhang B, Govindaraju RS (2003) Geomorphology-based artificial neural networks (GANNs) for estimation of direct runoff over watersheds. J Hydrol 273:18–34

Chapter 13
Analysis of Urbanization and Assessment of Its Impact on Groundwater and Land Use/Land Cover Using GIS Techniques: A Case Study of Bhopal and Gurugram District

Shubham Bhardwaj, Prerit Machiwar, Chander Kant, Shivukumar Rakkasagi, Ray Singh Meena, and Manish Kumar Goyal

Abstract Gurugram and Bhopal are among the top-developed cities in India. Gurugram is surrounded by the concrete jungle, i.e., it has massive urbanization surrounding; it comes in the National Capital Region (NCR) region; and it shares its border with the capital of India. It also has proximities with Noida and Faridabad, which is an industrial area and has already seen massive urbanization before Gurugram, and all the lakes in the surrounding region of Gurugram have been dried and overexploited, while Bhopal is the capital of Madhya Pradesh, which has scenic beauty and is surrounded by a complex web of lakes and rivers. But one thing that is common between these two urban centers is that they have witnessed an unprecedented increase in population in the last two decades, and there is a pattern of migration to these places from throughout India. Being in a suitable location and connected with national highways, Gurugram and Bhopal have seen massive urbanization during the last two decades. Because of this, land use and land cover of the area have drastically changed in the last 20 years. Groundwater level, water quality, and water resources of the district have been rapidly depleting. These districts are moving toward a major water crisis, and there is a high need for sustainable planning and preservation of this precious resource, water. For analysis satellite images, groundwater data and rainfall data are obtained for 20 years. Land use types in different periods, viz., 2000–2001, 2010–2011, 2015–2016, and 2020–2021, were analyzed using Landsat images and further implicated in the geographic information system (GIS). Groundwater potential zone maps, drought-prone area maps, and

S. Bhardwaj (✉) · P. Machiwar · C. Kant · R. S. Meena
Department of Civil Engineering, National Institute of Technology, Hamirpur, India

S. Rakkasagi · M. K. Goyal
Department of Civil Engineering, Indian Institute of Technology Indore, Indore, Madhya Pradesh, India
e-mail: mkgoyal@iiti.ac.in

A. K. Gupta et al. (eds.), *Ecosystem Restoration: Towards Sustainability and Resilient Development*, Disaster Resilience and Green Growth,
https://doi.org/10.1007/978-981-99-3687-8_13

flood-prone area maps have been developed using remote sensing and GIS techniques. Further, groundwater quality data of Bhopal of years 2000, 2011, and 2019 are obtained from the other sources. The results obtained from the analysis of groundwater for different years are compared with those of the World Health Organization (WHO) and the Bureau of Indian Standards (BIS). Furthermore, water quality has also been analyzed. The results obtained from the study will be helpful for local authorities and the government in the planning and sustainable development of the Gurugram district.

Keywords Climate · Groundwater · Land use/land cover · Urbanization

13.1 Introduction

Water is the most precious element on earth; without it, human's life can't be imagined on earth. Although the Earth is known as a blue planet, it has an abundance of water, but it does not have much freshwater that can be used for drinking purposes. The majority of freshwater is not accessible to human beings because of its presence in glaciers and beneath the earth. Humans are sending missions to other planets for this search. It is a good thing, but including this, we need to take steps to preserve the water that is already on Earth. The population is increasing at a very rapid rate, and it causes stresses on already depleting natural resources. According to the World Health Organization (WHO), it is expected that population will rise to 9.8 billion by 2050. Because of this rising population, relocation is taking place from rural areas to urban areas, and 70% of global population will shift to cities by the time population will reach to 10 billion according to the United Nations Organization (UNO). Water is so precious that it is said that the next world war could be due to water. In India also there is scarcity of water, and India is moving toward a major water crisis. In India, water resources are depleting at a very fast rate (Goyal and Ojha 2010, 2012; Poonia et al. 2021).

Every year, the population is increasing, but the number of natural resources available with which to sustain this population, improve the quality of human lives, and eliminate mass poverty remains limited (Brundtland et al. 1987). This means that the rapid urbanization process also affects the management and distribution of available natural resources to the masses, and it cannot generate as fast as the human population. As cities develop, population increases and land cover use changes, with formerly nonurban regions, such as agricultural grounds, other vegetative regions, water bodies, and so on, being replaced by building structures and roadways (Strahler 1975). Furthermore, urban settlement does not increase uniformly in all areas of a city, even throughout the same time period. It is perhaps more active in certain sections of the city, while it displays development or expansion activity in others (Deng and Huang 2004). Urban sprawl results in intensive demands on the environmental resources and poses problems by penetrating the valuable natural habitats of their hinterlands. As the cities expand in areas of population growth, the land use changes with the hitherto nonurban areas, like water bodies, agriculture, etc., getting replaced by concrete jungle (Strahler 1975). Also, urban settlement is not growing in an unplanned manner or evenly in all direction. In some areas of the city, it is dynamically more active, and, in other parts, it shows less increment or expansion activities (Deng and Huang 2004). In many metropolitan cities, it has been observed that during urban growth of cities, the surrounding areas of urban built-up, which had natural vegetation cover or agriculture or water bodies, etc., are coming under encroachment. Spatial expansion of cities results in to change in earlier land use type cities (Goyal et al. 2022). Such phenomenon of urban sprawl is characterized by chaotic mix of development, which leads to an improper development in cities. To prevent this kind of sprawl in the future, it is necessary to monitor the growth of the city (Vinoth Kumar et al. 2007). Mapping and monitoring the urban sprawl, because of urban decay within the centers of cities, at regular intervals is very essential for urban planners to understand the development trend on the urban periphery and subsequently to regulate it (Lv et al. 2012; Goyal et al. 2018; Das et al. 2020; Jha et al. 2008).

The urban environment quality is decreasing continuously as the cities have reached their saturation point and are unable to cope with the increasing rate of the infrastructure projects. There is huge amount of stress on the existing water resources in many developing cities, which is going through water crisis, and the responsible factors for water shortages in the face of rapid urbanization include population explosion, unplanned growth of urban areas, change in lifestyle, and increasing per capita water consumption (Mohr et al. 2011). So, the positive aspects of urbanization have often been overshadowed in Asian megacities by deterioration in the physical environment and quality of life caused by the widening gaps between supply and demand for essential services and infrastructure (Rahman 2007). The unprecedented growth in urbanization will lead to many health and environment problem (Kanta Kumar et al. 2011). There are already a lot of disputes going on between various regions related to sharing of water, so this a high time to take steps toward sustainable development, preserving, and wise utilization of water resources. Remote sensing and GIS (geographic information system) techniques make a powerful tool available for analyzing land change and growth in urban areas by studying satellite images.

13.2 Study Area

Gurugram district (Fig. 13.1) is one of the 22 districts of Haryana. It is situated near Delhi and shares its border with the National Capital of India. It is known as the industrial and technological hub of Haryana. Its area is 1,258 km^2. Being the industrial hub of Haryana and being at the strategically location, Gurugram has many multinational companies. The district offers a huge amount of employment because of economic development and employment opportunities; there is an annual precedence shift of population from rural areas to urban area.

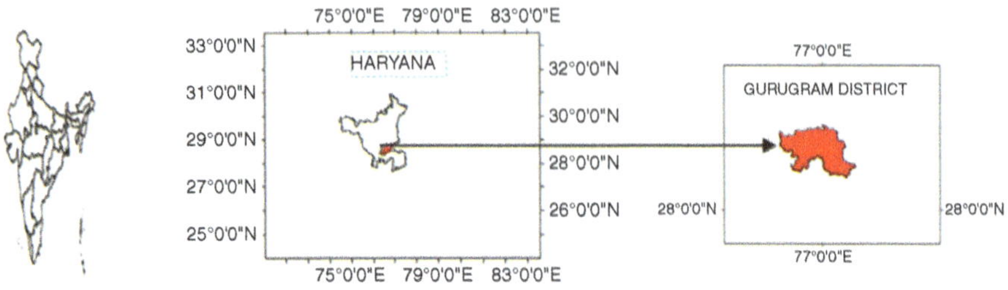

Fig. 13.1 Location of Gurugram district in India

Its population in 2011, according to the 2011 census, was 15, 14,432, while it was 8, 74,695 in 2001 according to the 2001 census. The population growth rate over the period 2001–2011 was 73.92%, and the population density of the district is 1241 inhabitants per square kilometer.

Its literacy rate is 84.4%. Gurugram is well-connected with many national highways. It is surrounded by major other cities like Delhi, Faridabad, and Noida. This district has witnessed massive urbanization in the last two decades because of its suitable location for industries and multinational companies.

Bhopal is the headquarters of the Madhya Pradesh government and is a beautiful city in the center of India. The Bhopal is popularly derived from Bhojpal or Bhojtal (i.e., Bhopal Lake); the dam, which is almost 35 km^2 area, now holds up the Bhopal city lakes and is said to have been built by Raja Bhoj. Bhopal is also known as the *City of Lakes* for its various natural and artificial lakes and for being one of the greenest cities in India for the last 5 years. The district is divided into three tehsil and two blocks. It has one city (Bhopal), one town (Berasia), and 512 villages. The total area of Bhopal district is approximately 2724 km^2. It is situated between the districts of Vidisha to the northeast, Guna to the north, Raisen to the east and southeast, Rajgarh to the northwest, and Sehore to the southwest and west. The city of Bhopal city situated in the south side of the district, and the maximum of the population resides within Bhopal Municipal Corporation. The city lies between north latitude 23o05′ and 23o54′ and east longitude 77o10′ and 77o40′ (Fig. 13.2), falling in Survey of India Topo Sheet No. 55 E.

13.2.1 Data Used

The present study uses data on rainfall information, the digital elevation model (DEM), soil types, and land use/land cover Landsat data in the chosen area. For the setting up of the geographical information system, the primary data required are the Landsat images for the different years, rainfall data, soil data, groundwater level information, and groundwater quality data. All the data sets were acquired from different sources, as shown in Table 13.1. These data were analyzed and altered for appropriate use as input for further processing.

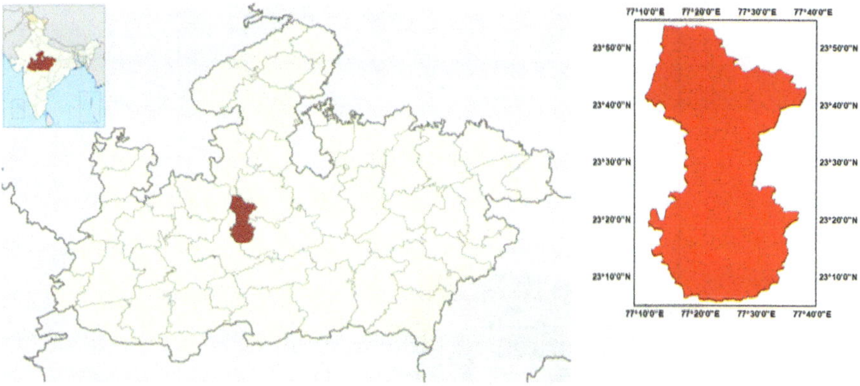

Fig. 13.2 Location of Bhopal district in India

Table 13.1 Types of data and the sources of their collection

Sr. No.	Source	Data type
1.	United States Geological Survey (https://earthexplorer.usgs.gov)	Landsat 8 Operational Land Imager/Landsat 4–5 Thematic Mapper
2.	Food and Agriculture Organization (FAO) (http://www.fao.org/soils-portal)	Soil map
3.	Climate Research Unit (https://crudata.uea.ac.uk/cru/data)	Rainfall data
4.	Climate-Data.org	Climate data
5.	Central Ground Water Board (https://indiawris.gov.in)	Groundwater data
6.	BIS, Bureau of Indian Standards (2012): IS 10500: 2012 WHO (2011): Guidelines of Drinking-water Quality	Water quality standards

13.3 Methodology

13.3.1 Land Use Classification

Gurugram has seen massive urbanization. Its infrastructure has completely changed in the last two decades. It is very important to know how land use and land cover in the region have changed for proper analysis of groundwater potential zones, drought-prone areas, and flood risk zones. For land use land change analysis, Landsat 8 images are used for observation in GIS. Land use is divided into majorly five categories, i.e., agriculture, urban Areas, thick vegetation (forest), water bodies, and barren land. A change in categories is observed for the years 2000, 2010, 2015, and 2021 for proper analysis (Figs. 13.3, 13.4, 13.5, 13.6).

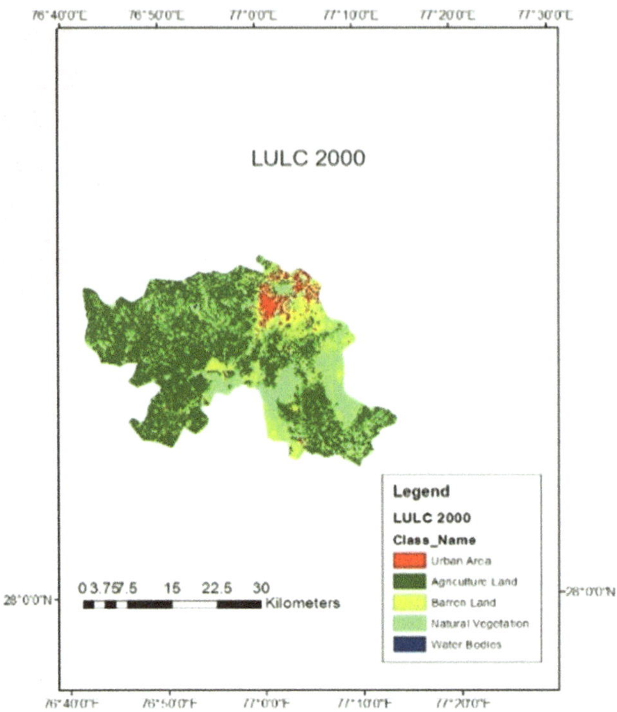

Fig. 13.3 Year 2000

13.3.2 Rainfall

Rainfall is another essential factor for analysis. Due to urbanization and global warming, climate change is taking place. Rainfall is an important factor for groundwater recharge. If rainfall is higher, water recharge will be higher, and there will be less groundwater recharge if precipitation is lower. Rainfall data for the years 2000, 2010, 2015 and 2021 is obtained. The Climatic Research Unit (CRU) monthly dataset is downloaded and converted into netCDF file in a usable format. Required data is separated into bands, and rainfall maps are prepared.

13.3.3 Drainage Density

Drainage density is the formation of surface and subsurface characteristics. It is an important factor in preparation for flood risk zone, drought-prone areas, and groundwater potential zones. Runoff is dependent on drainage density. If drainage density is greater than runoff, water will be less infiltrated in the area, and if drainage density is

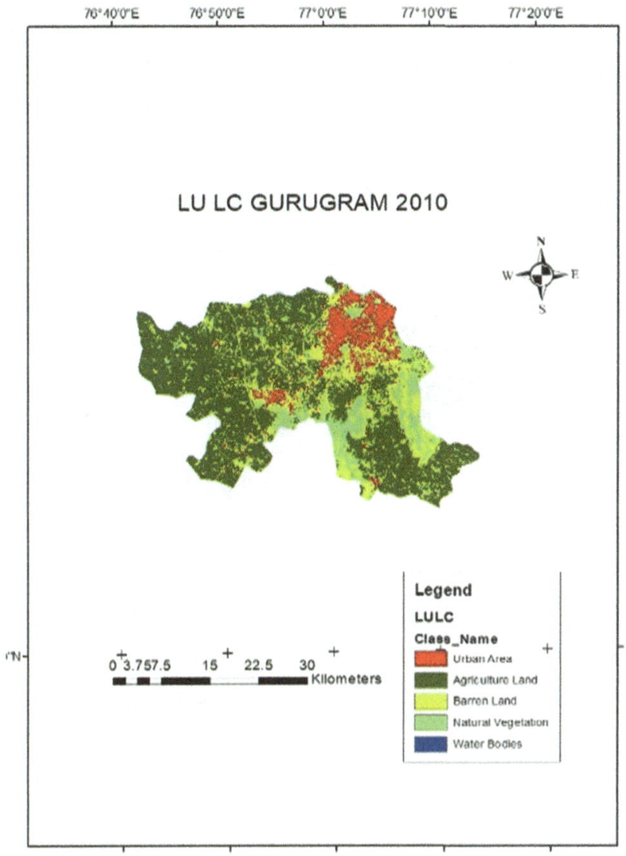

Fig. 13.4 Year 2010

less, then water infiltration will be greater, so there may be a groundwater potential zone (Fig.13.7).

13.3.4 Slope

The slope is an important feature for the analysis. The slope map is prepared from the digital elevation model of Gurugram district. If the slope is high, then runoff will be higher and water will have less time to infiltrate. Similarly, if the slope is less, then water will get maximum time to infiltrate, and the runoff will be less. If slope is low, then there are high chances that water will accumulate in that area, infiltrate the soil, and recharge the water table easily. This factor is used only for groundwater potential zones and flood risk zones (Fig. 13.8).

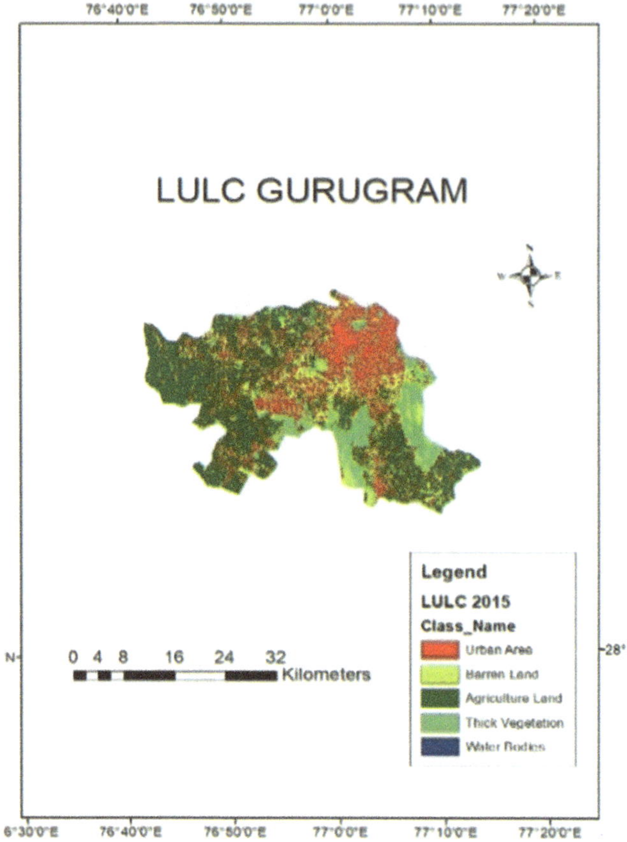

Fig. 13.5 Year 2020

13.3.5 Soil

Soil is an essential factor that governs the infiltration capacity of the area. Different types of soil have different infiltration capacities, like clay, loam, etc. Clay has very poor permeability. Gurugram district has four types of soil, and all have different kinds of permeability (Fig. 13.9).

13.4 Results and Discussion

13.4.1 Ground Potential Zones of Gurugram

It is very significant to identify the groundwater potential zones for planning for sustainable development. There are various factors that affect the groundwater

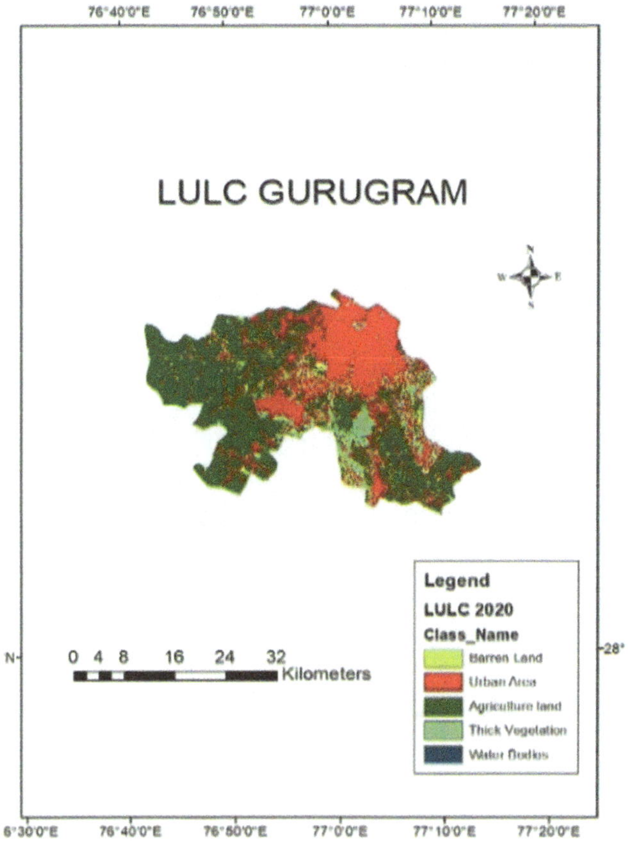

Fig. 13.6 Year 2015

potential zones, like drainage density, digital elevation model, rainfall map, slope, and land use/land cover maps (Table 13.2)

Thematic maps of the abovementioned factors are prepared, and overlay is done through the weightage overlay process in ArcGIS (Figs. 13.10, 13.11, 13.12 and 13.13). Different weightages are given to these factors rendering to their influence on the preparation of groundwater potential zones (Table 13.3).

13.4.2 Flood Risk Zones of Gurugram Region

A flood risk map helps identify flood risk zone areas and classifies areas according to their potential risk (Tables 13.4 and 13.5). Various factors that influence the flood-prone areas are drainage density, rainfall, slope, DEM, soil, and land use/land cover (Figs. 13.14, 13.15, 13.16, and 13.17).

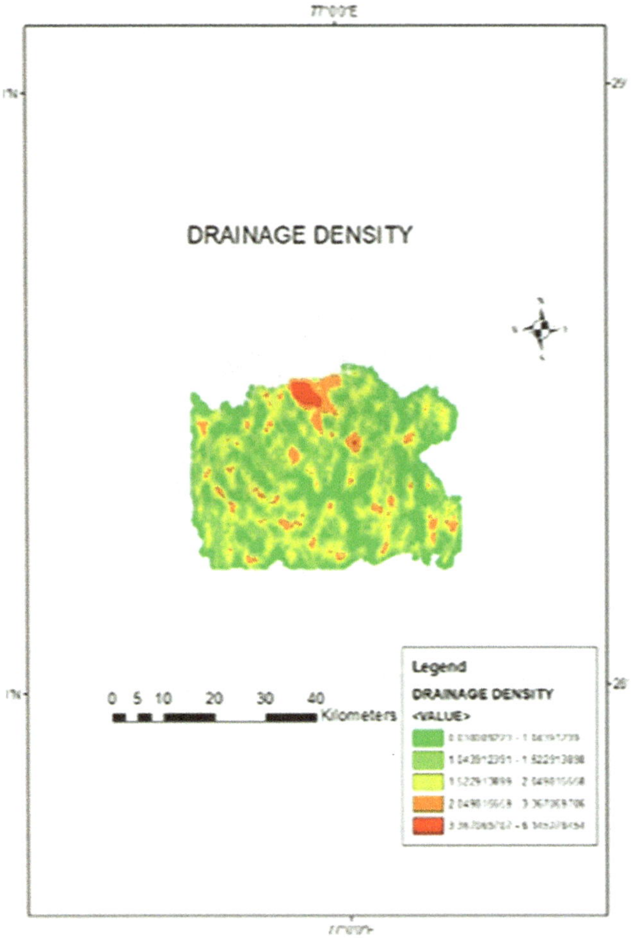

Fig. 13.7 Drainage density

13.4.3 Drought-Prone Area of Gurugram

Factor that influences drought-prone areas are drainage density, rainfall, digital elevation model, and land use (Tables 13.6 and 13.7). Drought is less dependent on the type of soil, as drought can occur in any type of soil, irrespective of its characteristics (Figs. 13.18, 13.19, 13.20 and 13.21)

The region is divided into different zones according to their preferences. It is observed that zone 4 is of maximum importance, and the highest chances of drought occurring are in these zones. Precautionary steps should be taken to prevent drought in these areas and to replenish groundwater.

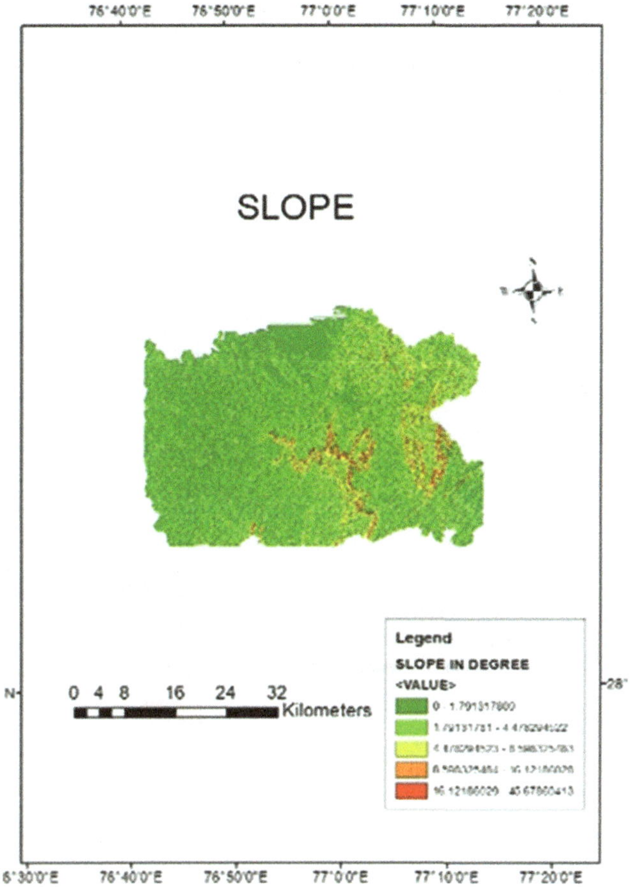

Fig. 13.8 Slope

13.4.4 Assessment of Land Use/Land Cover Changes of Gurugram

13.4.4.1 Assessment of Land Use/Land Cover Changes

Spatial patterns of LULC changes in the Bhopal district for the years 2000, 2011, and 2021 are shown in Figs. 13.22, 13.23 and 13.24. The total area of every land use/land cover category and its changes during 2000–2021 are presented in Tables 13.8 and 13.9. It is determined that the area under different LULC classes changed over the years.

The data on changes in LULC shows expansion of urban land or built-up area at the cost of forest, vegetation, water bodies, and agricultural land. As evaluated by the

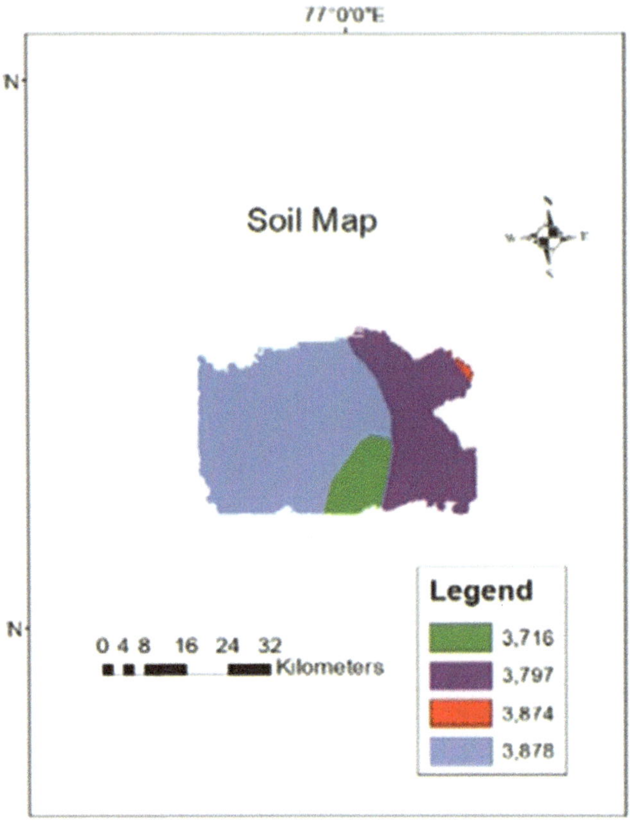

Fig. 13.9 Soil

Table 13.2 Weightage according to their influence on groundwater potential zones

Factors	Weightage
Drainage density	30
DEM	30
Rainfall	15
Slope	5
Soil	10
LULC	10

land use classes, over the last 20 years, an urban built-up area has increased approximately, as shown in Table 13.9, from 48 km^2 to 360 km^2. And the dynamic degree of the urban area was obtained at −31.28%, which shows the land use for this category increased.

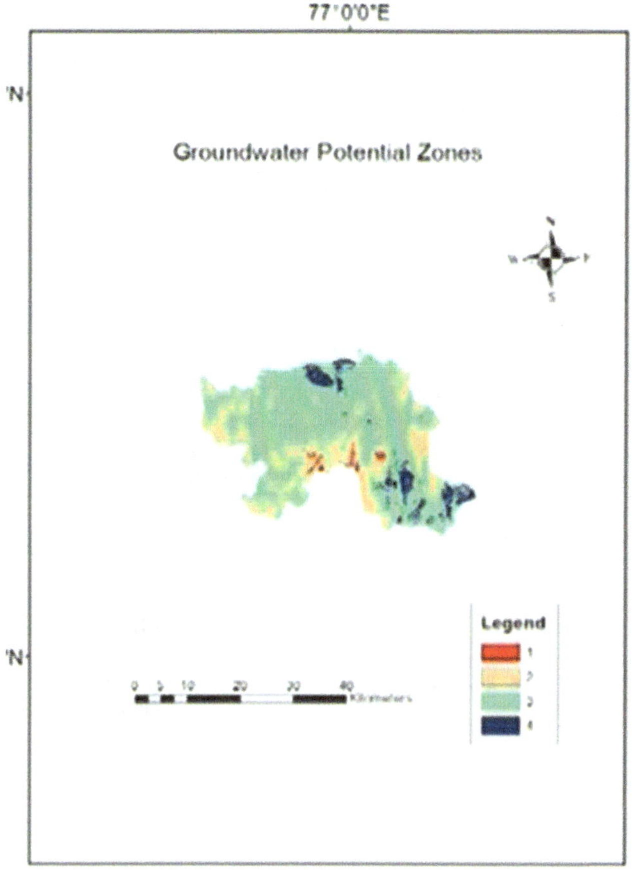

Fig. 13.10 Year 2000

13.4.5 *Effect on Groundwater Resources*

13.4.5.1 Groundwater Quality

Both the quality and quantity of groundwater in the Bhopal district are severely affected by urbanization. Based on the water quality analysis during the years 2001, 2011, and 2018, different locations in Bhopal are explained in Table 13.10. It is further compared with the water quality standards of the Bureau of India and the World Health Organization. The physicochemical parameters were analyzed by the CGWB, and the data were obtained and further analyzed.

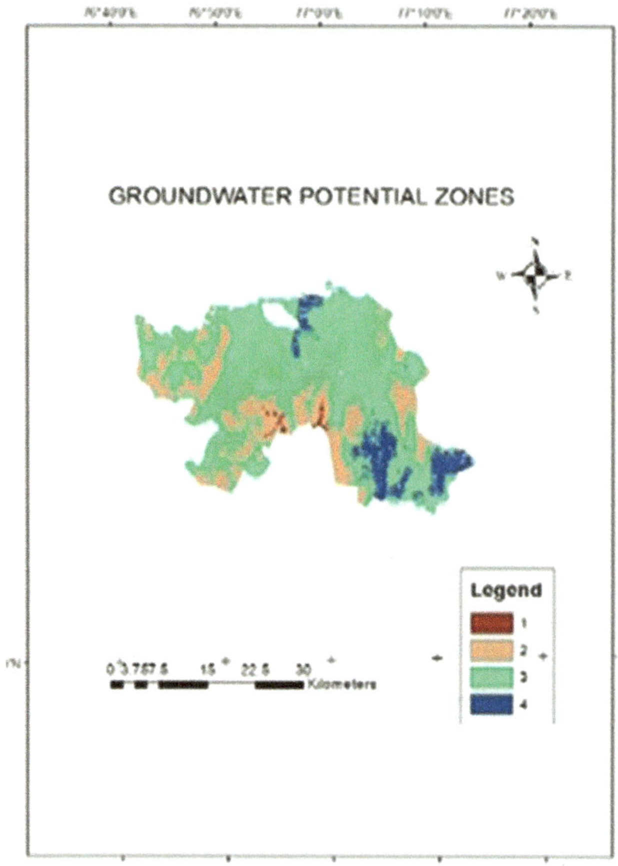

Fig. 13.11 Year 2010

13.4.5.2 Groundwater Level

One of the objectives for this research was to identify the impact of urban land use on the groundwater level. For this analysis, firstly, the changes in groundwater levels occurred in the last 20 years were analyzed. From the results, it has been determined that there are some areas where groundwater level increased and, in some areas, decreased. So, both kinds of effects are being identified in the Bhopal region due to urban growth. The groundwater level data were obtained by CGWB and analyzed in this report, as shown in Table 13.11 and Figs. 13.25, 13.26 and 13.27 shows the groundwater scenario in the Bhopal region for the different years 2000, 2011, and 2019. These maps are generated using ArcGIS tools.

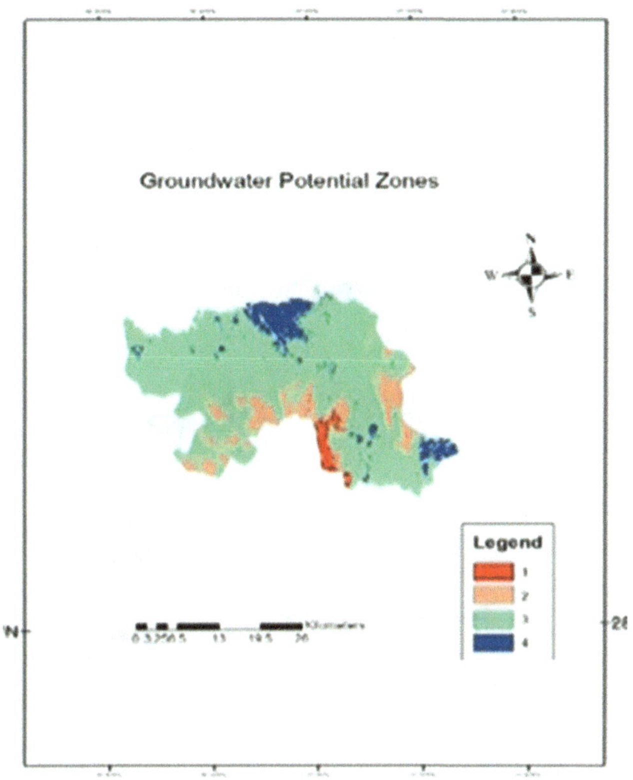

Fig. 13.12 Year 2015

13.4.5.3 Effect on Rainfall Pattern

The rainfall thematic maps of the years 2000, 2011, and 2020 were generated by Kriging methods with the help of the ArcGIS interpolation tool and are shown in the Figs. 13.28, 13.29 and 13.30 below; the figure represents a continuous rainfall surface. The green color represents low rainfall, and the peach color shows high rainfall areas in the Bhopal region. The rainfall data for the Bhopal region over the last 20 years is shown in Table 13.12 below.

13.5 Conclusion

There is a need to minimize the impact of urbanization on the environment. There has been massive degradation in the environment, and climate change is taking place. Rainfall has already decreased in the last 20 years. People need to be

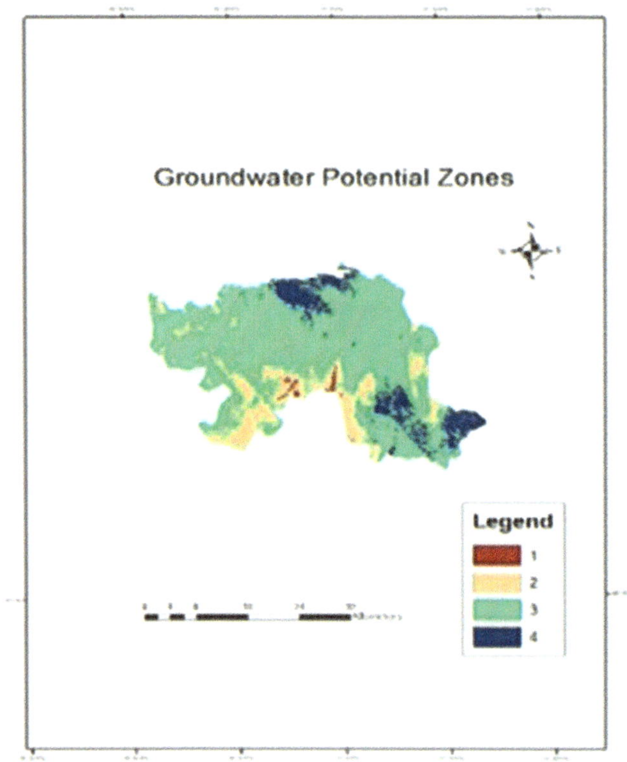

Fig. 13.13 Year 2020

Table 13.3 Types of zones in the region and their importance

Zones	Preference
Zone 1	Very poor (least important)
Zone 2	Poor (not important)
Zone 3	Good (important)
Zone 4	Very good (most important)

Table 13.4 Weightage according to their influence on Flood Risk Zones

Factors	Weightage
Drainage density	20
DEM	18
Rainfall	20
Slope	10
Soil	8
LULC	24

Table 13.5 Types of zones in the region and their importance

Zones	Preference
Zone 1	Very poor (least important)
Zone 2	Poor (not important)
Zone 3	Good (important)
Zone 4	Very good (most important)

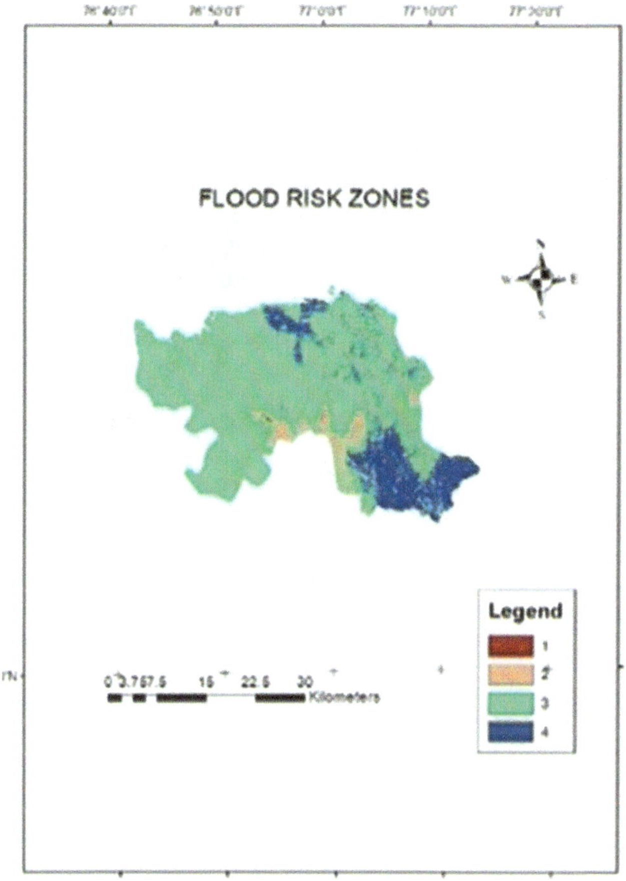

Fig. 13.14 Year 2000

encouraged to do afforestation, to use public transport, and to use water wisely. A policy is needed to be framed to conserve water and regard rainwater harvesting.

Groundwater potential zones need to be identified, and these zones need to be incorporated into future planning to extract water. Artificial recharge can also take place in these zones.

Flood-prone areas are identified. These are the areas where maximum flooding will occur. This water should be utilized properly. Flood spreading should be done.

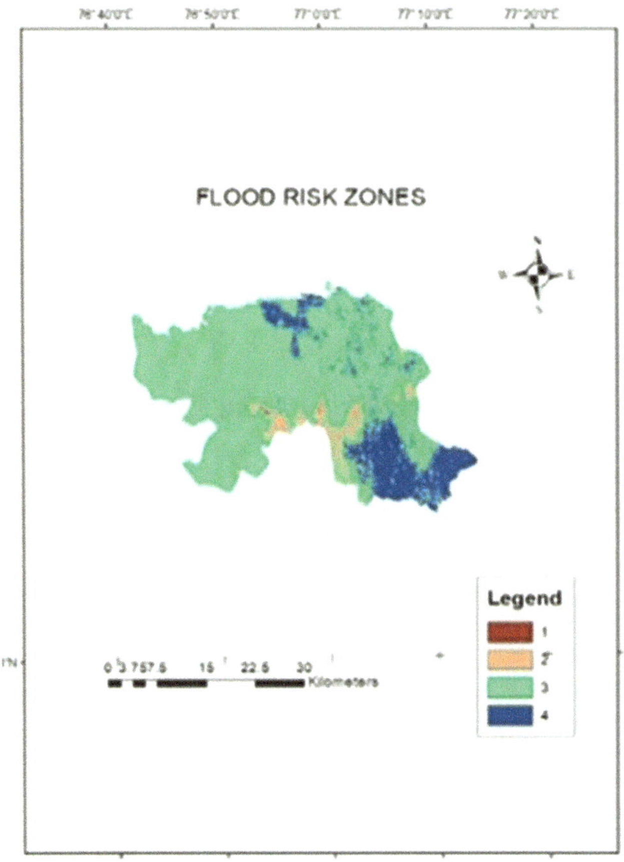

Fig. 13.15 Year 2010

Flood waters need to be transported to drought-prone areas or to groundwater potential zones, where they can be injected into the ground through injecting wells (artificial recharge should be done) in these areas.

There are lot of areas that are drought-prone. Drought should be prevented in these areas. These require the maximum attention of the authorities. Water conservation practices and rainwater harvesting should be implemented in these regions. Floodwater from flood zones should be transferred to these zones.

Urbanization is taking place at a very rapid rate, and water bodies are very low in this region. It is high time to move toward sustainable development, and proper planning is required for further development. Traditional and advanced technologies need to be used by incorporating groundwater, flood risk zones, and drought-prone area studies.

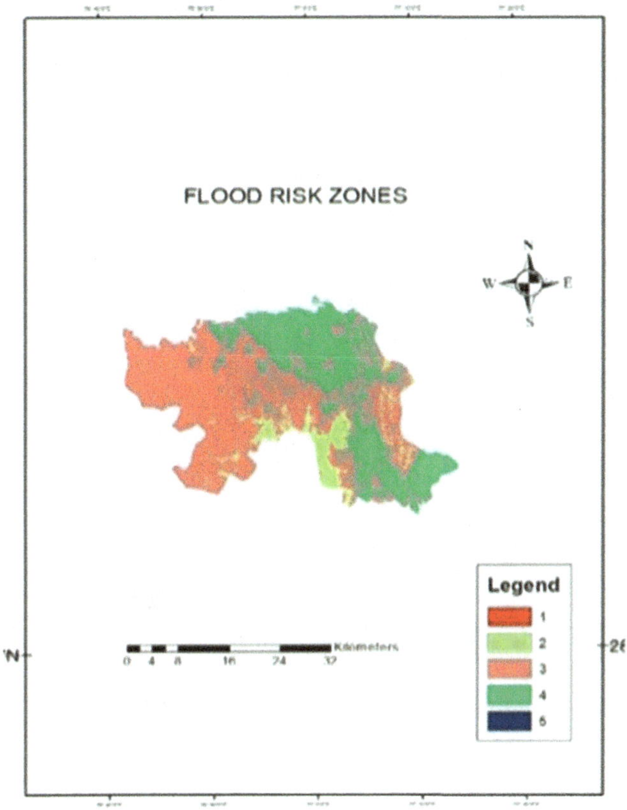

Fig. 13.16 Year 2015

13.5.1 For Bhopal Region

- The overall urban built-up area expanded from 47.47 km^2 in 2001 to 359.30 km^2 in 2021, according to the results. There was a total increase of 311.87 km^2 in built-up area.
- Agriculture and barren land areas are the classes most affected by urbanization in Bhopal during the previous 20 years
- Since 2000, many surface water bodies have vanished, and the surface area of most water bodies has reduced due to urbanization in the previous 20 years.
- Some water quality parameters, like calcium, nitrate, electrical conductivity, potassium, and total hardness of $CaCO_3$, cross the desirable limit of standard water quality parameters in this time period.
- For the result of groundwater level, it seems that in some regions of Bhopal, like Ahmedpur, Misrod, Balampur ghati, Bairagarh, and Berasia, the water level decreased at a high rate.

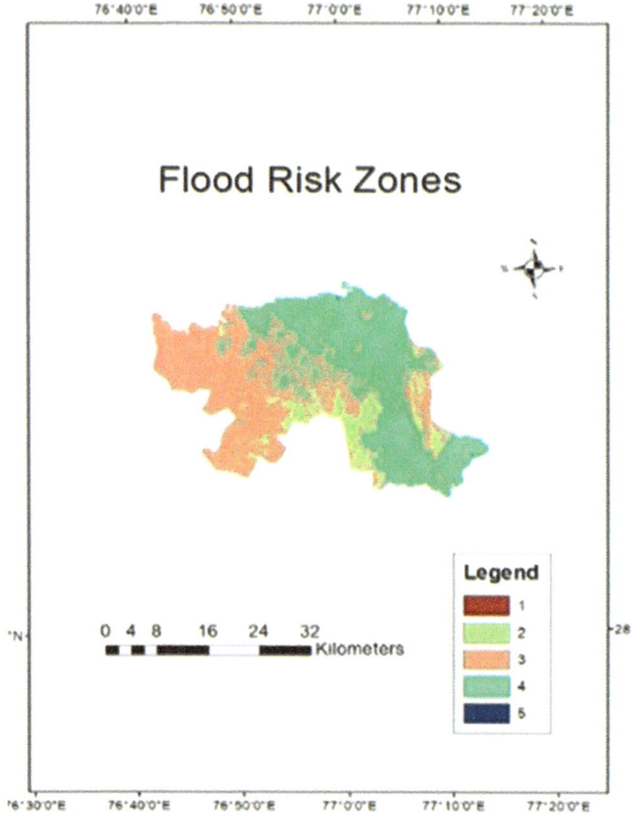

Fig. 13.17 Year 2020

Table 13.6 Weightage according to their influence on drought-prone areas

Factors	Weightage
Drainage density	30
Rainfall	20
DEM	10
LULC	40

Table 13.7 Types of zones in the region and their importance

Zones	Preference
Zone 1	Very poor (least important)
Zone 2	Poor (not important)
Zone 3	Good (important)
Zone 4	Very good (most important)

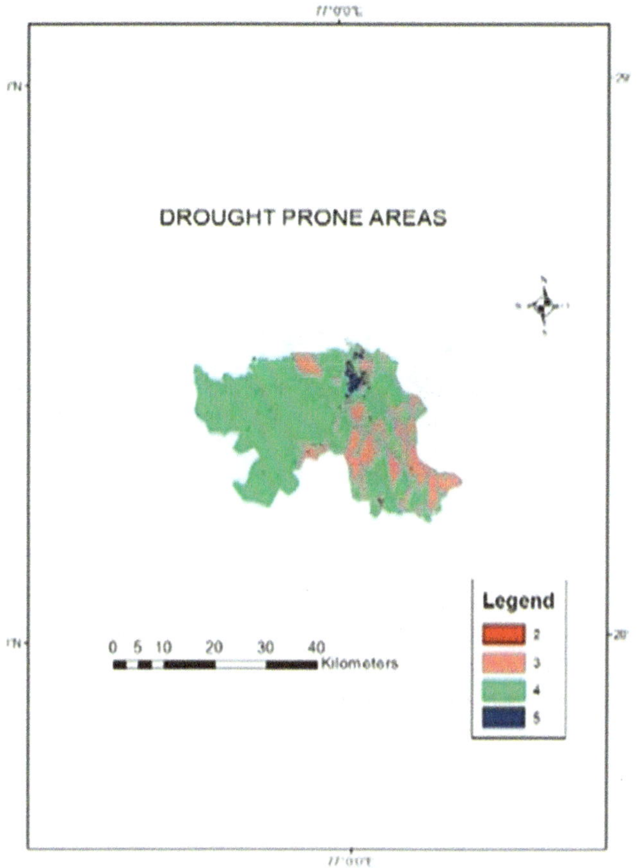

Fig. 13.18 Year 2000

13.6 Action Plan

The mapping of groundwater prospects is most important for two aspects, viz., (1) to classify areas for detecting sources of groundwater inside the potential area, near a residence to be provided with drinking water, and (2) to establish the location for building a suggested groundwater recharge structure. Likewise, the mapping of groundwater prospects, in addition to groundwater prospect regions, indicates suitable sites for certain recharge structures. There are two features—(1) recharge structure sites and (2) recharge structure type—that happen to be considered through applying the recharge structure mapping. Supplying drinking water from groundwater is a difficult job that includes numerous methodological problems, such as identifying the areas of groundwater prospect and recharge structures, evaluating its sustainability and quality, expanding wells, etc. The map shown in Fig. 13.31 can be

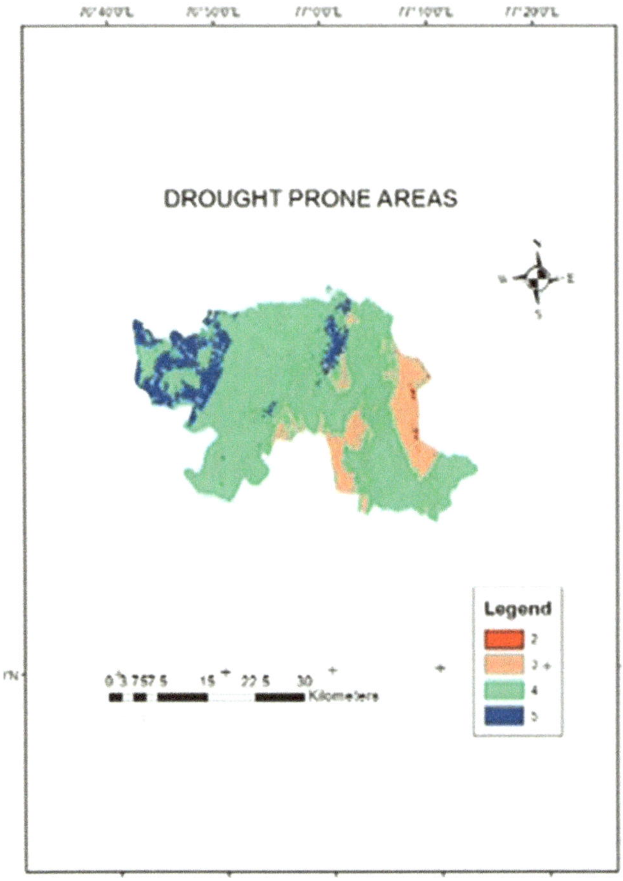

Fig. 13.19 Year 2010

effectively utilized for the detection of the location of groundwater prospect areas and recharge structures.

The standards for the choice of some sites for diverse types of groundwater and recharge structures are provided below for implementation in and around urbanizing areas.

- **Check dam**: Check dams are typically constructed on the first- and second-order streams along the bottom hill areas with a 0–5% slope.
- **Percolation tank**: These structures are typically constructed on the first- to third-order streams situated in the valleys as well as on plains having adequate weathered region/loose soil/ruptures.
- **Nala bund**: These structures are typically constructed on the first- to fourth-order streams flow across the valleys and plains, where the attainment of land for the

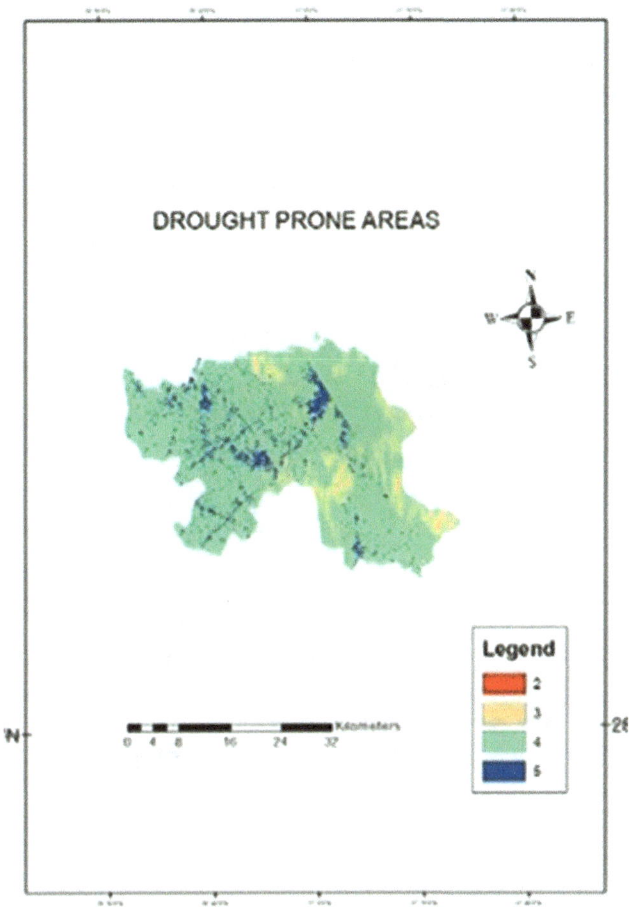

Fig. 13.20 Year 2015

inundation of larger regions is not feasible. The restricted water can be accumulated in the riverbed for several days, which improves recharge.

- **Recharge well**: These structures are typically constructed in regions where spreading of the upper strata is weak, for instance, in schists lying beneath granites or in interred pediplains with topsoil with less permeability.
- **Tanks de-silting**: The de-silting must be performed in tanks of small size that are moderately silted up.
- **Recharge pit**: These structures are typically constructed near houses where drainage cannot exist, for instance, water division regions, hill peaks, etc. The recharge pits are also selected in the prevailing tanks.
- **Dyke**: These structures are typically constructed on stream routes that flow in unsaturated areas, for instance, eroded/ruptured stones like basalt, terrains of laterite, etc., where the discharge of groundwater is substantial.

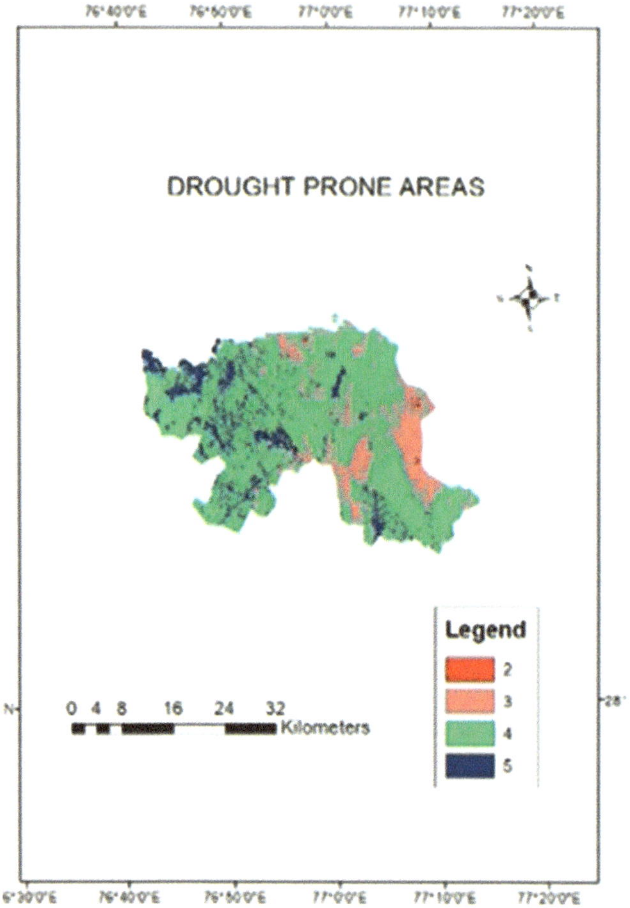

Fig. 13.21 Year 2020

13.7 Recommendations

Groundwater accessibility in Bhopal and Gurugram districts is inadequate, and now it is excessively exploited, resulting in decreased water table levels. Because of significant urban and industrial expansion, the majority of storm runoff flows into the sewers or drains, reducing the recharge from rainfall. As a result, there is a serious need to conserve this critical resource by implementing the following actions.

- **Upgrading integrated three-dimensional urban planning:** Spatial urban planning is a critical tool for achieving methodical urbanization in the cities of both districts. The major objective of urban planning is to integrate multiple LULC, construct tactical and city infrastructure, and improve local governance. Such an

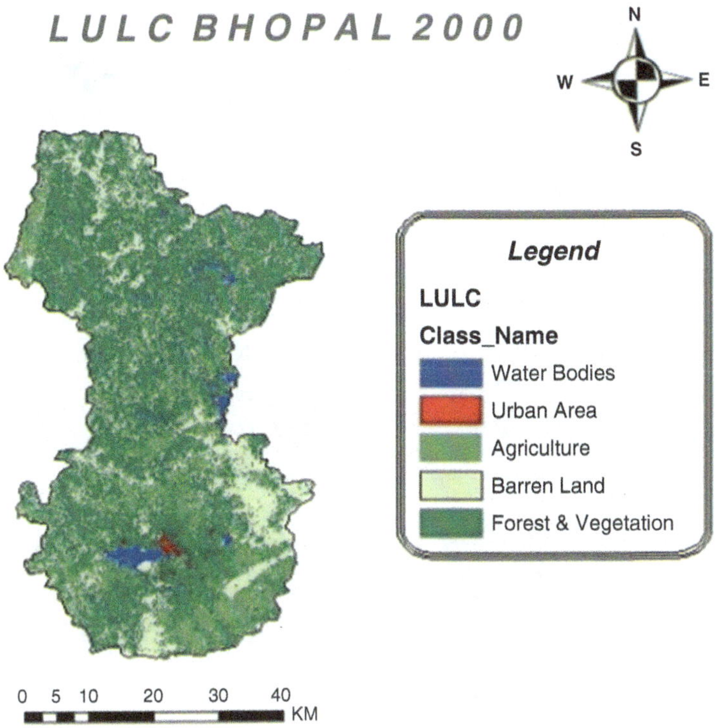

Fig. 13.22 LULC of year 2000

attempt is required in the long run to allow the sensible use of land and water resources.

- **Rainwater harvesting and artificial recharging:** To reverse the district's diminishing water level trend, rooftop rainwater harvesting must be implemented, and recharge structures should be built in decline regions where water accumulates in the rainy period to improve groundwater recharge. Water harvesting and artificial recharge structures must be constructed in the urban areas of both districts. The runoff must be redirected to deserted quarrying pits, and small check dams can be built in hilly regions to rejuvenate surplus runoff.
- **Control of industrial effluents:** To reduce groundwater pollution, industrial effluents that pollute groundwater must be treated prior to discharge. Groundwater pumping requires rigorous regulatory procedures, specifically for industrial use. All tube wells must include a water meter that can withdraw a certain amount of groundwater. Industries must be encouraged to recycle the waste matter to reduce water consumption.

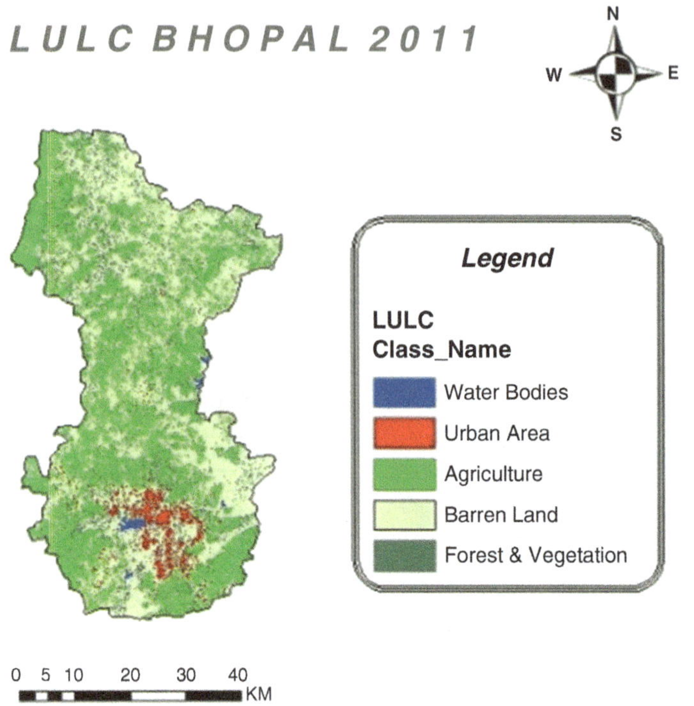

Fig. 13.23 LULC of year 2011

- **Expand greenery in urban areas (toward green infrastructure):** Flora is a biological cooling system because it promotes evapotranspiration, and energy is dispersed more due to latent heating instead of prudent heating. It will also be a sink for CO_2 and likely improve groundwater recharge and soil conservation. But the type of plant has a significant role in this; the authorities and agencies should advise the people on the proper mix of vegetables to farm and different inputs to employ.
- **Management of waterbodies and other marshlands:** Water retention in the urban environment increases evaporation. Likewise, the flora and marshlands have a tendency to absorb heat, and the ambient air temperatures (particularly during extreme temperatures in the daytime) are possible to lessen. As a result, water-penetrating urban planning is required to support climate change. There must be an adequate regulatory system in place for marshland and waterbody protection and restoration.
- **Supporting planning for sustainable groundwater management:** Sustainable planning must aim to alleviate an "alarming" act of groundwater condition at a town level. This planning strategy can accentuate on prudent usage of groundwater, supporting conservation of surface water, improving public awareness, systematic planning for water supply, and helping household and industrial water conservation methods.

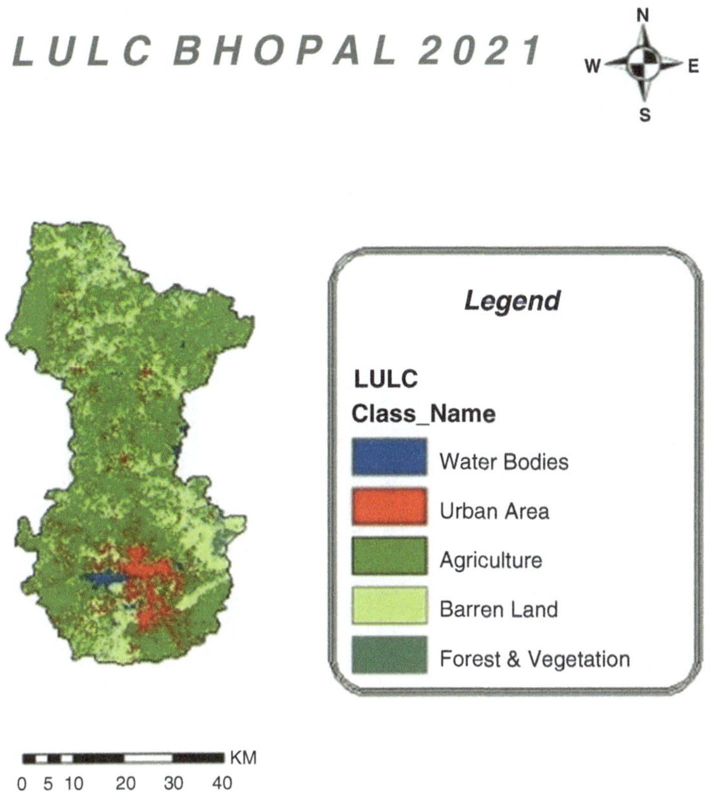

Fig. 13.24 LULC of year 2021

Table 13.8 Land use/cover classification statistics from 2000 to 2020 (area is in approximate)

Year: 2000 area (%)	2010 area (%)	2015 area (%)	2020 area (%)
Urban area – 3.784	15.687	25	36.6
Agriculture land –61.36	53.6	48	39
Barren land –8.32	10	9	9
Thick vegetation –25.9	20	18	15
Water bodies – < 1	<1	<1	<1

- **Education and awareness program:** Ultimately, citizens must be made aware of the reality that fast, unplanned, and disorganized urban expansion has primarily produced socioeconomic and ecological variations, resulting in incompetent use of land and water. The governments of state and local governments as well as NGOs should undertake proper capacity development and awareness programs to convey knowledge regarding the negative consequences of rapid urbanization.

Table 13.9 Land use/cover classification statistics from 2000 to 2021

	2000		2011		2021		2000–2021	
Land use type	Area (km^2)	%	Area (km^2)	%	Area (km^2)	%	Change of area (km^2)	Dynamic degree (%)
Agriculture	1209.1545	44.395523	1182.348	43.41129	1096.2133	40.24875	−112.941	0.444786
Barren land	619.6545	22.75134	593.2333	21.78126	492.68148	18.08938	−126.973	0.975759
Forest and vegetation	785.418	28.837541	737.8745	27.09193	717.33687	26.33786	−68.0811	0.412769
Urban /built-up area	47.4786	1.7432324	149.4023	5.485481	359.30564	13.19233	311.827	−31.2749
Water bodies	61.89	2.2723638	60.7374	2.230045	58.058307	2.131679	−3.83169	0.294816
Grand total	**2723.5956**	**100**	**2723.595**	**100**	**2723.5956**	**100**	–	–

Table 13.10 Groundwater quality comparison

Parameter	Unit	Year			BIS (IS 10500:2012)[a]		WHO
		2001	2011	2018	Desirable limit	Permissible limit	Max. allowable concentration
Bicarbonate as $HCO3^-$	mg/L	159	305	352	–	–	–
Calcium as Ca^{+2}	mg/L	22	92	87	75	200	–
Chlorides as Cl^-	mg/L	64	28	37	250	1000	250
Electrical conductivity	µmhos/cm	520	657	902			
Fluoride as F^-	mg/L	0.77	0.43	0.3	1	1.5	1.5
Potassium as K^+	mg/L	0.4	0.4	0.9	–	–	3
Magnesium as mg^{+2}	mg/L	5	13.48	21.87	30	100	–
Nitrate as NO_3^-	mg/L	26	46	19	45	No relaxation	50
pH	–	7.3	7.5	7.34	6.5 to 8.5	No relaxation	6.5 to 8.5
Residual sodium carbonate	meq/L	1.1	1	–0.37	<1.25	2.5	–
Sodium adsorption ration (SAR)	–	2.4	0.72	1.42	–	–	–
Sodium as Na^+	Mg/L	48	28	57	–	–	200
Sulfate as SO_4	Mg/L	25	6	45	200	400	–
Total hardness as $CaCO_3$	Mg/L	175	285	307	200	600	500

[a]BIS Bureau of Indian Standards (2012): IS10500:2012
[b]WHO (2011): Guidelines of Drinking-water Quality

Table 13.11 Groundwater level of Bhopal district

Sr. No.	Station	Ground level (m bgl)		
		2000	2011	2019
1	Ahmedpur	7.85	17.76	–
2	Bairagarh	–	–	8.74
2	Balampur ghati	3.44	2.4	9.55
3	Berasia	10.15	12	10.24
4	Berasia(D)	9.5	32.55	20.92
5	Bilkhiria	3.04	4.9	4.52
6	Govindpura(D)	10.62	–	–
7	Gunga(S)	7.29	–	–
8	Gunga	6.87	17.6	10.19
9	Islamnagar	12.39	12.4	12.4
10	Misrod	5.53	–	10.16
11	Phanda	–	18.89	20.95
	Average	**7.968**	**13.171**	**15.497**

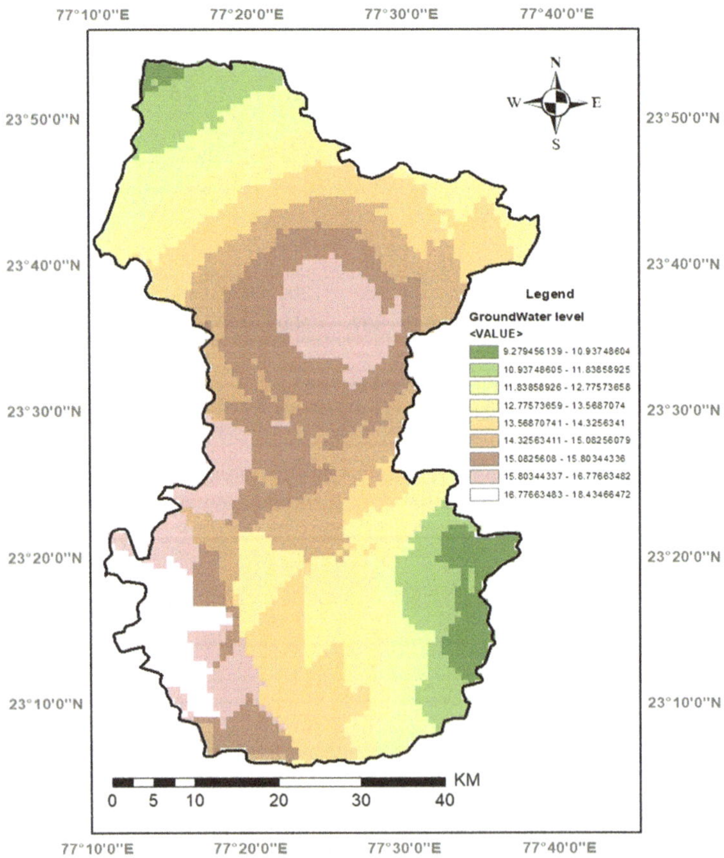

Fig. 13.25 GW level of year 2000

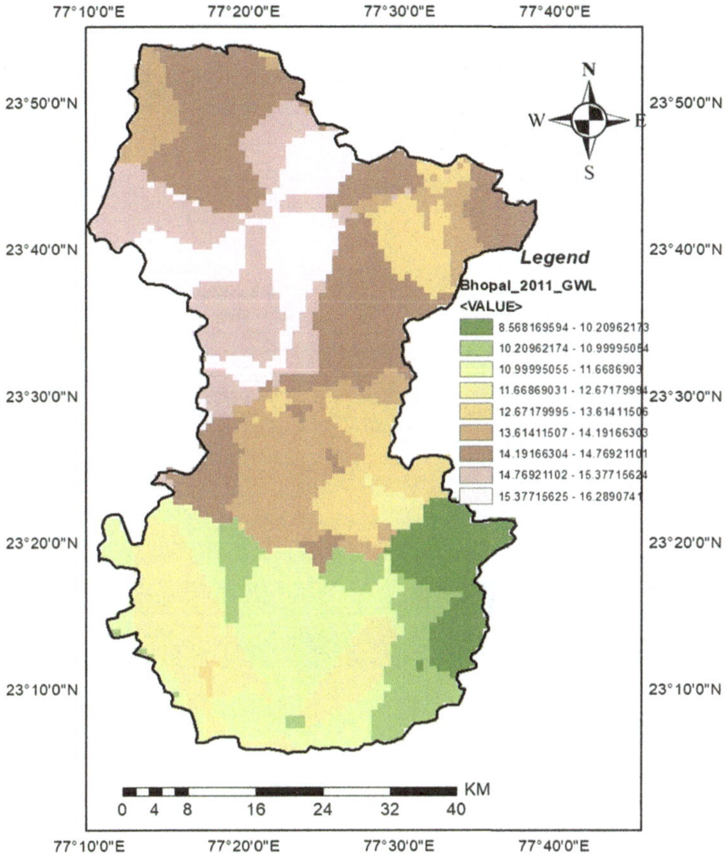

Fig. 13.26 GW level of year 2011

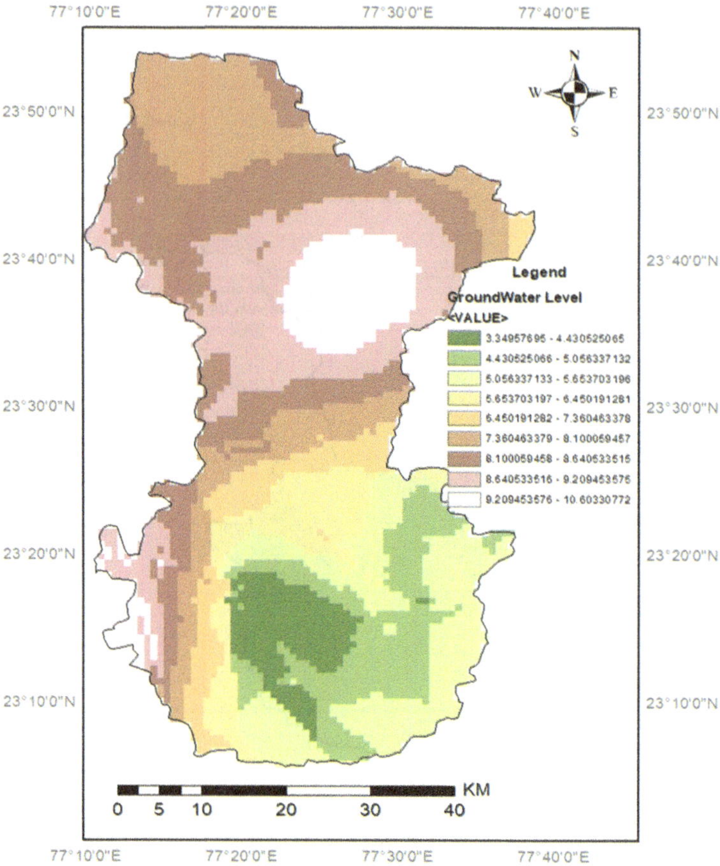

Fig. 13.27 GW level of year 2019

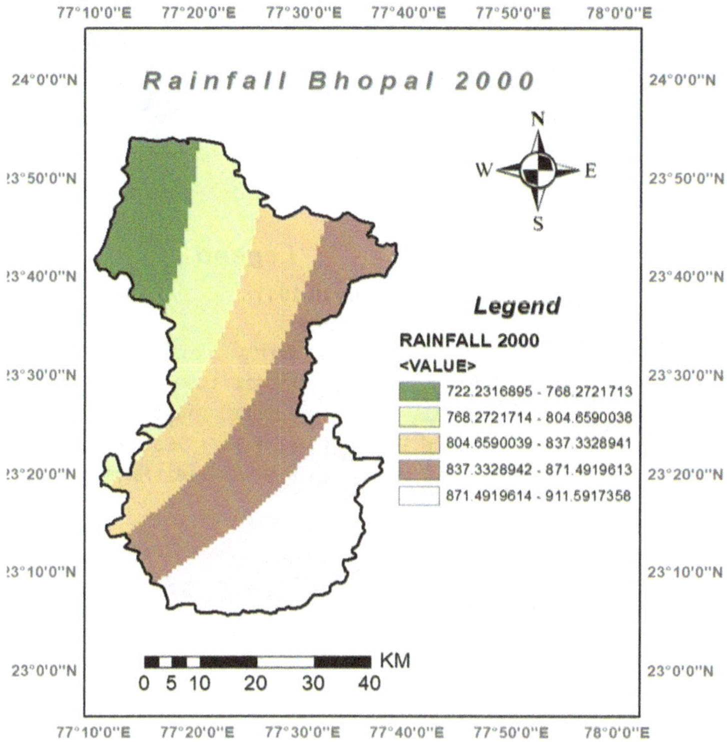

Fig. 13.28 Rainfall of Year 2000

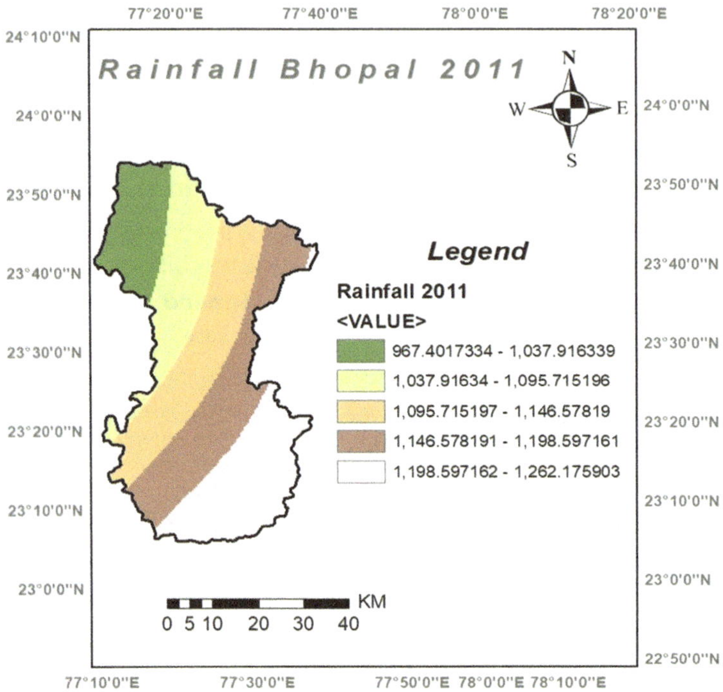

Fig. 13.29 Rainfall of Year 2011

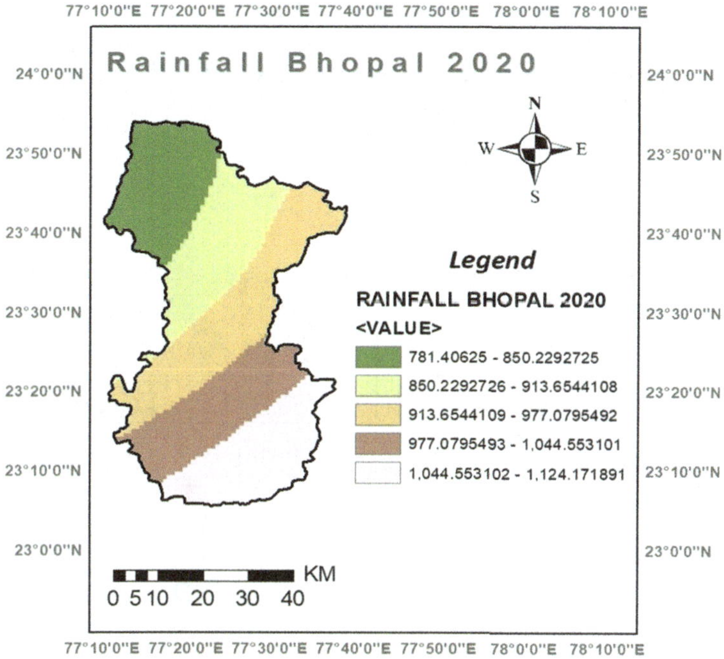

Fig. 13.30 Rainfall of Year 2020

Table 13.12 Rainfall data (*Data source: India WRIS*)

Year	Rainfall (mm)	Year	Rainfall (mm)	Year	Rainfall (mm)
2000	847.92	2007	816.61	2014	980.26
2001	1010.98	2008	839.25	2015	1126.88
2002	830.33	2009	1128.5	2016	1536.62
2003	1109.51	2010	771.26	2017	815.33
2004	1036.49	2011	1140.66	2018	872.8 5
2005	865.76	2012	1129.95	2019	1795.17
2006	1433.71	2013	1423.66	2020	1159.33

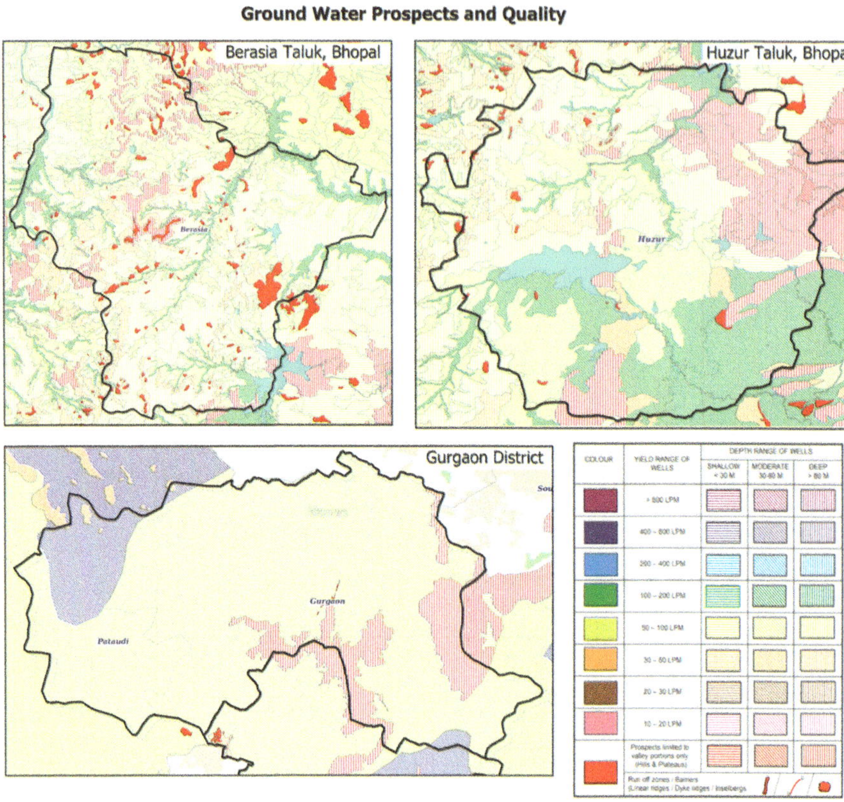

Fig. 13.31 Map of groundwater prospects and quality for both districts

References

Brundtland G et al (1987) Our common future: Report of the 1987. World Commission on Environment and Development

Das J, Jha S, Goyal MK (2020) On the relationship of climatic and monsoon teleconnections with monthly precipitation over meteorologically homogenous regions in India: wavelet &; global coherence approaches. Atmos Res 238:104889. https://doi.org/10.1016/j.atmosres.2020.104889

Deng FF, Huang YQ (2004) Uneven land reform and urban sprawl in China: the case of Beijing. Prog Plann 61:211–236. https://doi.org/10.1016/j.progress.2003.10.004

Goyal MK, Ojha CSP (2010) Evaluation of various linear regression methods for downscaling of mean monthly precipitation in arid Pichola watershed. Nat Resour 01(01):11–18. https://doi.org/10.4236/nr.2010.11002

Goyal MK, Ojha CSP (2012) Downscaling of precipitation on a lake basin: evaluation of rule and decision tree induction algorithms. Hydrol Res 43(3):215–230. https://doi.org/10.2166/nh.2012.040

Goyal MK, Panchariya VK, Sharma A, Singh V (2018) Comparative assessment of SWAT model performance in two distinct catchments under various DEM scenarios of varying resolution, sources and resampling methods. Water Resour Manag 32(2):805–825. https://doi.org/10.1007/s11269-017-1840-1

Goyal MK, Gupta AK, Jha S, Rakkasagi S, Jain V (2022) Climate change impact on precipitation extremes over Indian cities: non-stationary analysis. Technol Forecast Soc Change 180:121685. https://doi.org/10.1016/j.techfore.2022.121685

Jha R, Singh VP, Vatsa V (2008) Analysis of urban development of Haridwar, India, using entropy approach. KSCE J Civ Eng 12:281–288. https://doi.org/10.1007/s12205-008-0281-z

Kanta Kumar NL, Sawant NG, Kumar S (2011) Forecasting urban growth based on GIS, RS and SLEUTH model in Pune metropolitan area. Int J Geomat Geosci 2:568–579

Lv ZQ, Dai FQ, Sun C (2012) Evaluation of urban sprawl and urban landscape pattern in a rapidly developing region. Environ Monit Assess 184:6437–6448. https://doi.org/10.1007/s10661-011-2431-x

Mohr J, Baier K, Jha R, Azzam R (2011) Urbanization and the social impact on water resources in India. Fourth international conference on water resources and renewable energy development in Asia, Chiang Mai, Thailand

Poonia V, Goyal MK, Gupta BB, Gupta AK, Jha S, Das J (2021) Drought occurrence in Different River basins of India and blockchain technology based framework for disaster management. J Clean Prod 312:127737. https://doi.org/10.1016/j.jclepro.2021.127737

Rahman A (2007) Application of remote sensing and GIS technique for urban environmental management and sustainable development of Delhi, India. Springer, Cham

Strahler AN (1975) In: Strahler AN (ed) Physical geography, 4th edn. Wiley, New York

Vinoth Kumar JA, Pathan SK, Bhanderi RJ (2007) Spatio-temporal analysis for monitoring urban growth - a case study of Indore City. J Indian Soc Remote Sens 35:11–20

WHO (2011) Guidelines for drinking-water quality. WHO Library Cataloguing-in-Publication Data, Geneva

Chapter 14
Application of Water Accounting Plus Framework for the Assessment of the Water Consumption Pattern and Food Security

Pooja Patle, Pushpendra Kumar Singh, Shivukumar Rakkasagi, Ishtiyaq Ahmad, and Manish Kumar Goyal

Abstract India's major issue is water scarcity and security. To deal with these issues, proper management plans and strategies are required. To diminish the adverse impacts and improve the benefits of water depletion for citizens, a strong knowledge of hydrological processes at basin level is required. Water professionals currently lack a mutual framework that connects reduction to water user groups and their profits. The lack of ground-based data drives the application of remotely sensed data in the WA+ framework. The water accounting framework assists decision-makers in understanding and implementing policies to address water scarcity and security. Water accounting requires precise input data to provide accurate explanations of water allocation and depletion in river basins. This framework improves understanding of the basin's complex hydrological processes by separating non-manageable, manageable, reserved, and committed flow for downstream demands and environmental flows with the interaction of land use on the basis of the water management classes. The WA+ evapotranspiration sheet depicts the basin's consumption pattern by categorizing evaporation, transpiration, and interception. The set of indicators applied to assess the basin's overall water resource condition. The Water Accounting Plus (WA+) framework provides an agricultural services sheet that connects the productivity of land and water of rainfed and irrigated areas using the green and blue water concepts of the Budyko framework. To understand the applicability of the framework, we discussed a recent study of the

P. Patle (✉) · I. Ahmad
Department of Civil Engineering, National Institute of Technology, Raipur, India
e-mail: iahmad.ce@nitrr.ac.in

P. K. Singh
Division of Water Resources, National Institute of Hydrology, Roorkee, India

S. Rakkasagi · M. K. Goyal
Department of Civil Engineering, Indian Institute of Technology Indore, Indore, Madhya Pradesh, India
e-mail: mkgoyal@iiti.ac.in

A. K. Gupta et al. (eds.), *Ecosystem Restoration: Towards Sustainability and Resilient Development*, Disaster Resilience and Green Growth,
https://doi.org/10.1007/978-981-99-3687-8_14

Krishna basin, which is used to investigate the conditions of water resources applying the WA+ framework formed by IHE Delft, FAO, and IWMI.

Keywords Water accounting plus · Evapotranspiration · Budyko framework · Water productivity · Land productivity · Krishna basin

14.1 Introduction

The goal of water accounting plus is to track inflows and outflows and assess liabilities, stocks, and resources for a specific area over a period of time (Karimi et al. 2013). Water accounting plus framework has been developed with the collaboration of IHE DELFT, FAO, IWMI, CGIAR and UNESCO. Recently, many organizations developed their water accounting frameworks, which had some sort of limitations. The water accounting framework of the International Water Management Institute (IWMI) (Molden and Sakthivadivel 1999), System of Environmental and Economic Accounting (SEEA) UN Statistics Division' Accounting for Water (SEEAW) (Vardon et al. 2012), Australian Water Accounting Conceptual Framework (Merz et al. 2006), and UNEP' Water Footprint are samples of existing water accounting systems. However, none of those programs of water accounting frameworks was taken as a broad normal (Dost et al. 2013). The water accounting system developed by AQUASTAT does not give detailed information regarding the interface between the use of land and water. The system emphases on water withdrawals only and will not differentiate between consumptive and non-consumptive use. The United Nations Statistics Division proposed a WA framework called System of Environmental Economic Accounting for Water (SEEA-WATER); however, important necessary data are not likely to be accessible (Perry 2012). The vital change in green and blue water resources of Australian Water Accounting Standard (AWAS) formed by the Water Accounting Standards Board (WASB). The framework accounts for water withdrawals instead of consumptive use (refer to Table 14.1), and it consists of irrigated agriculture, industrial, and domestic users. However, it will not offer any information on rainfed systems and natural evapotranspiration (ET) processes. A WA procedure was developed by the International Water Management Institute (Molden 1997) with the purpose of stalking water depletion instead of withdrawals to prevent errors when ignoring recycling and to account for ET. Agriculture is the world's largest water user, and India is even more so. According to recent research by the National Bank for Agriculture and Rural Development (Sharma et al. 2021), agriculture consumes 78% of available water resources. The idea of WP (linking engineering, agronomical, and economic factors) can be one of the most effective tools for addressing India's current water scarcity and food security issues. Molden (1997), for example, coined the term WP for the first time to emphasize the advantages of water use in terms of productivity. As a result, the agricultural WP can be used to improve agricultural water management in India (Zwart and Bastiaanssen 2004; Brauman et al. 2013; Mekonnen and Hoekstra 2014). A greater WP signifies either a higher crop yield from the same water supply

Table 14.1 WA+ sheets and their description

S No.	WA+ sheets	Purposes
1.	Resource-based sheet	Represent overexploitation, unmanageable, manageable, exploitable, reserved, utilized, and utilizable flows at river basin scale that are discussed in general. Examine the many water sources that influence net inflow. Determine the difference between landscape ET (due to rain) and incremental ET (due to natural and artificial withdrawals)
2.	Evapotranspiration sheet	Assess water consumption patterns (in terms of volume of water) by land use classes and water user groups, explaining the human influence on ET, beneficial and non-beneficial consumption, and breaking down beneficial consumption in agricultural, ecological, economic, energy, and leisure
3.	Agricultural sheet	Assess productivity of agricultural (kg/ha) and water (kg/m^3) and water consumption in terms of crops, timber, and fish product. Assess the productivity of land and water from the rain-fed and irrigated regions
4.	Utilized flow sheet	Report water shortage based on water needs and supplies, as well as an overview of all man-made withdrawals. Make strategies for water allocation. Make volumetric water entitlements, and assess non-authorized use and monitor compliance with water withdrawals. Calculate natural withdrawals (e.g., seasonal floods, shallow groundwater tables, and groundwater dependent ecosystems)
5.	Surface water sheet	Provide a summary of the basin's surface water and estimate the flow of the river in different reaches (even ungauged). Govern the availability of surface water and the amount of water that can be withdrawn. Calculate the amount of surface water that can be stored
6.	Ground water sheet	Assess the role of groundwater in renewable water resources, specifically for dry season base flow, making reliable groundwater withdrawal strategies (i.e., avoid decreasing groundwater tables) and map groundwater withdrawals for irrigation

or the same crop productivity from a low water supply (Goyal et al. 2018; Das et al. 2020; Poonia et al. 2021).

P. Karimi and Bastiaanssen (2015) had presented only four sheets, which were (1) resource-based sheet (2) evapotranspiration sheet, (3) productivity sheet, and (4) withdrawal sheet. The resource-based sheet gives a broad summary on overexploitation, unmanageable, manageable, exploitable, reserved, utilized, and utilizable flows at river basin scale, whereas the total ET sheet provides a thorough understanding on how, where, and when water is consumed in river basins and designs ET management principles to define a cap on consumptive use from withdrawals and inundations and classify beneficial and non-beneficial water consumptions. The productivity sheet presents relations between water depletion and biomass production, production of crop and water, and the withdrawal sheet delivers information on water withdrawals and reuse. With the passage of time, due to increasing demands, these sheets have been updated, the withdrawal sheet has been replaced with utilized flow sheet, and more sheets has been included to understand the hydrological processes (Table 14.1). Each sheet is associated with the set of indicators used to describe the situation of water resources of the basin.

14.2 Land Use in the WA+ Framework

Land use land cover is an essential component in water accounting since, it establishes whether water is manageable or non-manageable. Land use in water accounting plus has been classified centered on the water management classes, which are as follows: (1) protected land use (PLU), (2) utilized land use (ULU), (3) modified land use (MLU), and (4) managed water use (MWU). Protected land uses are the areas where no interferences are allowed; these are secured by the government and international NGOs, including national parks, RAMSAR sites, tropical rainforests, wetlands, etc. (Karimi et al. 2013; Dembele 2020; Food and Agriculture Organization of the United Nations and IHE Delft Institute for Water Education 2019). Utilized land use (ULU) is the land where vegetation is not managed on a regular basis and human influences are limited. It includes forests, woodlands, shrublands, pastures, savannas, etc. Modified land uses are the regions where vegetation and soils are managed, but water supply is not disturbed (rainfall). It includes rainfed agriculture, biofuel crops, forest plantations, etc. Managed water uses are the classes where human interventions are present and water supply is also not natural and withdraws from surface and groundwater resources, for instance, irrigation canals, hydropower schemes, urban water supply, treatment plants, etc. Figure 14.1 represents the water accounting land use pattern of Meghalaya.

14.3 Total Evapotranspiration Sheet

The WA+ total evapotranspiration sheet illustrates water depletion and recognizes components of water use that can be managed, manageable, and non-manageable on the basis of the LULC classifications. Total evapotranspiration (ET) is divided into soil and water evaporation (E) and vegetation transpiration (T) and interception (I), and then helpful and non-beneficial water consumption is distinguished (Karimi et al. 2013; Dembele 2020; Food and Agriculture Organization of the United Nations and IHE Delft Institute for Water Education 2019) (Fig. 14.2). ET is classified as beneficial or non-beneficial based on a value judgment based on case studies that must be up-to-date (Dembele 2020). For example, soil evaporation is seen as non-beneficial, but transpiration is regarded as a helpful ET contributing to food production and the economy. The formula developed by Von Hoyningen-Hüne (1983) and Braden (1985) is used to estimate interception losses.

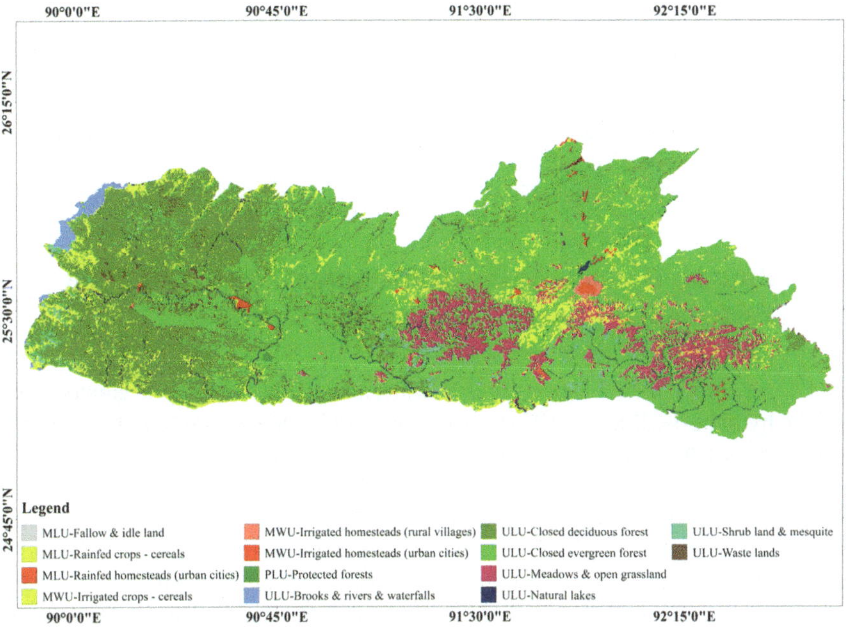

Fig. 14.1 WA+ land use classes of Meghalaya state, India

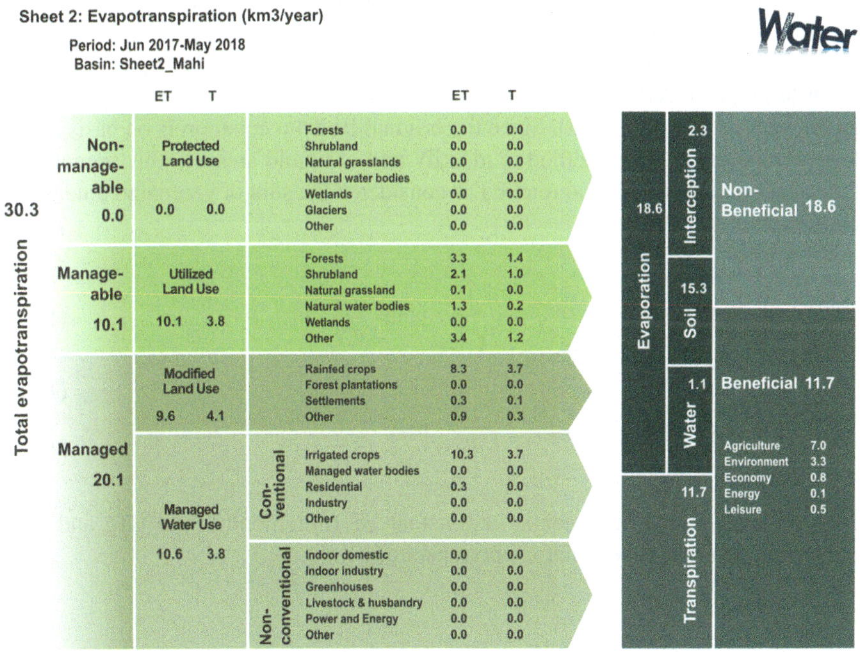

Fig. 14.2 Total ET sheet showing water consumption from different water management classes

$$\mathrm{Im} = \left(1 - \frac{1}{1 + \frac{\mathrm{Pm}}{\mathrm{Nm}} \times \left(1 - e^{(-0.5 \times \mathrm{LAIm})}\right) \times \frac{1}{\mathrm{LAIm}}}\right) \times \mathrm{LAIm} \times \mathrm{Nm} \tag{14.1}$$

Here, LAI represents leaf area index (m^2/m^2), Nm denotes the no. of rainy days in a month, and Pm denotes monthly precipitation.

14.4 Budyko Hypothesis for Estimation of Green and Blue Water ET

For ET separation into ETgreen and ETblue, the WA+ framework employs the Budyko hypothesis (Budyko 1974). The water held in the soil is referred to as green water, whereas the water accumulated in rivers, ponds, lakes, other bodies of surface water, and aquifers is referred to as blue water (Singh et al. 2021; Goyal and Ojha 2010, 2012; Falkenmark and Rockström 2006). The Budyko hypothesis determines an experiential relationship between AET, reference evapotranspiration (PET), and P (Sposito 2017; Singh et al. 2021) and hence offers first-order estimations of evaporation employing only P and PET (Mianabadi et al. 2019). The Budyko hypothesis is based on the combination of two approaches: (a) water balance and (b) energy balance (Singh et al. 2021) For each green and blue pixel, the water balancing is performed separately. The aridity index (PET/P) and the evaporative index (AET/P) are used to explain the Budyko curve. A pixel-based analysis has been adopted in this framework to identify the rainfed and irrigated pixels. Pixels falling over the water limit are considered blue water pixels, and those falling below are considered green water pixels (Fig. 14.3). Since the original Budyko equation is on basis of the long-term water balance method to identify arid or humid areas, in this framework, the equation is modified, and total ET is considered the sum of green and blue water consumption. ·

$$\frac{\mathrm{ETg}*}{\mathrm{P}} = \mathrm{f}(\phi) = \sqrt{\phi \cdot \tanh\left(\frac{1}{\phi}\right) \cdot \left(1 - e^{-\phi}\right)} \tag{14.2}$$

$$\mathrm{ETg} = \min(\mathrm{ETa}, f(\phi) \cdot \mathrm{P}) \tag{14.3}$$

$$\phi = \frac{\mathrm{PET}}{\mathrm{P}} \tag{14.4}$$

PET and AET represent potential and actual ET (mm/month), and ETg and ETb represent green and blue water evapotranspiration.

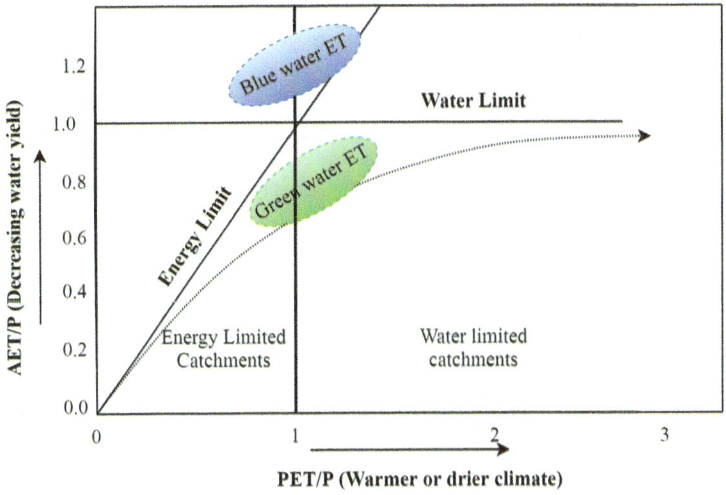

Fig. 14.3 Budyko curve for assessing blue water ET and green water ET

14.5 Agricultural Services Sheet

Water scarcity and food security are the two major threats due to changing climatic conditions, inefficient water usage, and improper management plans; hence, it's important to focus on them. The agricultural services sheet distinguishes between the productivity of land (Kg/ha) and water (Kg/m^3) (Fig. 14.4). Productivity measurement in WA+ is on biomass production base (Karimi et al. 2013). This sheet indicates the possibilities for saving water in agriculture and making agricultural water management more efficient. Water productivity (WP) is a basic indicator in the performance evaluation of river basins, and it has huge food and water security consequences (Molden 2007; Sharma et al. 2021). Land and water productivity are calculated on the basis of green and blue water ET calculations. The water productivity concept mainly focuses on the *More Crop Per Drop*, which represents how much volume of water is consumed to generate per *Kg* of crop. The agricultural sheet reflects light on the type of crop grown in the area and whether it is suitable for that area to produce it (in terms of water consumption). The main purpose of this sheet is to plan future rainfed and irrigated cropping methods using rainfall, exploitable, and available water and show potentials for conserving water in agriculture and making agricultural water management more efficient.

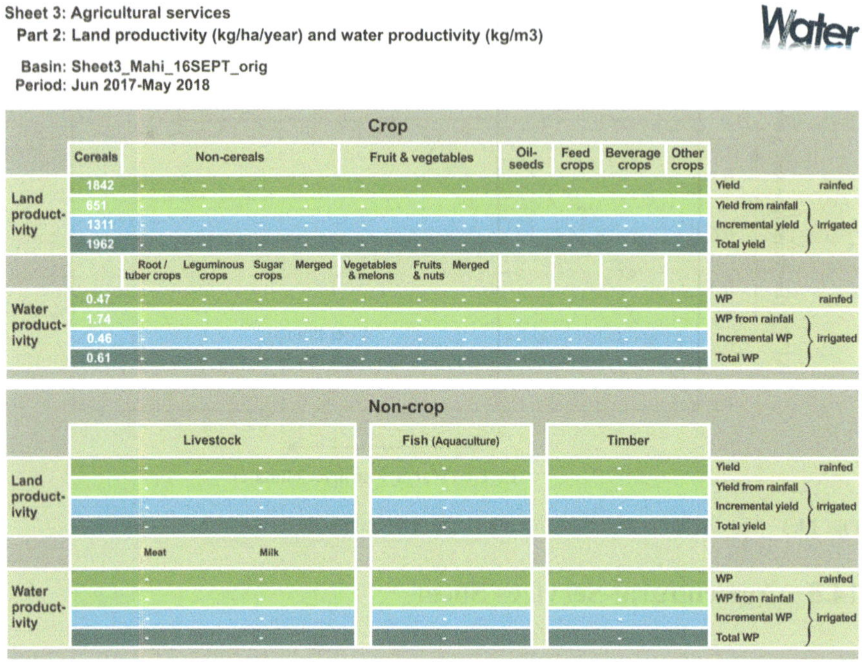

Fig. 14.4 Agricultural services sheet showing land and water productivity of irrigated and rainfed areas

14.6 Key Indicators

WA+ summarizes the overall water resources and their consumption with the help of some indicators associated with each sheet, enabling the understanding of a common man. These indicators are summarized in Table 14.2.

Harvest index can vary crop to crop (Murray et al. 2021). For example, rice and wheat have the harvest index value of 0.44 and 0.37, respectively.

14.7 An Example of Water Accounting Study of Krishna Basin, India

One of the recent studies (ABD and IHE DELFT 2020) used water accounts to investigate the conditions of water resources in the Krishna basin, which consists of three subbasins, applying the water accounting plus (WA+) framework formed by IHE Delft, FAO, and IWMI. In this example of study, authors examined the Krishna basin in Karnataka state which consists of three subbasins, namely, Middle Krishna (K2), Ghatprabha (K3), and Malaprabha (K4; Fig. 14.5). This study considered

Table 14.2 Key indicators of total ET (sheet 2) and agricultural services sheet (sheet 3)

Sheet-2 Indicators		
1.	Transpiration ET fraction	$\frac{\text{Transpiration}}{\text{ET}}$
2.	Beneficial ET fraction	$\frac{\text{Ebeneficial} + \text{Tbeneficial}}{\text{ET}}$
3.	Managed ET fraction	$\frac{\text{ET managed}}{\text{ET}}$
4.	Agricultural ET fraction	$\frac{\text{Agricultural ET}}{\text{ET}}$
5.	Irrigated ET fraction	$\frac{\text{Irrigated ET}}{\text{Agricultural ET}}$
Sheet-3 Indicators		
1.	Land productivity (Kg/ha)	$\frac{\text{Biomass production} \times \text{Harvest index}}{\text{Crop area}}$
2.	Water productivity (Kg/m^3) (rainfed crops)	$\frac{\text{Rainfed crop production} \times \text{Harvest index}}{\text{Rainfed crops ET}}$
3.	Water productivity (Kg/m^3) (irrigated crops)	$\frac{\text{Irrigated crop production} \times \text{Harvest index}}{\text{Irrgated crop ET}}$
4.	Food-irrigation dependence	$\frac{\text{Irrigated agriculture ET}}{\text{Total agriculture ET}}$

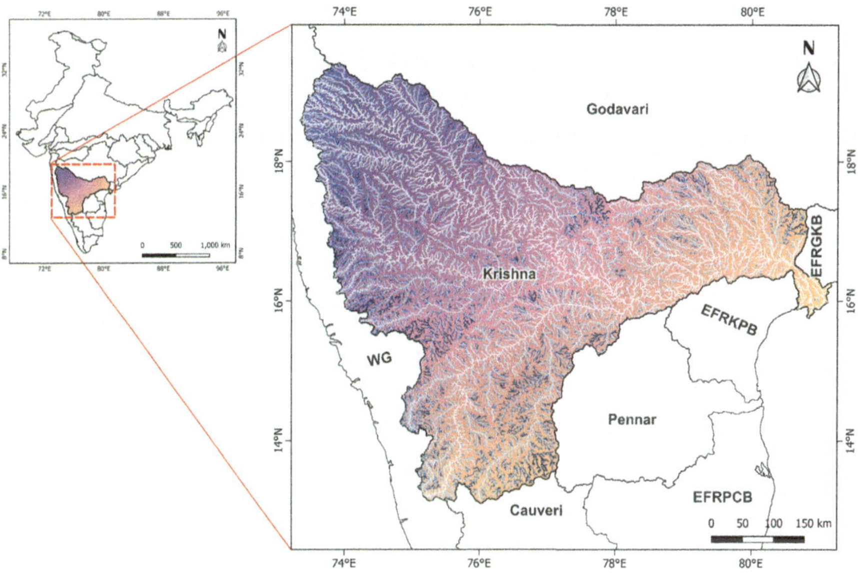

Fig. 14.5 Location of the Krishna basin boundary with stream network overlaid

subbasins of the Krishna basin in Karnataka state for the analysis, and inflows are assessed by means of available in situ measures. This study used the distinction between precipitation and evapotranspiration, which were derived from remote sensing data across the upstream region in locations where inflows are not calculated. The Krishna basin includes surface water reservoirs of over 40,000. Water resource development is ongoing, and currently, 76 large and 135 medium irrigation projects are proposed in the basin. The basin is under significant environmental strain due to a rising population (now above 66 million), increased need for agricultural production, and intensive water resource development.

The most appropriate datasets were chosen in this study based on the following: (a) inter-relationship of data products, (b) validation with making use of in situ measurement, (c) annual water balance measurement and evaluation with in situ discharge quantities, and (d) accessibility of data in recent times. The CHIRPS dataset was used for precipitation, while the SSEBop dataset was used for actual ET estimations. Remotely sensed ET data exhibits less marked month-to-month and seasonal variation than precipitation, with greater ET values during the monsoon season, when ample amounts of water and energy are available, and lower values throughout the winter season. The basic rationale for categorizing these four land use types is that their management options range from maintaining pristine conditions to managing hourly water flows. The PLU regions account for barely 1% of K3, whereas natural lands (ULU) account for 4–5% in all three subbasins. The MLU and MWU have comparable coverage (about 50%) in K2 and K3, while irrigated agriculture is less established in K4, and the MWU only reports 63%.

Figure 14.6 illustrates a flowchart of the central computational stages in the water accounting method, including input datasets and data type applied in the study.

Key findings of this study (Krishna basin) are:

- The quantity of consumption of non-beneficial water in agriculture is higher throughout the subbasins, accounting for 42% in rainfed and 48% in irrigated cultivations. The observed soil evaporation was high in all three subbasins, particularly in paddy crops; however, other crop varieties also have high evaporation values. This demonstrates that there is substantial potential to boost agricultural output without increasing overall water usage by reducing wasteful soil evaporation. Measures to reduce soil evaporation, such as advanced irrigation planning and effective management of farm field water, should also be explored in the agriculture sector.
- The pixel balancing model results show a cumulative storage loss of 0.2 km^3/year for all three subbasins, which is below 1% of precipitation. There are spatial changes, with K2 and K3 losing and K4 gaining. This shows that there is no evidence of overexploitation of water resources over the time period studied. The changes in interannual storage are substantially larger and are closely related to the monsoon climate. Changes in storage for both surface and groundwater must be closely observed on a regular/seasonal basis.
- The water accounting statistics are generated using an underlying pixel balance model, which is derived by remotely sensed datasets with constrained calibration settings. The present version excludes routing and dam operations. Furthermore, the development of the water accounts might be achieved by model improvement or the use of outputs of locally calibrated model.

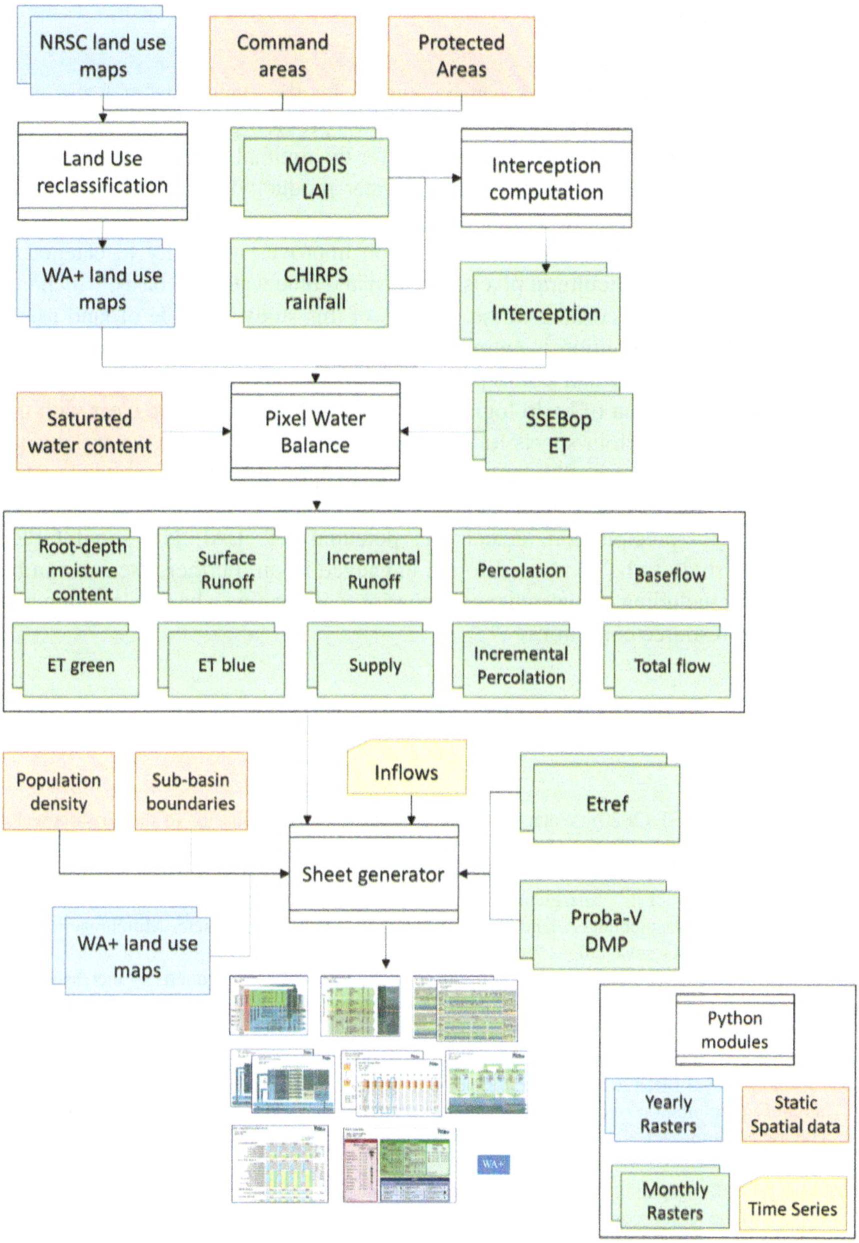

Fig. 14.6 Water accounting plus flow chart (Source: ABD and IHE DELFT 2020)

14.8 Conclusion

The use of the WA+ framework can be studied for the evaluation of water consumption patterns and land and water productivity, providing insights into developing adequate management plans and schemes for the optimal use of water resources for increased agricultural output. WA + 's water productivity approach promotes effective water utilization. This study also suggests the amount of water required by a specific crop. Budyko framework integration improves its ability to categorize rainfed and irrigated agricultural pixels on green and blue water use basis, which is a unique feature of it. According to the findings of this study, the role of land use is critical because consumption is entirely dependent on it, as it becomes feasible to check the water consumption from a particular land use class, and can be managed or not. The comprehension of hydrological processes will become more clear with the assistance of the other four sheets for the assessment of downstream water demand, surface and groundwater withdrawals, and available, utilizable, and exploitable water. Another advantage of utilizing WA+ is that it allows for the use of satellite-driven datasets (precipitation, actual ET, potential ET, LAI, NPP, and GPP), allowing the study to be conducted for an ungauged basin. Further research can be conducted by including climate change scenarios to determine a basin's existing and future water resource conditions.

References

ABD and IHE DELFT (2020). Water accounting study for three sub-basins of the Krishna water system in Karnataka, India 152; https://www.wateraccounting.org/files/projects/adb/phase2/IHE_WA_India_Karnataka_Final_Report.pdf

Braden H (1985) Ein Energiehaushalts- und Verdunstungsmodell for Wasser und Stoffhaushaltsuntersuchungen landwirtschaftlich genutzer Einzugsgebiete. Mittelungen Dtsch Bodenkundliche Geselschaft 42:294–299

Brauman KA, Siebert S, Foley JA (2013) Improvements in crop water productivity increase water sustainability and food security—a global analysis. Environ Res Lett 8(2):024030. https://doi.org/10.1088/1748-9326/8/2/024030.

Budyko MI (1974) Climate and Life. Academic, Orlando, Fla.

Das J, Jha S, Goyal MK (2020) On the relationship of climatic and monsoon teleconnections with monthly precipitation over meteorologically homogenous regions in India: Wavelet & global coherence approaches. Atmos Res 238:104889. https://doi.org/10.1016/j.atmosres.2020.104889

Dembele M (2020) Spatially explicit hydrological modelling for water accounting under climate change in the Volta river basin in West Africa. Ph.D. University of Lausanne

Dost R, Obando EB, Bastiaanssen W, Hoogeveen J (2013) Water accounting plus (WA+) in the awash river basin

Falkenmark M, Rockström J (2006) The new blue and green water paradigm: breaking new ground for water resources planning and management. J Water Resour Plan Manag 132(3):129–132

Food and Agriculture Organization of the United Nations and IHE Delft Institute for Water Education (2019) (rep.). Water accounting in the Litani River Basin. Rome. p 1–63

Goyal MK, Ojha CSP (2010) Evaluation of various linear regression methods for downscaling of mean monthly precipitation in arid Pichola watershed. Nat Res 01(01):11–18. https://doi.org/10.4236/nr.2010.11002

Goyal MK, Ojha CSP (2012) Downscaling of precipitation on a lake basin: evaluation of rule and decision tree induction algorithms. Hydrol Res 43(3):215–230. https://doi.org/10.2166/nh.2012.040

Goyal MK, Panchariya VK, Sharma A, Singh V (2018) Comparative assessment of SWAT model performance in two distinct catchments under various DEM scenarios of varying resolution, sources and resampling methods. Water Resour Manag 32(2):805–825. https://doi.org/10.1007/s11269-017-1840-1

Karimi P, Bastiaanssen WG (2015) Spatial evapotranspiration, rainfall and land use data in Water Accounting – Part 1: Review of the accuracy of the remote sensing data. Hydrol Earth Syst Sci 19(1):507–532. https://doi.org/10.5194/hess-19-507-2015.

Karimi P, Bastiaanssen W, Molden D, Cheema M (2013) Basin-wide water accounting based on remote sensing data: an application for the Indus Basin. Hydrol Earth Syst Sci 17(7):2473–2486

Mekonnen MM, Hoekstra AY (2014) Water footprint benchmarks for crop production: A first global assessment. Ecol Indicators 46:214–223. https://doi.org/10.1016/j.ecolind.2014.06.013.

Merz R, Blöschl G, Parajka J (2006) Spatio-temporal variability of event runoff coefficients. J Hydrol 331(3–4):591–604. https://doi.org/10.1016/j.jhydrol.2006.06.008.

Mianabadi A, Coenders-Gerrits M, Shirazi P, Ghahraman B, Alizadeh A (2019) A global Budyko model to partition evaporation into interception and transpiration. Hydrol Earth Syst Sci 23: 4983–5000. https://doi.org/10.5194/hess-23-4983-2019

Molden D (1997) Accounting for water use and productivity. SWIM paper no. 1 International irrigation management institute, Colombo, Sri Lanka

Molden David (2007) Water for food, Water for Life. Comprehensive Assessment of Water Management in Agriculture. London. www.earthscan.co.uk

Molden D, Sakthivadivel R (1999) Water accounting to assess use and productivity of water. Int J Water Res Dev 15(1–2):55–71

Murray U, Jeff B, Matthew F (2021) Variability in harvest index of grain crops and potential significance for carbon accounting: examples from Australian agriculture. In: Advances in agronomy. Elsevier, Amsterdam, p 47

Perry CJ (2012) Accounting for water; stocks, flows, and values. In: Inclusive wealth report 2012 measuring progress toward sustainability, UNU-IHPD and UNEP. Cambridge University Press, Cambridge, pp 215–230

Poonia V, Goyal MK, Gupta BB, Gupta AK, Jha S, Das J (2021) Drought occurrence in different river basins of India and blockchain technology based framework for disaster management. J Clean Prod 312:127737. https://doi.org/10.1016/j.jclepro.2021.127737

Sharma B, Gulati A, Mohan G, Manchanda S, Ray I, Amarasinghe U (2021) Water productivity mapping of major Indian crops. NABARD & ICRIER, p 1–212.

Singh P, Jain S, Mishra P, Goel M (2021) An assessment of water consumption patterns and land productivity and water productivity using WA+ framework and satellite data inputs. Physics and Chemistry of the Earth, Parts A/B/C, p 103053.

Sposito G (2017) Understanding the Budyko equation. Water 9(4):236

Vardon M, Lagunes-Martinez R, Gan H, Nagy M (2012). The system of environmental-economic accounting of water: development, implementation and use.

Von Hoyningen-Hüne J (1983) Die Interception des Niederschlags in landwirtschaftlichen Beständen. Schriftenr. des DVWK, 57, p 1–53

Zwart SJ, Bastiaanssen WGM (2004) Review of measured crop water productivity values for irrigated wheat, rice, cotton and maize. Agric Water Manag 69(115–133):2004. https://doi.org/10.1016/j.agwat.2004.04.007

Chapter 15
Comparison of Probability Distributions for Extreme Value Analysis and Predicting Monthly Rainfall Pattern Using Bayesian Regularized ANN

N. Vivekanandan, Shivam Singh, and Manish Kumar Goyal

Abstract The estimation of rainfall for a given return period is of utmost importance for the planning and design of minor and major hydraulic structures. This can be accomplished using an extreme value analysis (EVA) of rainfall, which involves fitting a series of annual 1-day maximum rainfall data to probability distributions including the 2-parameter normal, 2-parameter log normal, Pearson type 3, log Pearson type 3, extreme value type 1 (EV1), and generalized extreme value (GEV). The method of moments (MoM), maximum likelihood method (MLM), and L-moments (LMO) are used to determine the distributional parameters depending on the intended applications and the variable under consideration. The six probability distributions used in the EVA of rainfall for the Afzalpur, Aland, and Kalaburagi sites are adequately fitted when measured quantitatively by the goodness-of-fit and diagnostic tests (chi-square and Kolmogorov-Smirnov) and qualitatively by the fitted curves of the estimated rainfall. According to the study's findings, the GEV (LMO) is the most suited among the six distributions tested in EVA for estimating rainfall for Afzalpur and Kalaburagi, while the EV1 (MLM) is more suited for Aland. An artificial neural network with Bayesian regularization has been implemented to model and predict monthly rainfall patterns from all three sites. The model performance analysis shows significant correlation of approximately 0.95 for the training dataset and 0.40 for validation.

Keywords Chi-square · D-index · Extreme value type 1 · Generalized extreme value · Kolmogorov-Smirnov · L-moments · Maximum likelihood method · Rainfall · Artificial neural network · Bayesian regularization

N. Vivekanandan (✉)
Central Water and Power Research Station, Pune, Maharashtra, India
e-mail: vivek.n@cwprs.gov.in

S. Singh · M. K. Goyal
Department of Civil Engineering, Indian Institute of Technology Indore, Indore, Madhya Pradesh, India

A. K. Gupta et al. (eds.), *Ecosystem Restoration: Towards Sustainability and Resilient Development*, Disaster Resilience and Green Growth,
https://doi.org/10.1007/978-981-99-3687-8_15

15.1 Introduction

Assessment of extreme rainfall for a desired return period is of utmost importance for the planning and design of minor and major hydraulic structures, viz., dams, bridges, barrages, and storm water drainage systems. Such information can be applied to the planning and design of water resources projects linked to reservoir design, river bank protection works, soil and water conservation, etc., as well as to the prevention of floods and droughts (CWC 2010). In the post-commissioning stage, where it is necessary to analyze the risk of hydraulic structures failing, extreme rainfall occurrences are crucial (Baratti et al. 2012). Extreme rainfall with a desirable return duration is used, depending on the design life of the structure. Extreme value analysis (EVA), which involves fitting a probability distribution to the series of annual 1-day maximum rainfall (AMR) data, can be used to achieve this (Abida and Ellouze 2008). The robust forecasts of precipitation patterns at a significant lead time can be of great importance in managing water resources as well as mitigating the risk associated with prolonged and flash flooding (Das et al. 2020; Floods 2019; Goswami et al. 2018). Numerical weather prediction models are conventionally used to forecast these extremes by mathematically modeling the governing physical processes (Bauer et al. 2015; Dueben et al. 2021). Being inspired by the learning capability of neural networks, an advancement in machine learning from nonlinear complex data (Pisner and Schnyer 2020; Scher and Messori 2018), we have incorporated artificial neural networks (ANNs) to model and predict the monthly rainfall of all three considered stations with Bayesian regularization to overcome the issue of overfitting in the data.

15.2 Literature Review

For analyzing the rainfall data, a variety of probability distributions from the normal, gamma, and extreme value families of distributions will be provided (Arvind et al. 2017; Esberto, 2018; Sasireka et al. 2019). The distributions that are most frequently applied and employed in EVA are the two-parameter normal (N2) and log normal (LN2), Pearson type 3 (P3), log Pearson type 3 (LP3), extreme value type 1 (EV1), and generalized extreme value (GEV) distributions. From this, it is clear that the N2 and LN2 belong to the family of normal distributions, the P3 and LP3 to the family of gamma distributions, and the EV1 and GEV to the family of extreme value distributions (Bhuyan et al. 2010; Mujere 2011; Olumide et al. 2013; Haberlandt and Radtke 2014; Sharma and Sharma 2019; Singh et al. n.d.; Tank et al. 2021). The parameter estimation algorithms method of moments (MoM), maximum likelihood method (MLM), and L-moments (LMO) are used to determine the distribution's parameters based on the planned applications and the variate under consideration (Acar et al. 2008; Malekinezhad et al. 2011; Vivekanandan 2020).

Using Gumbel (also known as EV1), LN2, and LP3, AlHassoun (2011) conducted a study on establishing an empirical formula to estimate rainfall intensity in the Riyadh region. He came to the conclusion that among the three distributions examined for the assessment of rainfall intensity, the LP3 provides greater accuracy. In order to calculate the extreme rainfall depths at several rain gauge stations in southeast United Kingdom, Esteves (2013) used the EV1 distribution. For the purpose of creating intensity-duration-frequency curves for seven divisions in Bangladesh, Rasel and Hossain (2015) used the EV1 distribution. In the Bamenda Mountain region of Cameroon, Afungang and Bateira (2016) used the EV1 distribution to estimate the maximum quantity of rainfall for various times. Eight probability distributions were used by Baghel et al. (2019) to analyze the frequency of daily maximum rainfall data in the Udaipur district. For the Anakapalli, Atchutapuram, Kasimkota, and Parvada sites, Vivekanandan and Srishailam (2020) examined the MoM and MLM estimators of the EV1, LN2, and LP3 distributions utilized in the EVA of rainfall. The question of which distribution model best fits a given collection of data comes frequently when multiple probability distributions are used in the EVA of rainfall. Both quantitative and qualitative analyses may be able to provide an answer, and the conclusions are quantifiable and trustworthy. The effectiveness of fitting the selected probability distributions is assessed quantitatively using chi-square (χ^2) and Kolmogorov-Smirnov (KS), and diagnostic (viz., D-index) tests, as well as qualitatively using fitted curves for the estimated extreme rainfall. The methods used in the EVA of rainfall and the evaluation of EVA results using GoF and diagnostic tests are succinctly discussed with an example, and the outcomes of the study are reported in the paper.

There have been numerous attempts to predict rainfall using different physical, empirical, and physio-empirical models. In many parts of the world, these models have been used to predict rainfall on an annual, seasonal, monthly, and daily scale (Bauer et al. 2015; Goyal et al. 2018; Nardi et al. 2018). Physical models are created taking into account the synoptic climate and the physical processes by which various climatic variables interact to produce rainfall in a location. In order to anticipate rainfall in physical models, numerical models are created to describe ingrained physical processes (Goyal and Ojha 2010). These models require data and knowledge on numerous land-ocean-atmospheric variables and are typically quite complicated. In spite of this, physical models frequently fail to provide accurate rainfall predictions, because they frequently use crude approximations of complicated physical events (Coles 2001; Su et al. 2012). Physical models have been replaced with empirical models based on statistical techniques. These models are typically created with a specific climate variable in mind, such as rainfall prediction, based on the connection defined by statistics between that variable and its antecedent variables. Statistical models are always region-specific; thus, the model created using the statistical relationship between local rainfall and other atmospheric factors is only applicable for the region (Goyal et al. 2012; Katz 2013). These models are also referred to as empirical models for this reason. In most instances, empirical models are more accurate than physical models and are easier to construct and apply. The primary flaw of empirical models is their total reliance on the historical data utilized to construct them. As a result, they are unable to replicate rainfall in an unknown environment. For instance, empirical models cannot accurately predict the abrupt variations in rainfall that the region has experienced recently and is expected to experience in the future since the models were not constructed with such a wide range of data (Hewitson et al. 2014). In this circumstance, physical models are used as a forecasting option. The development of physical-empirical models, in which empirical models are based on the variables physically responsible for the region's climate, has recently received attention in an effort to address the limitations of both types of modeling approaches (Vergés et al. 2016).

Regression analysis, including linear and higher-order polynomial (nonlinear) regressions, is typically used to create statistical models (Chen and Zhang 2022; Su et al. 2012). However, the relationship between rainfall and the climatic variables that cause it is frequently enigmatic and cannot be captured using commonly used statistical approaches. Complex regression analysis is frequently created using machine learning (ML) methods (Hinge et al. 2018; Sharma and Goyal 2020). Climate extremes in changing climate are getting worse, and robust prediction of these extremes is one of the most essential aspects in managing water resources and disaster management (Poonia et al. 2021a, b). As a result, the focus of meteorologists has been shifted to the development of machine learning (ML)-based physical-empirical forecasting models in recent years. ANNs are one of the numerous possibilities explored and presented in the literature for prediction. They are a flexible and rich concept that may be used to tackle difficulties with clustering, time series, and function approximation in addition to classification tasks (Rautela et al. 2022). The adaptability of ANNs encouraged researchers to look into their suitability for classification and regression problems (Vu et al. 2019). Studies reveal that ANNs and other artificial intelligence techniques are capable of outperforming conventional statistical techniques. The Bayesian assessment of an ANN for prediction indicates that the performance of the neural architecture is critical to its configuration, because it strongly influences the estimate efficacy of the framework (Burden and Winkler 2008). To avoid over fitting, however, in this study we concentrate on Bayesian regularization (BR) of the ANN employed for predicting monthly rainfall (Okut 2016). The huge amount of complex nonlinear meteorological data from various sources necessitate great care to avoid over fitting ANN algorithm (Ye et al. 2021). We have implemented BR-ANN to explore the nonlinear relationships associated with monthly rainfall at three stations and predict it for the next.

15.3 Methodology

The cumulative distribution function (CDF) and quantile estimator of six distributions (viz., N2, LN2, P3, LP3, EV1, and GEV) adopted in EVA is presented in Table 15.1. The empirical equations involved in determining the MoM, MLM, and

Table 15.1 CDF with quantile estimator six probability distributions

Distribution	CDF (F(x) or F(y))	Quantile estimator x(T)
N2 (μ(x), σ(x))	$F(x)=\varphi\left(\frac{x-\mu(x)}{\sigma(x)}\right)$	$x(T)=\mu(x)+K(T)\sigma(x)$
LN2 (μ(y), σ(y))	$F(y)=\varphi\left(\frac{y-\mu(y)}{\sigma(y)}\right)$ wherein y = ln(x)	$x(T)=\exp(\mu(y)+K(T)\sigma(y))$
P3 (ξ, α, β)	$F(x)=\begin{cases} G\left(\beta,\frac{x-\xi}{\alpha}\right) & ,\alpha>0 \\ 1-G\left(\beta,\frac{x-\xi}{\alpha}\right) & ,\alpha<0 \end{cases}$	No explicit expression of the quantile function is available
LP3 (ξ, α, β)	$F(x)=\begin{cases} G\left(\beta,\frac{\ln(x)-\xi}{\alpha}\right) & ,\alpha>0 \\ 1-G\left(\beta,\frac{\ln(x)-\xi}{\alpha}\right) & ,\alpha<0 \end{cases}$	No explicit expression of the quantile function is available
EV1 (ξ, α)	$F(x)=e^{-e^{-\left(\frac{x-\xi}{\alpha}\right)}},\ \alpha>0$	$x(T)=\xi+\alpha[-\ln(-\ln(F(x)))]$
GEV (ξ, α, β)	$F(x)=e^{-\left(1-\frac{\beta(x-\xi)}{\alpha}\right)^{1/\beta}},\ \alpha>0,\beta>0$	$x(T)=\xi+\frac{\alpha\left[1-(-\ln(F(x)))^{\beta}\right]}{\beta}$

Wherein

ξ	:	Location parameter
α	:	Scale parameter
β	:	Shape parameter
T	:	Return period (in year)
F(x)	:	CDF of a variable x (i.e., AMR)
K(T)	:	Frequency factor of a return period (T) corresponding to the coefficient of skewness (CS), say CS = 0.0 for N2 and LN2
x(T)	:	Estimated extreme rainfall for a return period (T)
G(...)	:	Incomplete gamma integral
ϕ(...)	:	CDF of standard normal distribution
μ(x)	:	Average of observed data (x)
σ(x)	:	Standard deviation of observed data (x)
μ(y)	:	Average of log transformed (y = ln(x)) series of observed data
σ(y)	:	Standard deviation of log transformed series of observed data

A relation between F(x) and T is given by F(x) = 1–1/T

LMO estimators of the distributions are presented in Table 15.2 (Rao and Hamed 2000).

15.3.1 MoM of P3 Distribution

The MoM estimators of P3 distribution (Bobee and Askhar 1991) can be determined by solving the system of equations, which is given as below:

Table 15.2 Determination of parameters of six distributions using MoM, MLM, and LMO

Distribution	Parameters of the distribution		
	MoM	MLM	LMO
N2	$\mu(x)=\frac{1}{N}\sum_{i=1}^{N}x(i)-\frac{1}{2N}\sum_{i=1}^{N}x(i)^2+\frac{1}{2}\left(\frac{1}{N}\sum_{i=1}^{N}x(i)\right)^2$ $\sigma(x)=\left[\left(\frac{1}{N}\sum_{i=1}^{N}x(i)^2\right)-2\left(\frac{1}{N}\sum_{i=1}^{N}x(i)\right)\right]^{1/2}$	$\mu(x)=\frac{1}{N}\sum_{i=1}^{N}x(i)$ $\sigma(x)=\left(\frac{1}{N}\sum_{i=1}^{N}(x(i)-\mu(x))^2\right)^{1/2}$	$\lambda_1=\mu(x)$ $\lambda_2=\sigma(x)/\sqrt{N}$
LN2	$\mu(y)=\frac{1}{N}\sum_{i=1}^{N}y(i)-\frac{1}{2N}\sum_{i=1}^{N}y(i)^2+\frac{1}{2}\left(\frac{1}{N}\sum_{i=1}^{N}y(i)\right)^2$ $\sigma(y)=\left[\frac{1}{N}\sum_{i=1}^{N}y(i)^2-2\left(\frac{1}{N}\sum_{i=1}^{N}y(i)\right)\right]^{1/2}$	$\mu(y)=\frac{1}{N}\sum_{i=1}^{N}y(i)$ where $y(i)=\ln(x(i))$ $\sigma(y)=\left(\frac{1}{N}\sum_{i=1}^{N}(y(i)-\mu(y))^2\right)^{1/2}$	$\lambda_1=\mu(y)$ $\lambda_2=\sigma(y)/\sqrt{N}$
P3	Can be determined by solving (Eqs. 15.1 and 15.2)	Can be determined by solving Eq. (15.3)	No explicit expression is available
LP3	Can be determined by solving (Eqs. 15.1 and 15.2) by replacing x(i) with ln(x(i))	Can be determined by solving Eq. (15.3) by replacing x (i) with ln(x(i))	No explicit expression is available
EV1	$\xi=\mu(x)-(0.5772157)\alpha$ $\alpha=\left(\frac{\sqrt{6}}{\pi}\right)\sigma(x)$	$\xi=-\alpha\ln\left[\sum_{i=1}^{N}\exp(-x(i)/\alpha)/N\right]$ $\alpha=\mu(x)-\left[\sum_{i=1}^{N}x(i)\exp(-x(i)/\alpha)/\sum_{i=1}^{N}\exp(-x(i)/\alpha)\right]$	$\xi=\lambda_1-(0.5772157)\alpha$ $\alpha=\frac{\lambda_2}{\ln(2)}$
GEV	$\mu(x)=\xi+\frac{\alpha(1-\Gamma(1+\beta))}{\beta}$ $\sigma(x)=\frac{\alpha}{\beta}\left(\Gamma(1+2\beta)-\Gamma(1+\beta)^2\right)^{1/2}\mathrm{CS}=(\mathrm{sign}\,\beta)\frac{\Gamma(1+3\beta)+3\Gamma(1+\beta)(1+2\beta)-2\Gamma^3(1+\beta)}{\{\Gamma(1+\beta)-\Gamma^2(1+\beta)\}^{1/2}}$	Can be determined by using the modified Newton-Raphson algorithm	$z=(2/(3+\tau_3))-(\ln(2)/\ln(3))$; $\tau_3=(2(1-3^{-\beta})/(1-2^{-\beta}))-3$; $\alpha=(\lambda_2\beta)/(1-2^{-\beta})\Gamma(1+\beta)$; $\xi=\lambda_1+(\alpha(\Gamma(1+\beta)-1)/\beta)$; $\beta=7.817740z+2.930462z^2+13.641492z^3+17.206675z^4$

$$M'_r = \frac{\exp(r\xi)}{(1 - r\alpha)^\beta} \tag{15.1}$$

where in 1-rα > 0, $r = 1, 2, 3$

$$\mu(x) = M'_1, \quad \sigma(x)^2 = M'_2 - (M'_1)^2 \text{and} \quad \gamma(x) = M'_3 - 3M'_2M'_1 + 2(M'_1)^3 \tag{15.2}$$

where M'_r is the rth moment of x about the origin and μ(x) is the coefficient of skewness of the observed data.

15.3.2 MLM of P3 Distribution

The MLM estimators of P3 distribution (Bobee and Askhar 1991) can be determined by solving the following system of equations:

$$\left.\begin{aligned} &\sum_{i=1}^{N} (x(i) - \xi) = N\alpha\beta \\ &\sum_{i=1}^{N} \frac{1}{\alpha}(x(i) - \xi) = N\psi(\beta) \\ &\sum_{i=1}^{N} \frac{1}{(x(i) - \xi)} = \frac{N}{\alpha(\beta - 1)} \end{aligned}\right\} \tag{15.3}$$

Here, ψ(β) is the digamma function of estimator of the scale parameter (β).

In Table 15.2, λ_1, λ_2, and λ_3 are the first, second, and third, respectively, LMOs (Hosking 1990) that can be determined in Eq. (15.4), which is given as below:

$$\lambda_1 = b_0, \quad \lambda_2 = 2b_1 - b_0, \quad \lambda_3 = 6b_2 - 6b_1 + b_0 \quad \text{and} \quad \tau_3 = \lambda_3/\lambda_2 \tag{15.4}$$

wherein λ_{r+1} is the r + 1th LMO (Hosking and Wallis 1993), which is defined by:

$$\lambda_{r+1} = \sum_{k=0}^{r} \frac{(-1)^{r-k}(r+k)!}{(k!)^2(r-k)!} b_k \tag{15.5}$$

wherein b_k is an unbiased estimator (Saf 2009; Gubareva and Gartsman 2010) and given by:

$$b_k = N^{-1} \sum_{i=k+1}^{N} \frac{(i-1)(i-2)\ldots\ldots(i-k)}{(N-1)(N-2)\ldots\ldots(N-k)} x(i) \tag{15.6}$$

where x(i) is the observed data of ith sample and N is the total number of samples.

15.3.3 Goodness-of-Fit Tests

GoF tests are essential for checking the adequacy of probability distributions to the AMR series in rainfall estimation. Out of a number GoF tests available, the widely accepted GoF tests are χ^2 and KS (Zhang 2002), which are used in the study.

χ^2 test statistic is defined by:

$$\chi^2 = \sum_{j=1}^{NC} \frac{(O_j(x) - E_j(x))^2}{E_j(x)} \tag{15.7}$$

where $O_j(x)$ is the observed frequency value of x for jth class, $E_j(x)$ is the expected frequency value of x for jth class, and NC is the number of frequency classes (Charles Annis 2009). The rejection region of χ^2 statistic at the desired significance level (η) is given by $\chi_C^2 \geq \chi_{1-\eta,NC-m-1}^2$. Here, m denotes the number of parameters of the distribution, and χ_C^2 is the computed value of χ^2 statistic by the probability distribution.

KS test statistic is defined by:

$$KS = \underset{i=1}{\overset{N}{Max}} \left| F_e(x(i)) - F_c(x(i)) \right| \tag{15.8}$$

where x(i) is the observed data for ith sample, $F_e(x(i)) = r/(N + 1)$ is the empirical CDF of x(i) of ith sample, "r" is the rank assigned to sample values arranged in ascending order (i.e., $x(1) < x(2) < \ldots..x(N)$), and $F_c(x(i))$ is the computed CDF of x(i) of ith sample.

Test criteria: If the computed values of GoF tests statistic given by the distribution are less than that of the theoretical values at the desired level of significance, then the distribution is considered to be acceptable for EVA at that level.

15.3.4 Diagnostic Test

Sometimes the GoF test results would not offer a conclusive inference, thereby posing a bottleneck for the user in selecting the suitable distribution for the application. In such cases, a diagnostic test in adoption to GoF is applied for making inference. The selection of the most suitable distribution is performed through the D-index test (United States Water Resources Council (USWRC) 1981), which is defined as:

$$\mathrm{D-index} = (1/\mu(x)) \sum_{i=1}^{6} |x(i) - x(i)^*| \tag{15.9}$$

Here, x(i) (i = 1 to 6) and $x(i)^*$ are the six highest observed and the corresponding estimated values of ith sample. The probability distribution having the least D-index is considered as a better suited for rainfall estimation.

15.3.5 Bayesian Regularized Artificial Neural Network

BR-ANN (Burden and Winkler 2008; Okut 2016) is a more robust variant of ANNs than the standard ANN (Sasireka et al. 2019). This robustness of BR-ANN is obtained through the BR of the ANN parameters. A popular error function (E_D) of ANN is as follows:

$$\mathrm{E_D}(\mathrm{D}|w, \mathrm{M}) = \sum_{i=1}^{n} (\mathrm{t}_i - \widehat{\mathrm{t}}_i)^2 \tag{15.10}$$

where w denotes weight, M denotes ANN structure, n denotes size of training data, t_i is the ith target output, and $\widehat{t}_i$is the ith model (BR-ANN) output.

Regularizing ANN with the Bayesian technique aids in optimizing the ANN parameter by utilizing prior ANN parameter values. In order to do this, an additional term (E_w) is added to the BR-ANN's target function as follows:

$$\mathrm{E_D}(\mathrm{D}|\mathrm{w}, \mathrm{M}) = \sum_{i=1}^{\mathrm{n}} (\mathrm{t}_i - \widehat{\mathrm{t}}_i)^2 + \mathrm{E}_w \tag{15.11}$$

In order to improve generalization and gradual conversion, E_w is employed to compensate the unrealistic weights. The function is minimized using an optimization technique based on gradients:

$$\mathrm{F} = \beta\ \mathrm{E_D}(\mathrm{D}|\mathrm{w}, \mathrm{M}) = \sum_{i=1}^{\mathrm{n}} (\mathrm{t}_i - \widehat{\mathrm{t}}_i)^2 + \alpha\ \mathrm{E_w}(\mathrm{w}|\mathrm{M}) \tag{15.12}$$

where the hyper-parameters that need to be optimized are represented by α and β and $E_w(w|M)$ represent sum of square of the ANN architecture. BR-ANN is an efficient predictive model, because it can uncover theoretically complex input-output relationships.

15.4 Application

In this paper, a study on the comparison of six probability distributions (viz., N2, LN2, P3, LP3, EV1, and GEV) adopted in the EVA of rainfall is carried out. MoM, MLM, and LMO determine the distribution's characteristics, which are also employed in the estimation of rainfall. The daily rainfall data recorded at the Afzalpur, Kalaburagi, and Aland locations from 1970 to 2018 and 1970 to 2017, respectively, are used. From the daily rainfall data, the AMR series is taken out and used for EVA CWPRS (2021) do not have data for the intermittent period, according to a review of the daily rainfall statistics. Additionally, it is highlighted that the observed rainfall at Kalaburagi, which was 1.5 mm in 1993 and 15.6 mm in 2015, is inconsistent and was not taken into account while analyzing the data. However, the data for the missing years are ignored and not taken into account in EVA because of the significance of the hydrological extremes. The AMR descriptive data are provided in Table 15.3. The index map shown in Fig. 15.1 shows the locations of the rain gauge stations taken into consideration for the investigation.

From the descriptive statistics (Table 15.3), it is noted that the higher order moments (CS and CK) of the AMR series behave differently for Afzalpur when compared to the values for Aland and Kalaburagi. The CV of the AMR series of Afzalpur, Aland, and Kalaburagi varies between about 27% and 67%, as shown in Table 15.3.

Table 15.3 Descriptive statistics of the observed AMR

Site	Average (mm)	SD (mm)	CS	CK	Minimum (mm)	Maximum (mm)
Afzalpur	79.1	52.6	5.061	30.614	37.1	400.3
Aland	83.9	35.6	0.935	1.368	30.1	190.0
Kalaburagi	81.7	22.3	0.367	−0.474	41.6	129.0

SD standard deviation, *CS* coefficient of skewness, *CK* coefficient of kurtosis

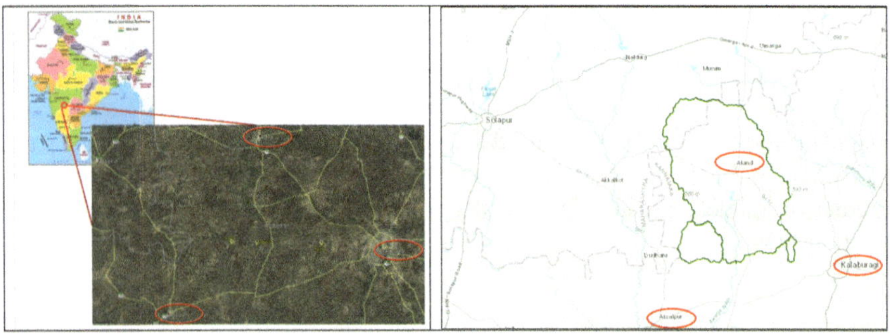

Fig. 15.1 Index map of the study area with locations of rain gauge stations

15.5 Results and Discussion

By applying the procedures as described above, a computer code was developed and used in the EVA of rainfall. The code computes the (1) parameters of N2, LN2, P3, LP3, EV1, and GEV (using MoM, MLM, and LMO) distributions; (2) extreme rainfall estimates for different return periods; and (3) GoF test statistic and D-index values.

15.5.1 *Estimation of Extreme Rainfall*

The MoM, MLM and LMO were used to determine the characterisitcs of N2, LN2, P3,LP3, EV1 and GEV distributions adopted in EVA, wherever applicable. These parameters are used in the following study. Tables 15.4, 15.5, and 15.6 provide estimates of the 1-day maximum rainfall for Afzalpur, Aland, and Kalaburagi for various return periods. The EVA and GoF test results of P3 (LMO) and LP3 (LMO) are not shown in Tables 15.4, 15.5, 15.6, and 15.7 due to the absence of LMO in P3 and LP3 distributions. From the EVA results, it can be seen that, when compared to the values of other distributions for the return periods ranging from 20 to 1000 years, the LP3 (MLM) offered higher estimates for Afzalpur and Aland and the EV1 (MLM) for Kalaburagi.

15.5.2 *Analysis of Results Based on GoF Tests*

Six distributions were used to compute the GoF test values for the AMR series of Afzalpur, Aland, and Kalaburagi. The results are shown in Table 15.7. In the current study, the number of frequency classes (NC) is taken into account to be six, and as a result, the degree of freedom (NC-m-1) is taken into account to be two for distributions with three parameters (m), namely, P3, LP3, and GEV, and three for distributions with two parameters (m), namely, N2, LN2, and EV1, when computing the two statistic values. According to the degree of freedom, the theoretical values at the 5% level of significance are observed to be 5.99 for P3, LP3, and GEV and 7.815 for N2, LN2, and EV1. Likewise, the theoretical values of the KS statistic at 5% level of significance with reference to the number of samples considered in EVA are observed as 0.196 for Afzalpur, 0.203 for Aland, and 0.200 for Kalaburagi. From the GoF test results, some of the observations drawn from the study were summarized and presented below:

- χ^2 test results didn't support the use of MoM, MLM, and LMO estimators of five distributions (viz., LN2, P3, LP3, EV1, and GEV) for the EVA of rainfall for Afzalpur and Aland.

Table 15.4 Estimated 1-day maximum rainfall for different return periods by six probability distributions for Afzalpur

Distribution	Method	Estimated 1-day maximum rainfall (mm) for a return period (in year)									
		2	5	10	20	25	50	100	200	500	1000
N2	MoM	79.1	123.3	146.4	165.5	171.1	187.0	201.4	214.5	230.4	241.5
	MLM	79.1	122.8	145.7	164.6	170.1	185.9	200.1	213.1	228.8	239.8
	LMO	79.1	107.5	122.3	134.6	138.1	148.4	157.6	166.0	176.2	183.3
LN2	MoM	65.8	109.5	143.0	178.1	189.9	228.1	269.1	312.9	375.7	427.2
	MLM	71.4	99.6	118.5	136.7	142.6	160.7	178.9	197.4	222.4	241.9
	LMO	71.4	97.2	114.1	130.3	135.4	151.3	167.1	183.1	204.4	218.4
P3	MoM	59.3	81.6	120.2	171.7	190.3	252.5	320.0	391.3	489.8	566.7
	MLM	60.8	83.7	123.2	176.0	195.0	258.8	328.0	401.1	502.0	580.9
	LMO	NA	NA	NA	NA	NA	NA	NA	NA	NA	NA
LP3	MoM	64.3	93.2	121.2	156.6	169.9	218.2	279.5	357.2	492.9	627.8
	MLM	80.2	111.0	144.1	188.2	205.3	269.5	354.6	467.6	675.4	893.1
	LMO	NA	NA	NA	NA	NA	NA	NA	NA	NA	NA
EV1	MoM	70.4	116.9	147.7	177.2	186.6	215.4	244.0	272.6	310.2	338.7
	MLM	72.2	98.8	116.5	133.4	138.8	155.3	171.7	188.1	209.7	226.0
	LMO	73.3	104.4	125.0	144.8	151.1	170.4	189.6	208.7	233.9	252.9
GEV	MoM	66.0	102.9	133.5	168.5	180.9	223.9	274.3	333.7	429.0	516.4
	MLM	68.1	95.2	117.6	143.2	152.3	183.5	220.1	263.1	331.9	394.8
	LMO	66.4	93.8	119.3	151.2	163.2	207.1	263.2	335.3	463.0	592.0

Table 15.5 Estimated 1-day maximum rainfall for different return periods by six probability distributions for Aland

Distribution	Method	Estimated 1-day maximum rainfall (mm) for a return period (in year)									
		2	5	10	20	25	50	100	200	500	1000
N2	MoM	83.9	113.8	129.5	142.4	146.2	157.0	166.7	175.5	186.3	193.8
	MLM	83.9	113.5	129.0	141.8	145.5	156.1	165.7	174.5	185.1	192.6
	LMO	83.9	113.0	128.2	140.8	144.5	155.0	164.4	173.0	183.5	190.8
LN2	MoM	77.2	108.8	130.1	150.8	157.4	178.0	198.9	220.1	248.9	271.4
	MLM	76.9	110.0	132.6	154.8	161.9	184.2	206.9	230.1	261.7	286.4
	LMO	76.9	110.7	134.0	156.9	164.2	187.3	210.8	234.9	267.8	289.7
P3	MoM	78.4	111.1	131.6	150.2	156.0	173.3	189.9	206.0	226.6	241.8
	MLM	77.8	110.8	131.8	151.2	157.2	175.4	192.9	209.9	231.8	248.0
	LMO	NA	NA	NA	NA	NA	NA	NA	NA	NA	NA
LP3	MoM	76.9	110.4	133.4	156.0	163.3	186.1	209.2	233.0	265.4	316.9
	MLM	75.9	111.9	135.2	158.1	165.4	188.5	212.0	236.0	268.8	321.0
	LMO	NA	NA	NA	NA	NA	NA	NA	NA	NA	NA
EV1	MoM	78.1	109.5	130.3	150.3	156.6	176.2	195.5	214.8	240.3	259.5
	MLM	78.0	109.8	130.9	151.1	157.5	177.2	196.8	216.3	242.0	261.5
	LMO	78.0	109.9	131.0	151.3	157.7	177.5	197.2	216.8	242.6	262.2
GEV	MoM	78.7	110.5	130.9	149.9	155.8	173.7	191.1	207.9	229.5	245.3
	MLM	78.3	109.8	130.3	149.7	155.8	174.4	192.6	210.6	233.9	251.3
	LMO	79.5	111.2	130.9	148.8	154.2	170.6	186.0	200.6	218.7	231.7

Table 15.6 Estimated 1-day maximum rainfall for different return periods by six probability distributions for Kalaburagi

		Estimated 1-day maximum rainfall (mm) for a return period (in year)									
Distribution	Method	2	5	10	20	25	50	100	200	500	1000
N2	MoM	81.7	100.4	110.2	118.3	120.7	127.4	133.5	139.1	145.8	150.5
	MLM	81.7	100.2	109.9	117.9	120.3	126.9	132.9	138.4	145.1	149.8
	LMO	81.7	100.7	110.7	118.9	121.3	128.2	134.3	140.0	146.8	151.6
LN2	MoM	78.8	98.7	111.1	122.4	126.0	136.6	147.0	157.1	170.4	180.4
	MLM	78.7	99.3	112.1	123.9	127.6	138.8	149.6	160.3	174.2	184.7
	LMO	78.7	99.9	113.3	125.6	129.4	141.1	152.4	163.6	178.3	190.5
P3	MoM	80.3	99.9	111.0	120.5	123.4	131.7	139.4	146.7	155.8	162.3
	MLM	80.1	99.6	110.7	120.4	123.3	131.8	139.7	147.2	156.6	163.4
	LMO	NA	NA	NA	NA	NA	NA	NA	NA	NA	NA
LP3	MoM	77.9	99.2	113.2	126.7	131.0	137.8	145.5	152.8	161.9	168.4
	MLM	78.7	99.5	112.5	124.5	128.2	139.5	150.5	161.4	175.5	186.2
	LMO	NA	NA	NA	NA	NA	NA	NA	NA	NA	NA
EV1	MoM	78.0	97.7	110.8	123.3	127.2	139.5	151.6	163.7	179.7	191.7
	MLM	78.1	99.8	114.2	128.0	132.4	145.9	159.3	172.7	190.3	203.6
	LMO	77.8	98.7	112.5	125.8	130.0	142.9	155.8	168.6	185.5	198.3
GEV	MoM	80.1	100.1	111.5	121.1	123.9	131.9	139.0	145.2	152.4	157.2
	MLM	82.5	103.2	114.8	124.7	127.6	135.9	143.1	149.6	157.0	161.9
	LMO	79.6	100.2	112.2	122.6	125.7	134.7	142.9	150.3	159.1	165.0

Table 15.7 Computed and values of GoF and diagnostic tests statistic by six probability distributions for Afzalpur, Aland, and Kalaburagi

		Computed values of GoF (viz., χ^2 and KS) and diagnostic (viz., D-index) tests								
		Afzalpur			Aland			Kalaburagi		
Distribution	Method	χ^2	KS	D-index	χ^2	KS	D-index	χ^2	KS	D-index
N2	MoM	37.250	0.234	4.828	5.267	0.088	1.866	1.739	0.075	0.452
	MLM	25.250	0.218	4.793	6.734	0.091	1.426	1.739	0.075	0.479
	LMO	13.250	0.201	3.714	7.200	0.094	1.871	1.739	0.076	0.413
LN2	MoM	22.250	0.202	4.510	9.000	0.114	1.813	0.696	0.051	0.387
	MLM	23.250	0.205	3.379	10.100	0.120	1.799	0.698	0.053	0.340
	LMO	24.250	0.208	3.381	11.200	0.126	1.812	0.700	0.055	0.320
P3	MoM	36.250	0.277	3.200	11.133	0.095	1.810	0.957	0.074	0.375
	MLM	37.250	0.278	3.280	11.135	0.098	1.077	0.975	0.078	0.390
	LMO	NA	NA	NA	NA	NA	NA	NA	NA	NA
LP3	MoM	8.250	0.118	3.287	8.525	0.105	1.898	2.783	0.069	0.301
	MLM	9.125	0.120	4.335	8.750	0.108	1.955	2.825	0.071	0.323
	LMO	NA	NA	NA	NA	NA	NA	NA	NA	NA
EV1	MoM	29.750	0.189	4.818	8.467	0.108	1.813	3.565	0.075	0.415
	MLM	30.250	0.192	3.404	8.725	0.110	1.030	3.675	0.078	0.362
	LMO	8.750	0.137	3.708	9.800	0.103	1.810	3.043	0.048	0.369
GEV	MoM	12.250	0.171	3.971	10.333	0.101	1.813	0.957	0.044	0.346
	MLM	7.250	0.125	3.234	10.333	0.100	1.075	0.827	0.060	0.383
	LMO	6.250	0.084	3.216	10.333	0.098	1.816	0.696	0.075	0.319

- χ^2 test results supported the use of MoM, MLM, and LMO estimators of all six distributions adopted in EVA of rainfall for Kalaburagi.
- KS test results didn't support the use of MoM, MLM, and LMO estimators of N2, LN2, and P3 distributions for EVA of rainfall for Afzalpur.
- KS test results confirmed the applicability of MoM, MLM, and LMO estimators of all six distributions adopted in the EVA of rainfall for Aland and Kalaburagi.

15.5.3 Analysis of Results Based on Diagnostic Test

In addition to the GoF test, the D-index was used to determine which of the six distributions in the EVA model best fit the criteria for estimating rainfall. These values were calculated using the N2, LN2, P3, LP3, EV1, and GEV distributions and are shown in Table 15.7. Based on the results of the diagnostic tests, it can be deduced that the D-index values of P3 (MoM) for Afzalpur, EV1 (MLM) for Aland, and LP3 (MoM) for Kalaburagi are less than those of other distributions used in the EVA.

15.5.4 Selection of Probability Distribution

Based on the EVA results from the diagnostic tests and GoF quantitative assessment, it was determined that the analysis produced conflicting inferences, necessitating qualitative evaluation. As a result, the best fit for rainfall estimates was again evaluated using fitted curves of the estimated severe rainfall along with D-index values, and a decision was taken as a result.

- According to the results of the diagnostic tests, EVA could be performed using P3 (MoM) for Afzalpur, EV1 (MLM) for Aland, and LP3 (MoM) for Kalaburagi.
- The MoM estimators of the distributions, however, are frequently less precise than MLM and LMO, as were previously mentioned. As a result, while choosing the best fit for estimating rainfall in Afzalpur and Kalaburagi, the D-index values obtained from P3 (MoM) and LP3 (MoM) are not taken into account.
- In light of the foregoing, it is determined that the D-index value of GEV (LMO) is the second subsequent minimum for Afzalpur and Kalaburagi after excluding the D-index values obtained from MoM of P3 and LP3 distributions from the selection.
- The GEV (LMO) is more matched among the six distributions chosen in EVA for rainfall estimation for Afzalpur and Kalaburagi, whereas EV1 (MLM) is better suited for Aland, according to the qualitative assessment (plots of EVA results) of the outcomes. Figure 15.2 shows the plots of the estimated 1-day maximum rainfall with 95% confidence limits based on the chosen distribution and actual AMR data for the Afzalpur, Aland, and Kalaburagi sites.

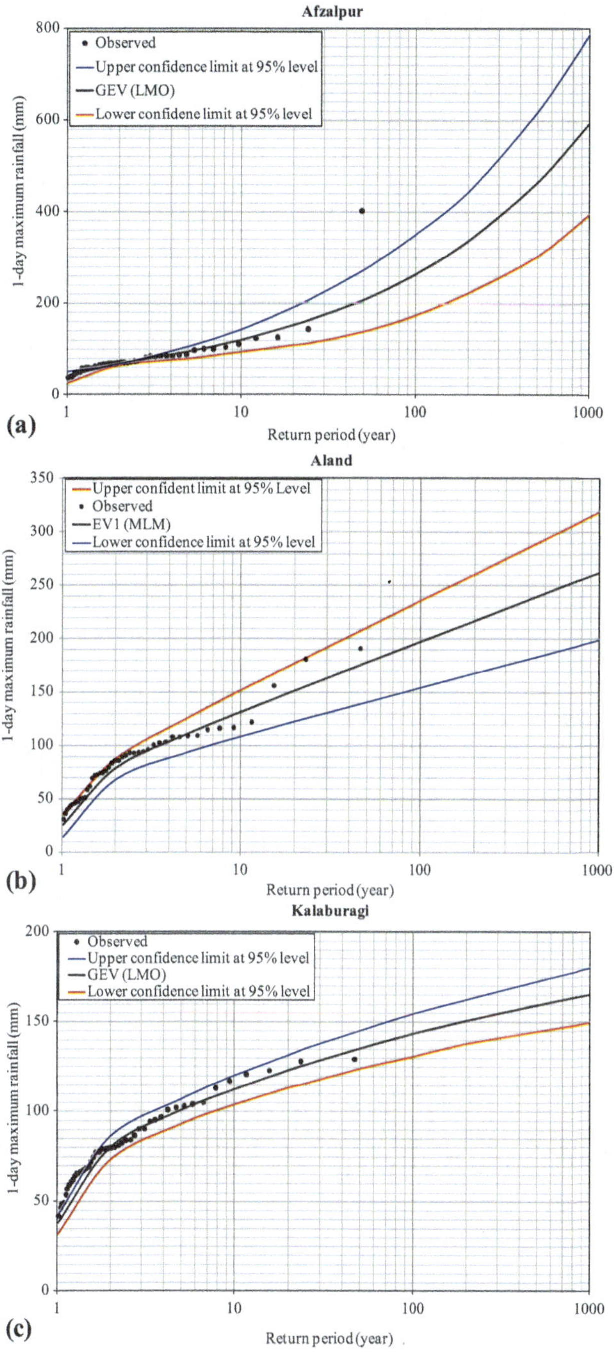

Fig. 15.2 Plots of estimated 1-day maximum rainfall by the selected distribution and observed AMR data for Afzalpur, Aland, and Kalaburagi

15.5.5 Efficiency Analysis of BR-ANN

Since a neural network demands comparatively more data to recognize the probable pattern (correlation) associated in a time series, the monthly rainfall dataset of all three sites, Afzalpur, Aland, and Kalaburagi, have been taken for 64 years, starting from January 1951 to December 2014. We have distributed this data for all three sites in the percentage of 70:30 for training and validation (testing) purposes to avoid overfitting in the performance evaluation of the predicting capability of ANN.

The performance of the ANN model with the BR technique used in predicting the targeted output has been expressed for training, validation, and the collective monthly rainfall data of all three sites, as shown in Fig. 15.3 (Afzalpur), Fig. 15.4 (Aland), and Fig. 15.5 (Kalaburagi). The efficiency of predicting outputs and targets through fit and errors has also been presented collectively for training and testing for all three sites. The model shows a high correlation between output and targeted input monthly rainfall data while training for all three stations. Since the model was trained for the training dataset, it has a higher correlation of approximately 0.95 between

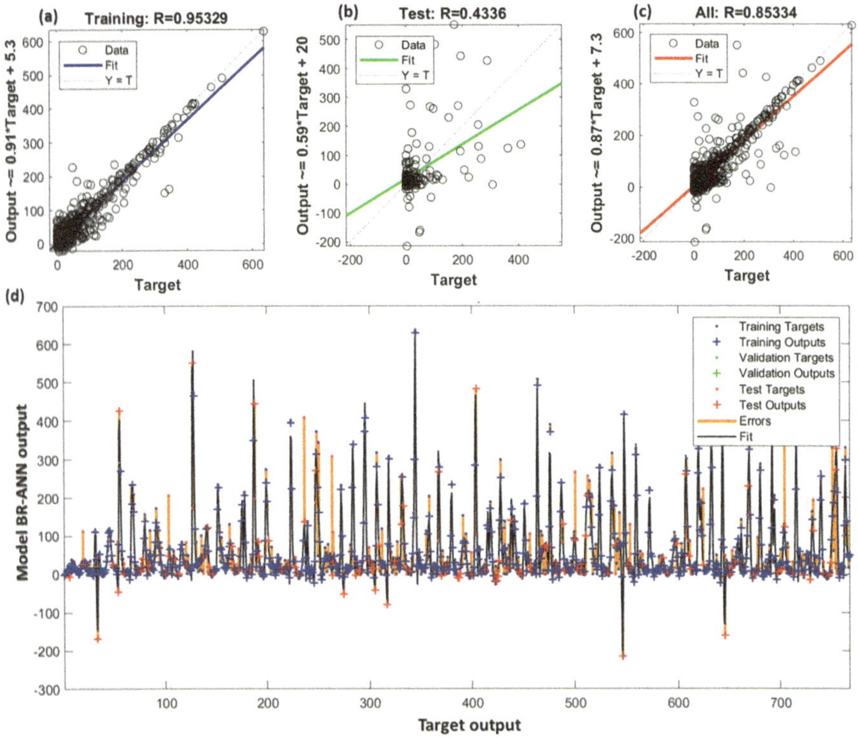

Fig. 15.3 Scatter plot showing correlation between the outputs of BR-ANN and input monthly data (target) of (**a**) training dataset; (**b**) testing dataset; and (**c**) all dataset of Afzalpur; (**d**) time series plot of predicted monthly rainfall using BR-ANN

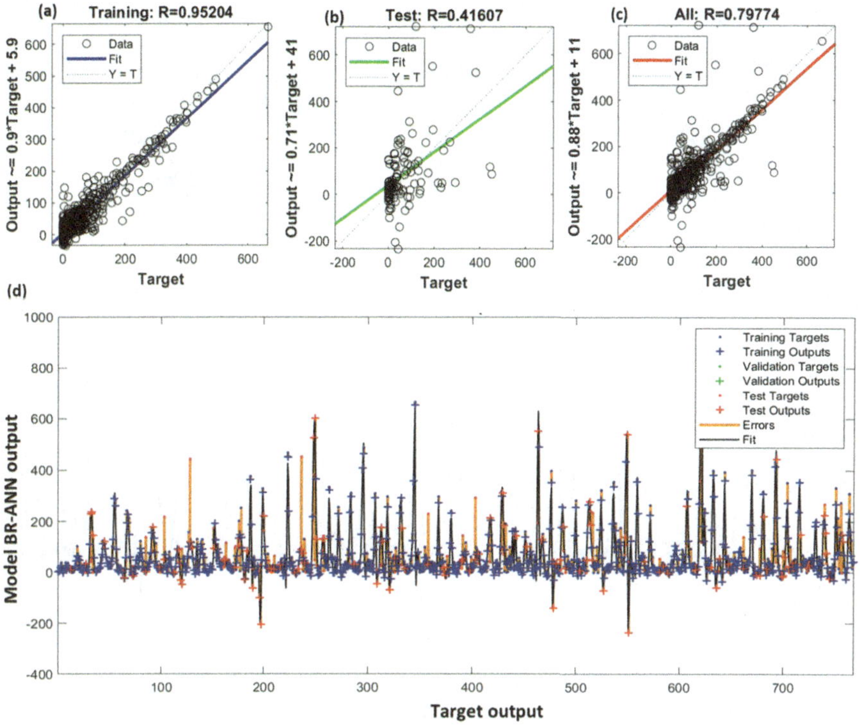

Fig. 15.4 Scatter plot showing correlation between the outputs of BR-ANN and input monthly data (target) of (**a**) training datasets; (**b**) testing dataset; and (**c**) all the dataset of Aland; (**d**) time series plot of predicted monthly rainfall using BR-ANN

model output and target, whereas it gets reduced to approximately 0.40 in the case of validation (testing).

15.5.6 Conclusions

This study compared the MoM, MLM, and LMO estimators of six probability distributions (N2, LN2, P3, LP3, EV1, and GEV) used in the EVA of rainfall for Afzalpur, Aland, and Kalaburagi with the aim of identifying the best distribution for rainfall estimation through quantitative (GoF tests using χ^2 and KS, and diagnostic test using D-index) and qualitative (fitted curves of the estimated rainfall) assessments. The monthly rainfall patterns of all three stations have also been proposed to model and predict using an ANN with Bayesian regularization technique. The inferences made from the study were condensed and are given below based on the EVA data and BR-ANN predictions:

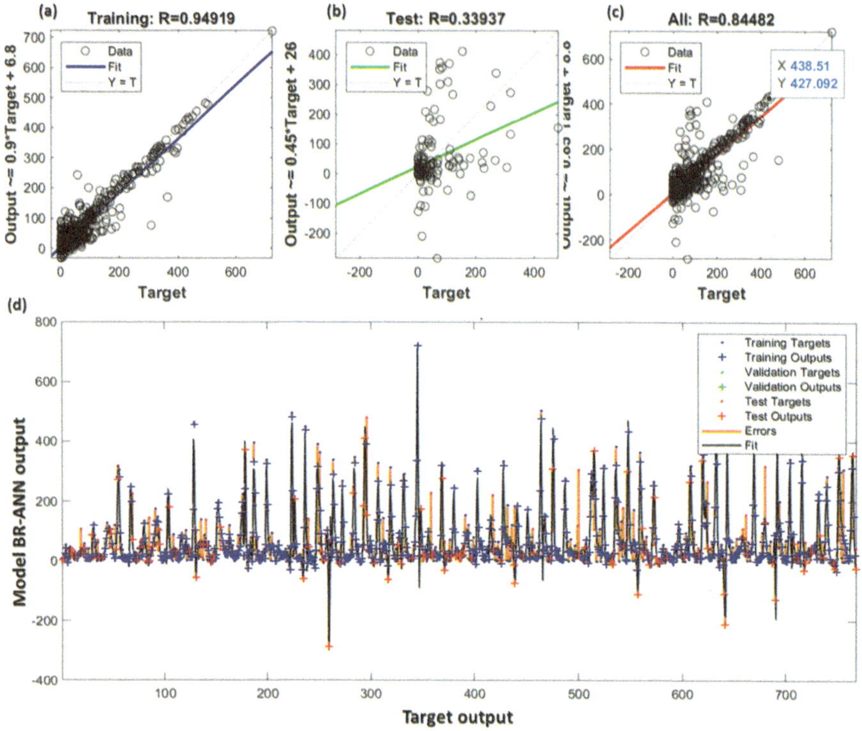

Fig. 15.5 Scatter plot showing correlation between the outputs of BR-ANN and input monthly data (target) of (**a**) training datasets; (**b**) testing dataset; and (**c**) all the dataset of Kalaburagi; (**d**) time series plot of predicted monthly rainfall using BR-ANN

- χ^2 test results didn't support the use of all six distributions for EVA of rainfall for Afzalpur. For Aland, the χ^2 test result indicates the N2 distribution is acceptable for EVA.
- χ^2 test results supported the use of all six distributions for EVA of rainfall for Kalaburagi.
- KS test results didn't support the use of N2, LN2, and P3 distributions for EVA of rainfall for Afzalpur.
- KS test results confirmed the applicability of all six distributions for EVA of rainfall for Aland and Kalaburagi.
- The GEV (LMO) is the superior choice among the six distributions chosen in EVA for rainfall estimation for Afzalpur and Kalaburagi, while EV1 (MLM) is better suited for Aland, according to the qualitative assessment (plots of EVA findings) of the outcomes, which was weighed with D-index values.
- The BR-ANN model has shown high correlation (approximately 0.95) in predicted output values and targeted input data at monthly scale from all three sites while training the datasets.

- The proposed model has shown comparatively lesser correlation between output and target values in validation, which means the model can be further optimized by hyper tuning the Bayesian regularization parameters.

By considering the data length (i.e., 48 years for Afzalpur, 45 years for Aland, and 46 years for Kalaburagi) of the AMR series available for the study, the study suggested that the estimated rainfall for the return period beyond 200 years may be cautiously used due to uncertainty in the higher-order return periods while designing the hydraulic structures in the respective sites. Some other machine learning algorithms, such as support vector machine (regression), Bayesian linear regression, etc., and some deep learning models, such as recurrent neural network, long short-term memory, etc., can also be explored to model nonstationary and complex rainfall patterns in a comparative analysis.

Acknowledgments The authors are grateful to the Director, Central Water and Power Research Station, Pune, for providing the research facilities to carry out the study. The authors are thankful to the India Meteorological Department for the supply of rainfall data used in the study.

References

Abida H, Ellouze M (2008) Probability distribution of flood flows in Tunisia. Hydrol Earth Syst Sci 12(3):703–714. https://doi.org/10.5194/hess-12-703-2008

Acar R, Celik S, Senocak S (2008) Rainfall intensity-duration-frequency (IDF) model using an artificial neural network approach. J Sci Ind Res 67(3):198–202

Afungang R, Bateira C (2016) Statistical modelling of extreme rainfall, return periods and associated hazards in the Bamenda Mountain, NW. Geogr Spat Plan J 1(9):5–19. https://doi.org/10.17127/got/2016.9.001

AlHassoun SA (2011) Developing empirical formulae to estimate rainfall intensity in Riyadh region. J King Saud Univ Eng Sci 23(1):81–88. https://doi.org/10.1016/j.jksues.2011.03.003

Arvind G, Kumar PA, Girish Karthi S, Suribabu CR (2017) Statistical analysis of 30 years rainfall data: a case study. Proceedings of IOP Conference Series: Earth and Environmental Science. 80, 012067. https://doi.org/10.1088/1755-1315/80/1/012067

Baghel H, Mittal HK, Singh PK, Yadav KK, Jain S (2019) Frequency analysis of rainfall data using probability distribution models. Int J Curr Microbiol Appl Sci 8(6):1390–1396. https://doi.org/10.20546/ijcmas.2019.806.168

Baratti E, Montanari A, Castellarin A, Salinas JL, Viglione A, Bezzi A (2012) Estimating the flood frequency distribution at seasonal and annual time scales. Hydrol Earth Syst Sci 16(12): 4651–4660. https://doi.org/10.5194/hess-16-4651-2012

Bauer P, Thorpe A, Brunet G (2015) The quiet revolution of numerical weather prediction. Nature 525:47. https://doi.org/10.1038/nature14956

Bhuyan A, Borah M, Kumar R (2010) Regional flood frequency analysis of north-bank of the river Brahmaputra by using LH-moments. Water Resour Manag 24(9):1779–1790. https://doi.org/10.1007/s11269-009-9524-0

Bobee B, Askhar F (1991) The gamma family and derived distributions applied in hydrology. Water Resources Publications, Littleton, CO

Burden F, Winkler D (2008) Bayesian regularization of neural networks, pp 23–42. https://doi.org/10.1007/978 1 60327-101-1_3

Central Water Commission (2010) Development of hydrological design aids (surface water): State of the Art Report (Report No.2009097/WR/REP-02). Central Water Commission, New Delhi
Charles Annis PE 2009 Goodness-of-fit tests for statistical distributions
Chen, Y., & Zhang, N. (2022). Optimal subsampling for large sample ridge regression
Coles S (2001) An introduction to statistical modeling of extreme values. Springer, London. https://doi.org/10.1007/978-1-4471-3675-0
CWPRS (2021) Desk Studies on Estimation of Flood at Various Railway Bridges on Daund-Kalaburagi Line Doubling Project for Railway Vikas Nigam Limited, Mumbai. Technical Report No. 5940, July 2021, Pune
Das J, Jha S, Goyal MK (2020) On the relationship of climatic and monsoon teleconnections with monthly precipitation over meteorologically homogenous regions in India: Wavelet & global coherence approaches. Atmos Res 238:104889. https://doi.org/10.1016/j.atmosres.2020.104889
Dueben PD, Bauer P, Adams S (2021) Deep learning to improve weather predictions. In: Deep learning for the earth sciences. Wiley, Hoboken, NJ, pp 204–217. https://doi.org/10.1002/9781119646181.ch14
Esberto MDP (2018) Probability distribution fitting of rainfall patterns in Philippine regions for effective risk management. Environ Ecol Res 6(3):178–186. https://doi.org/10.13189/eer.2018.060305
Esteves LS (2013) Consequences to flood management of using different probability distributions to estimate extreme rainfall. J Environ Manag 115(1):98–105. https://doi.org/10.1016/j.jenvman.2012.11.013
Floods E (2019) Extreme floods and droughts under future climate scenarios, pp 1–5
Goswami UP, Hazra B, Goyal MK (2018) Copula-based probabilistic characterization of precipitation extremes over North Sikkim Himalaya. Atmos Res 212:273–284. https://doi.org/10.1016/j.atmosres.2018.05.019
Goyal MK, Ojha CSP (2010) Evaluation of various linear regression methods for downscaling of mean monthly precipitation in arid Pichola watershed. Nat Res 01(01):11–18. https://doi.org/10.4236/nr.2010.11002
Goyal MK, Ojha CSP, Burn DH (2012) Nonparametric statistical downscaling of temperature, precipitation, and evaporation in a semiarid region in India. J Hydrol Eng 17(5):615–627. https://doi.org/10.1061/(ASCE)HE.1943-5584.0000479
Goyal MK, Panchariya VK, Sharma A, Singh V (2018) Comparative assessment of SWAT model performance in two distinct catchments under various DEM scenarios of varying resolution, sources and resampling methods. Water Resour Manag 32(2):805–825. https://doi.org/10.1007/s11269-017-1840-1
Gubareva TS, Gartsman BI (2010) Estimating distribution parameters of extreme hydrometeorological characteristics by L-moments method. Water Resour 37(4):437–445. https://doi.org/10.1134/S00978078100.40020
Haberlandt U, Radtke I (2014) Hydrological model calibration for derived flood frequency analysis using stochastic rainfall and probability distributions of peak flows. Hydrol Earth Syst Sci 18(1):353–365. https://doi.org/10.15488/602
Hewitson BC, Daron J, Crane RG, Zermoglio MF, Jack C (2014) Interrogating empirical-statistical downscaling. Clim Chang 122:539. https://doi.org/10.1007/s10584-013-1021-z
Hinge G, Surampalli RY, Goyal MK (2018) Prediction of soil organic carbon stock using digital mapping approach in humid India. Environ Earth Sci 77(5):172. https://doi.org/10.1007/s12665-018-7374-x
Hosking JRM (1990) L-moments: analysis and estimation of distributions using linear combinations of order statistics. J R Stat Soc 52(1):105–124. https://doi.org/10.1111/j.2517-6161.1990.tb01775.x
Hosking JRM, Wallis JR (1993) Some statistics useful in regional frequency analysis. Water Resour Res 29:271–281. https://doi.org/10.1029/92WR01980

Katz RW (2013) Statistical methods for nonstationary extremes. In: Extremes in a changing climate. Springer, Dordrecht, pp 15–37. https://doi.org/10.1007/978-94-007-4479-0_2

Malekinezhad H, Nachtnebel HP, Klik A (2011) Regionalization approach for extreme flood analysis using L-moments. Agric Sci Technol 13:1183–1196

Mujere N (2011) Flood frequency analysis using the Gumbel distribution. J Comput Sci Eng 3(7): 2774–2778

Nardi KM, Barnes EA, Ralph FM (2018) Assessment of numerical weather prediction model reforecasts of the occurrence, intensity, and location of atmospheric rivers along the West Coast of North America. Mon Weather Rev 146(10):3343–3362. https://doi.org/10.1175/MWR-D-18-0060.1

Okut H (2016) Bayesian regularized neural networks for small n big p data. In: Artificial neural networks—models and applications. InTech, Lahore. https://doi.org/10.5772/63256

Olumide BA, Saidu M, Oluwasesan A (2013) Evaluation of best fit probability distribution models for the prediction of rainfall—runoff volume (case study Tagwai dam, Minna-Nigeria). Int J Eng Technol 3(2):94–98

Pisner DA, Schnyer DM (2020) Chapter 6—support vector machine. In: Mechelli A, Vieira SBT-ML (eds) Machine learning. Academic Press, Cambridge, MA, pp 101–121. https://doi.org/10.1016/B978-0-12-815739-8.00006-7

Poonia V, Goyal MK, Gupta BB, Gupta AK, Jha S, Das J (2021a) Drought occurrence in different river basins of India and blockchain technology based framework for disaster management. J Clean Prod 312:127737. https://doi.org/10.1016/j.jclepro.2021.127737

Poonia V, Jha S, Goyal MK (2021b) Copula based analysis of meteorological, hydrological and agricultural drought characteristics across Indian river basins. Int J Climatol 41(9):4637–4652. https://doi.org/10.1002/joc.7091

Rao AR, Hamed KH (2000) Flood frequency analysis. CRC Publications, New York, NY

Rasel M, Hossain SM (2015) Development of rainfall intensity duration frequency equations and curves for seven divisions in Bangladesh. Int J Sci Eng Res 6(5):96–101

Rautela KS, Kumar D, Gandhi BGR, Kumar A, Dubey AK (2022) Application of ANNs for the modeling of streamflow, sediment transport, and erosion rate of a high-altitude river system in Western Himalaya, Uttarakhand. RBRH 27:e22. https://doi.org/10.1590/2318-0331.272220220045

Saf B (2009) Regional flood frequency analysis using L-moments for the west mediterranean region of Turkey. Water Resour Manag 23(3):531–551. https://doi.org/10.1007/s11269-008-9287-z

Sasireka K, Suribabu CR, Neelakantan TR (2019) Extreme rainfall return periods using Gumbel and gamma distributions. Int J Recent Technol Eng 8(4):27–29. https://doi.org/10.35940/ijrte.D1007.1284S219

Scher S, Messori G (2018) Predicting weather forecast uncertainty with machine learning. Q J R Meteorol Soc 144(717):2830–2841. https://doi.org/10.1002/qj.3410

Sharma A, Goyal MK (2020) Assessment of drought trend and variability in India using wavelet transform. Hydrol Sci J 65(9):1539–1554. https://doi.org/10.1080/02626667.2020.1754422

Sharma NK, Sharma S (2019) Frequency analysis of rainfall data of Dharamshala region. Int J Sci Res 8(2):886–892. https://doi.org/10.21275/ART20195211

Singh S, Goyal M, Jha S (n.d.) Role of large-scale climate oscillations in precipitation extremes associated with atmospheric rivers: nonstationary framework. Hydrol Sci J 68:395. https://doi.org/10.1080/02626667.2022.2159412

Su X, Yan X, Tsai C-L (2012) Linear regression. Wiley Interdiscip Rev Comput Stat 4(3):275–294. https://doi.org/10.1002/wics.1198

Tank G, Dongre P, Obi Reddy GP, Sen P (2021) Rainfall trend analysis—a review. Int Res J Eng Technol 8(4):4028–4030

United States Water Resources Council (USWRC) (1981) Guidelines for determining flood flow frequency. In: Bulletin 17A. U.S Geological Survey, Washington, D.C

Vergés A, Doropoulos C, Malcolm HA, Skye M, Garcia-Pizá M, Marzinelli EM, Vila-Concejo A (2016) Long-term empirical evidence of ocean warming leading to tropicalization of fish

communities, increased herbivory, and loss of kelp. Proc Natl Acad Sci 113:13791. https://doi.org/10.1073/pnas.1610725113
Vivekanandan N (2020) A comparative study on Gumbel and LP3 probability distributions for estimation of extreme rainfall. Int J Water Res Eng 6(1):21–33
Vivekanandan N, Srishailam C (2020) Selection of best fit probability distribution for extreme value analysis of rainfall. Water Energy Int 63(10):13–19
Vu HL, Ng KTW, Bolingbroke D (2019) Time-lagged effects of weekly climatic and socio-economic factors on ANN municipal yard waste prediction models. Waste Manag 84:129–140. https://doi.org/10.1016/j.wasman.2018.11.038
Ye L, Jabbar SF, Abdul Zahra MM, Tan ML (2021) Bayesian regularized neural network model development for predicting daily rainfall from sea level pressure data: investigation on solving complex hydrology problem. Complexity 2021:1–14. https://doi.org/10.1155/2021/6631564
Zhang J (2002) Powerful goodness-of-fit tests based on the likelihood ratio. J R Stat Soc 64(2): 281–294. https://doi.org/10.1111/1467-9868.00337

Chapter 16
An Indexing Method for Evaluating Managerial Effectiveness of a Watershed Project and Functional Involvement of Participant Organizations

Bhabesh Mahanta, Arup Kumar Sarma, and Sashindra Kumar Kakoty

Abstract Integrated Watershed Management Program (IWMP) envisages multi-stakeholder participation in different watershed management processes. Usually, a conglomeration of participant community-based organizations (CBOs) manages an IWMP project. Therefore, the managerial effectiveness of the organizations towards performing essential management functions is vital. In this study, we propose two indexing methods: (a) a project management effectiveness index (PMEI), to gauge the degree of overall managerial effectiveness of a watershed project by measuring what management functions covered in the project, and (b) an organizational involvement index (OII), to gauge the degree of managerial involvement of various participant organizations, by measuring the number of management functions performed by all organizational elements.

Firstly, we derived the universal management processes and functions for any human endeavor from the functional theory of management. Then, we undertook a structured open-ended questionnaire survey among the randomly selected inhabitants of an IWMP project area in Assam to determine the management functions of different participant organizations. The indexing system is built on an analysis of the quantified opinion of the respondents.

The project effectiveness index is 47.03%, and the organizational involvement index is 45.55% for the surveyed project. Notably, the management function share is unevenly distributed, and most surprisingly, two participant organizations have zero managerial involvement.

B. Mahanta (✉)
School of Agro and Rural Technology IIT Guwahati, Guwahati India

e-mail: bhabesh@iitg.ac.in

A. K. Sarma
Civil Engineering IIT Guwahati, Guwahati India

S. K. Kakoty
Mechanical Engineering IIT Guwahati, Guwahati India

A. K. Gupta et al. (eds.), *Ecosystem Restoration: Towards Sustainability and Resilient Development*, Disaster Resilience and Green Growth,
https://doi.org/10.1007/978-981-99-3687-8_16

This management function indexing system will help the watershed planners track and compare the management environment at different points of time or compare the functional status of various projects. In addition, it will facilitate the design of suitable interventions for the management process reengineering for more congruent function sharing.

Keywords Watershed indexing · Management · Process reengineering

16.1 Introduction

The Integrated Watershed Management Program (IWMP) has envisaged participatory watershed management by building and developing community-based organizations (CBO). These CBOs are an integral part of all IWMP projects, and therefore, the overall success of these programs depends on how efficiently CBOs are roped into the watershed organizational structure (Fawcett et al. 1995; Sreedevi et al. 2008; Kozlowski and Bell 2012; FAO 2017). However, a conglomerate of organizations with inefficient function share cannot induce management transfer of created assets, however lofty the objectives might be.

So, the question is: Are the present participant organizations getting involved in the process as expected?

16.1.1 Integrated Watershed Management Program (IWMP) Organizational Structure

Many states in India have been implementing watershed projects under IWMP with significant objectives for achieving sustainable community participation (NITI Aayog 2019). The Indian planning authority, NITI Aayog, has formed the State Level Nodal Agency (SLNA) for the overall planning, management, and monitoring of IWMP projects, emphasizing the indispensability of CBOs in the organizational infrastructure (SLNA 2010a). In addition, IWMP guidelines acknowledge collaboration among local government organizations like Zila Parishads, Gaon Panchayatas, schools, and voluntary organizations (Gaur and Milne 2015). Table 16.1 shows some details of the prescribed participant organizations.

16.1.2 What Do the Managers Supposed to Do?

The term "management" refers to the process of getting things done effectively and efficiently, through and with other people. By the functional approach of management theory, managers plan, organize, lead, and control (Robbins et al. 2013; Goyal

Table 16.1 Present organizational structure in IWMP

IWMP participant	Members	Defined activity
Watershed Development Team (WDT)	The team members are a government employee	It provides technical assistance to watershed activities and oversees project implementation
Watershed Development Committee (WDC)	Ten members (including one chairman, one secretary, and members from general/SC/ST and other CBO selected by Gram Sabha)	Project implementation activities
Self-Help Group (SHG)	The village producers cooperative committee, Women's Group	Implementation of livelihood schemes
Villages/Users	Users of the watershed activities	Use of project deliverables
Gram Sabha (GS)	Local government representatives	Formation of WDC in collaboration with the PIA
Project Implementation Agency (PIA)	Government departmental officers selected by SLNA	Project planning to implementation

et al. 2018; Poonia et al. 2021). In the early part of the twentieth century, French industrialist Henry Fayol mentioned five categories of essential management processes or functions: planning, organizing, commanding, coordinating, and controlling (Hannan et al. 2003). These processes have generally been condensed to primary four: planning, organizing, leading, and controlling (Robbins et al. 2011). In the definition of management, two essential and related terms are efficiency and effectiveness. Generally, efficiency refers to minimizing resource utilization, and effectiveness means doing the right task to maximize goal attainment. Goal articulation is an essential step in infrastructure planning (Parkin and Sharma 1999).

In management theory, the planning component encompasses defining goals, establishing strategies, and developing plans to coordinate activities. The organizing component includes determining what tasks are to be done, who is to do them, how the tasks are grouped, who reports to whom, and where decisions are to be made. The leading component includes motivating employees, directing the activities of others, selecting the most effective communication channel, and resolving conflicts. Finally, the controlling element monitors performance, compares it with goals, and corrects any significant deviations (Carpenter et al. 1986; Ali et al. 2001; Darnall and Preston 2010).

Here a question arises: Are management activities universally applicable? Management theories are based on standard human behavior. Explaining management as a generic activity, Robbins et al. (2011) observes that "what a manager does should be essentially the same regardless of whether he or she is a top-level executive or a first-line supervisor; in a business firm or a government agency; in a large corporation or a small business; or located in Salt Lake Kolkata, or Salt lake city, USA."

16.1.3 State of Affairs in Watershed Management

A watershed is a complex infrastructure designed for natural resources and environmental management involving many biophysical and anthropogenic factors (Easterby-Smith and Lyles 2003; Conservation Ontario 2010; Bach et al. 2011). Moreover, a watershed project is a multi-stakeholder initiative to collaboratively govern water management issues by constituting some structured stakeholders' groups (2008). Therefore, the watershed organizational environment is a network of groups acting as an individual managerial unit. Thus, watershed planners should know that each constituent group should define internal management functions (Devine et al. 1999; Daspit et al. 2013; Wang et al. 2016). Also, it is expected that the groups finally perform as a cross-functional team with a collaborative work ethic.

The present performance of many IWMP projects shows that many predefined critical management functions remain unattended by organizational components, due to which the scope for better coordination remains underutilized. There are two significant reasons for this: (a) CBOs are largely unstructured with minimum defined functionality. (b) Allotted functions are not well implemented (Goyal and Ojha 2010, 2012; Das et al. 2020).

In that context, scrutinizing watershed organizational involvement is essential to study the scope of critical management functions by watershed organizations and analyze their present state of functioning. The primary question is: What kind of management functions the watershed organizations are undertaking against what they are supposed to do? Moreover, can there be an indexing method for evaluating the managerial effectiveness of a watershed project and participant organizations? Unfortunately, there is a shortage of research regarding the adaptability of standard management functions or indexing managerial effectiveness in a watershed organizational environment.

Therefore, we studied the present state of affairs in the management environment of an IWMP project by opinion survey among the beneficiaries. After analyzing the result, we have framed an indexing system to determine the effectiveness in sharing management functions among the watershed organizational elements and the overall managerial effectiveness of the project. The system might apply to any IWMP project to redesign the present organizational model for improving stakeholders' participation.

16.2 Materials and Methods

16.2.1 Study Area and Data Collection

We have reviewed the applicability of the functional theory of management to watershed management. We have identified essential management functions of an organization from literature review, IWMP project documents, and local observation. In each IWMP project, there are six active organizational components, namely, PIA, WDT, WDC, GS, SHG, and villagers. These components are supposed to undertake different management functions under four management processes or function categories: planning, organizing, leading, and controlling. For examining the management functions carried out by the present watershed organizational components under IWMP, we selected four project areas in the Brahmaputra Valley, Assam, namely, Turkunijan IWMP Kaldia IWMP, Satpokholi IWMP, and Maloibari IWMP (SLNA 2010a, b, c, 2011). To give a broader base to our observation, we conducted an opinion survey among 120 watershed beneficiaries with a close-ended questionnaire having a two point scale. Each person can give his opinion by choosing "y" for yes and "n" for no (numerically, $y = 1$ and $n = 0$). The questionnaires are distributed to a random sample of stakeholders in the project area.

We have established a final response table (as presented below) after summing up individual opinions. In the last response table, management function-wise scores of each component organization are denoted as Y or N.

For a question, if Σy denotes total numbers of "yes" responses and R represents whole numbers of respondents,

then the final response table score is Y, when (Σy/R) > =0.5. The score in the final response table is N, when (Σy/R) < 0.5. A Y against a function implies that, by the majority opinion, the function is carried out. Similarly, N against a function implies that, by the majority opinion, the function is not carried out. We suppose that, numerically, $Y = 1$ and $N = 0$.

16.2.2 Organizational Involvement Index (OII)

The share of management functions an individual organization undertakes reflects its involvement.

So, for any organization "X_i,"

The organizational involvement score of $X = OIX_i =$ (total management functions undertaken by X)/total management functions allotted to X.

The total involvement score by all component organizations will show the present state of affairs regarding function sharing. Therefore, for "m" numbers of organizations, the overall organizational involvement index (OII) is

$OII = \frac{\Sigma OIX_i}{m}$ where $i = 1, 2, \ldots, m$.

16.2.3 Project Managerial Effectiveness Index (PMEI)

The number of management functions carried out by the organizations will reflect the managerial effectiveness. There may be two types of management effectiveness index:

(a) Process-wise effectiveness index (PEI) and (b) the overall project management effectiveness index (PMEI). Therefore, for a process "P_i,"

$PEIP_i$ = (total numbers of management functions covered by all organizations in process P_i)/total functions under P_i.

Therefore, for the "n" number of management processes,

PMEI = (total functions covered score in all management processes by all organizations)/numbers of management processes $= \Sigma PEIP_i/n$, $i = 1, 2, \ldots, n$.

We have shown the results as a percentage of scores based on obtained positive responses to the maximum positive score in an ideal case.

We invited three experienced watershed experts and three watershed beneficiaries from CBOs for a validation interview to justify the result.

16.3 Results

Table 16.2 shows the summarized results based on primary field data.

Management Processes	PMEI
Planning	50
Organising	36.7
Leading	45.8
Controlling	55.6
Project Management Effectiveness Index (PMEI)	47.03

Organisations	OII
PIA	100
WDT	93.3
WDC	66.7
GS	33.3
VILLAGE	0
SHG	0
Organisational Involvement Index (OII)	45.55

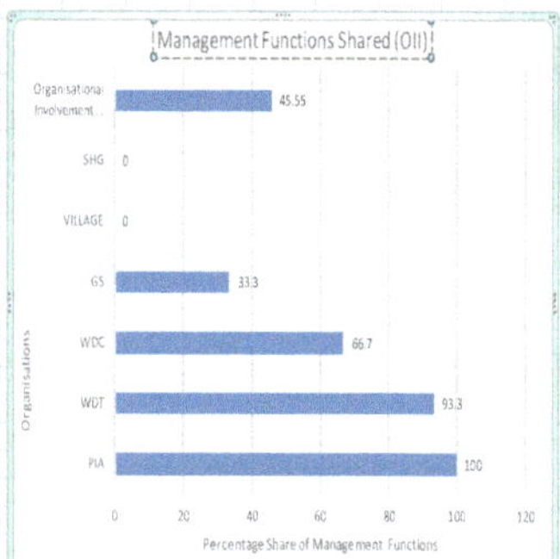

16.4 Discussion

16.4.1 Functional Involvement of Participant Organizations

Since IWMP focuses on people's participation in watershed management, there should be maximum numbers of organizations allotted with essential management functions. Here, the degree of organizational involvement given by the OII is only 45.55%. It means the participant organizations are either not allotted their functions share or slackness in their sides. It indicates a review of the present situation.

The result shows that the PIA, the watershed developer, involves a 100% share of management functions. The WDT also involves a good number of functions (66.7%). On the other hand, the 33.3% share of Gram Sabha, a statutorily elected

Table 16.2 Summarized response analysis

Management functions category	Management functions	PIA	WDT	WDC	GS	SHG	Villagers	Function-wise score	Management functions category-wise score
Planning	A1—defining goals	Y	Y	Y	N	N	N	3/6	$PEI_P = 9/18 = 50\%$ 
	A2—establishing strategies for achieving goals	Y	Y	Y	N	N	N	3/6	
	A3—developing comprehensive plans to integrate and coordinate	Y	Y	Y	N	N	N	3/6	
Organizing	B1—determining what tasks are to be done	Y	Y	Y	Y	N	N	4/6	$PEI_O = 11/30 = 36.7\%$ 
	B2—determining who is to do them	Y	Y	N	N	N	N	2/6	
	B3—determining how is to group the tasks	Y	Y	N	N	N	N	2/6	
	B4—determining who reports to whom	Y	Y	N	N	N	N	2/6	
	B5—determining where decisions to be made	Y	N	N	N	N	N	1/6	
Leading	C1—motivating employees	Y	Y	N	N	N	N	2/6	$PEI_L = 11/24 = 45.8\%$

(continued)

Table 16.2 (continued)

Management functions category	Management functions	PIA	WDT	WDC	GS	SHG	Villagers	Function-wise score	Management functions category-wise score
	C2—directing activities	Y	Y	Y	N	N	N	3/6	
	C3—selecting effective communication channels	Y	Y	Y	N	N	N	3/6	
	C4—resolving conflicts among members	Y	Y	Y	N	N	N	3/6	
Controlling	D1—monitoring the performance	Y	Y	Y	N	N	N	3/6	$PEI_C = 10/18 = 55.6\%$
	D2—comparing the results with goals	Y	Y	Y	N	N	N	3/6	
	D3—correcting	Y	Y	Y	Y	N	N	4/6	
Functions shared score by component organizations		$OI_{PIA} = 15/15 = 100\%$	$OI_{WDT} = 14/15 = 93.3\%$	$OI_{WDC} = 10/15 = 66.7\%$	$OI_{GS} = 2/15 = 13.3\%$	$OI_{SHG} = 0/15 = 0$	$OI_{VILL} = 0/15 = 0$	41/90	
Organizational involvement index (OII)		=Total functions shared score/no. of participant organizations = (100 + 93.3 + 66.7 + 13.3 + 0 + 0)/6 = 45.55%							
Project management effectiveness index (PMEI)		= Total functions covered score/no. of management processes = (50 + 36.7 + 45.8 + 55.6)/4 = 47.03%							

body, is not up to mark. Notably, the function shares decrease gradually toward community groups. Thus, it shows a top-heavy, spinning-top type of function share structure instead of the desired square one. This structure denotes a heavily centralized management environment.

Most surprisingly, villages and SHO groups are entangled with zero function shares. The probable reason for it may be that the planners' management scope is not well-defined or these are not percolated to the community level. Hence, it invites stringent reengineering interventions.

16.4.2 Managerial Effectiveness of the Watershed Project

For effective project management, it is expected that the project should cover all the management functions under all management processes. Here we see that all organizations' total functions for all management processes are only 47.03%. Therefore, we can expect a better value in the range of 70–80%.

The functions covered in the controlling process are maximum (55.6%). On the other hand, it is minimum for the organizing process (36.7%), which shows that the project gives a more crucial controlling process and lesser thrust in the organizing process. Practically significant project inefficiency occurs due to slackness in organizing. In such a case, a more powerful thrust on control cannot improve the project.

Primarily, watershed projects emphasize the participatory planning process. However, the coverage of management functions under the planning process is only 50%.

Many planners often downplay the leading process. It is also apparent here. Conflict resolution and the selection of effective communication channels are essential functions for a watershed project. A score of 45.8% is much less than expected.

16.5 Conclusion

This study assumes that the essential management functions are generic and apply to watershed management. In general, there may be a misconception that the management of a watershed project does not suit the purview of project management theory. Management functions are not adequately delineated for the cluster of participant organizations owing to such perception. This cluster acts like a cross-functional work team with complex inter-relationship. It should not deter the applicability of management functions to participant organizational units. Instead, watershed managers can improve project efficiency by focusing on managerial deficiencies.

In watershed projects, beneficiaries are project partners. Although the planners aim at improving their livelihood, assets are primarily created in build-operate-transfer mode. So, the diagnosis of managerial laxity will pave the way for an appropriate intervention designed for better involvement and sustainability. In this regard, the proposed indexing method will be a valuable tool for watershed managers.

References

Ali G Soliemani., Azadeh D, Reza M (2001) "An introduction to organizational behavior"

Bach H et al (2011) From local watershed management to integrated river basin management. In: Bach H, Taylor R (eds) Mekong river commission, p 48

Carpenter M, Bauer T, Erdogan B (1986) Management principles

Conservation Ontario (2010) Integrated watershed management—navigating Ontario's future: integrated watershed management in Ontario. Cinservation Ontario, Newmarket, Ont

Russell W. Darnall, John M. Preston (2010) Beginning project management

Das J, Jha S, Goyal MK (2020) On the relationship of climatic and monsoon teleconnections with monthly precipitation over meteorologically homogenous regions in India: Wavelet & global coherence approaches. Atmos Res 238:104889. https://doi.org/10.1016/j.atmosres.2020.104889

Daspit J, Tillman CJ, Boyd NG, Mckee V (2013) Cross-functional team effectiveness: an examination of internal team environment, shared leadership, and cohesion influences. Team Perform Manag 19:34–56. https://doi.org/10.1108/13527591311312088

Devine DJ, Clayton LD, Philips JL et al (1999) Teams in organizations: prevalence, characteristics, and effectiveness. Small Gr Res 30:678–711. https://doi.org/10.1177/104649649903000602

Easterby-Smith M, Lyles M (2003) Introduction: watersheds of organizational learning and knowledge management. Blackwell Handb Organ Learn Knowl Manag. p 1–15

FAO (2017) Watershed management in action: lessons learned from FAO field projects. FAO, Rome

Fawcett SB, Paine-andrews A, Lewis RK, et al. (1995) Work Group Evaluation Handbook: evaluating and supporting community initiatives for health and development "the art of research [is] the art of making difficult problems soluble by devising means of getting at them"

Gaur S, Milne G (2015) Improving operational effectiveness and impacts of the IWMP in India, Washington DC, p 20433

Goyal MK, Ojha CSP (2010) Evaluation of various linear regression methods for downscaling of mean monthly precipitation in arid Pichola watershed. Nat Res 01(01):11–18. https://doi.org/10.4236/nr.2010.11002

Goyal MK, Ojha CSP (2012) Downscaling of precipitation on a lake basin: evaluation of rule and decision tree induction algorithms. Hydrol Res 43(3):215–230. https://doi.org/10.2166/nh.2012.040

Goyal MK, Panchariya VK, Sharma A, Singh V (2018) Comparative assessment of SWAT model performance in two distinct catchments under various DEM scenarios of varying resolution, sources and resampling methods. Water Resour Manag 32(2):805–825. https://doi.org/10.1007/s11269-017-1840-1

Hannan MT, Pólos L, Carroll GR (2003) The fog of change: opacity and asperity in organizations. Adm Sci Q 48:399–432. https://doi.org/10.2307/3556679

Kozlowski SWJ, Bell BS (2012) Work groups and teams in organizations. Handb Psychol Second Ed. https://doi.org/10.1002/9781118133880.hop212017

NITI Aayog (2019) Composite water management index, pp 1–239

Parkin J, Sharma D (1999) Infrastructure planning, 1st edn. Thomas Telford Publishing, London

Poonia V, Goyal MK, Gupta BB, Gupta AK, Jha S, Das J (2021) Drought occurrence in different river basins of India and blockchain technology based framework for disaster management. J Clean Prod 312:127737. https://doi.org/10.1016/j.jclepro.2021.127737

Robbins SP et al (2011) Essentials of management, 6th edn. Pearson Education India, Bengaluru

Robbins SP et al (2013) Organizational behavior, 15th edn. Pearson Education India, Bengaluru
SLNA (2010a) Detailed project report for Turkunijan IWMP (3A2A6/a2), Barpeta District, Assam under the Department of Land Resources Ministry of Rural Development Government of India, New Delhi
SLNA (2010b) Detailed project report for Maloibari IWMP (3B2A2) of Kamrup district, Assam under the Department of Land Resources Ministry of Rural Development Government of India, New Delhi
SLNA (2010c) Detailed project report for Kaldia part III IWMP-upper (3A2A7/4), Barpeta District, Assam under the Department of Land Resources Ministry of Rural Development Government of India New Delhi
SLNA (2011) Detailed project report for Satpokholi IWMP(3B1C8) 2011–2012, Kamrup District, Assam under the Department of Land Resources Ministry of Rural Development Government of India, New Delhi
Sreedevi TK, Reddy TS V, Wani SP, et al. (2008) A comparative analysis of institutional arrangements in watershed development projects in India: global theme on agroecosystems report no 50
Wang G, Mang S, Cai H et al (2016) Integrated watershed management: evolution, development and emerging trends. J For Res 27:967–994. https://doi.org/10.1007/s11676-016-0293-3

Chapter 17
Pathways to Build Resilience Toward the Impact of Climate Change on the Indian Sunderban

Sweta Baidya, Pritha Chakraborty, Shivukumar Rakkasagi, Manish Kumar Goyal, and Anil Kumar Gupta

Abstract The Indian Sunderban is located in the southern part of West Bengal. Sunderban is known for its precious mangrove ecosystem. Sunderban holds the world's largest mangrove. Mangrove forest provides a huge amount of natural ecosystem services for the people of Sunderban. The mangroves of Sunderban have faced drastic losses, mainly from the impacts of climate change. Climate change plays a significant role, which makes the mangrove community vulnerable to its impacts. The people of Sunderban are highly dependent on the mangrove ecosystem; the loss of mangrove cover makes them vulnerable. The study found drivers and uncertainties of climate change that would destroy the mangrove community in future. The sea-level rise (cyclonic activity, erosion, and accretion), population rise (land cover change), and pollution (obstruct inflow of freshwater) are extreme threats for mangroves in Sunderban. The study shows that the changes need to be addressed through the government's policies and a restructure of

S. Baidya (✉)
ECDRM Division, National Institute of Disaster Management (NIDM), New Delhi, India

Mitigation Division, National Disaster Management Authority, Ministry of Home Affairs, New Delhi, India

P. Chakraborty
ESG Book, New Delhi, India

S. Rakkasagi · M. K. Goyal
Department of Civil Engineering, Indian Institute of Technology Indore, Indore, Madhya Pradesh, India
e-mail: mkgoyal@iiti.ac.in

A. K. Gupta
National Institute of Disaster Management (Ministry of Home Affairs, Government of India), New Delhi, India
e-mail: envirosafe2007@gmail.com

A. K. Gupta et al. (eds.), *Ecosystem Restoration: Towards Sustainability and Resilient Development*, Disaster Resilience and Green Growth,
https://doi.org/10.1007/978-981-99-3687-8_17

governance is required. In policymaking, the collaborative action of scientists, governments, and local people should be involved. The lack of local involvement created a gap in government policy, which increased the threat and risk to the mangrove ecosystem. The traditional knowledge of the local people of Sunderban needs to be incorporated into the management strategy of mangroves. Furthermore, in this study, the Sunderban wetland was studied based on the inundation patterns it has seen over the past 30 years. The study makes use of preprocessed Landsat images (1991–2020) to create annual composites from June to September. The composites were subjected to the short-wave infrared (SWIR) thresholding technique to produce inundation maps using Google Earth Engine (GEE).

Keywords Mangrove ecosystem · Climate change · Sea-level rise · Management strategy · Government policy · Governance · Traditional knowledge · Inundation mapping

17.1 Introduction

Sunderban's real meaning in the local language is "beautiful forest," where Sundar means beautiful and ban means forest. Sunderban holds the world's largest mangrove ecosystem; it was designated a "World Heritage Site" by UNESCO and was recognized as the Ramsar Site in 1985. The Sunderban is situated in India and Bangladesh, where 59% fall under Bangladesh. The Indian Sunderban is situated along the coast of West Bengal for around 260 km and extends over the districts of South 24 Parganas and the southern part of North 24 Parganas. The area of Sunderban was 20,000 km^2 in 1885, and now it has shrunk to 9600 km^2, out of which less than 4200 km^2 is mangrove forest (Bose 2004). The mangrove plant is classified into woody trees and thick shrubs that grow in groups in the intertidal zone of the coastal area. The mangroves have the capability to build the connection between ocean water and river water. The mangrove forest is also known as tidal forest; the Sunderban shares the world's largest and most interesting mangrove vegetation. There are three different types of mangrove: these are the black mangrove (*Avicennia germinans*), red mangrove (*Rhizophora mangle*), and white mangrove (*Laguncularia racemosa*), which are finite in tropical and subtropical areas due to their sensitivity to the temperature. The mangrove vegetation is a very adaptive nature in harsh conditions. They are highly salt-tolerant; it excretes the extra salt through the leaves. They have highly complex and shallow root systems, like prop, pneumatophores, and knee, which also help to bring in oxygen in a flooded condition. Each type of mangrove vegetation has distinctive characteristics that adapt to its environment. The mangrove has another surprising feature: it brings life through viviparous germination. Through viviparous germination, the propagule gets detached from the mother plant and grows independently with the contact of soil, and some propagule falls in the ocean, gets back into the soil through ocean current, and grows independently. Mangrove provides shelter to thousands of species found in mangrove areas. Some creatures, like oysters, crabs, mudskippers, and some sucker species, are most common; the mangrove provides a protective and harmless environment for coral fish to develop enough to travel into the ocean. Mangroves play an integral role in maintaining the coastal ecosystem, as they act as a sink for carbon dioxide (CO_2) and atmospheric pollutants, thereby helping the ecosystem cope with climate change-related disasters. It also protects the shoreline from natural hazards by working as a natural filter due to its unique and complex root system. It filters heavy metals, entraps sediments, and has the capability of absorbing excess nutrients to reduce water turbidity. The natural functionality of mangrove is decreasing gradually due to the contiguity of different types of toxic chemicals from agricultural fields, industries, and ports, which change the geochemical composition of the water. At the same time, the rising pollution from surrounding industries and the waste from the city through channels are increasing the risk and making the coastal region vulnerable. Climate change is one of the major influential weapons for destroying the mangroves in the Sunderban. Because the Sunderban is a low-lying coastal area, climate change has an adverse effect on mangroves. The loss of mangroves in the Sunderban area will erase many endemic species from the face of the earth. According to the report, some areas of the western part of Sundarban are most vulnerable to climate change, like Sagar Island, Namkhana, Ghoramara, Dakshin Surendranagar, and Jambudwip. The southwestern island Ghoramara and Supribhanga had been flooded and submerged (Panda 2010). Those displaced people, due to climate change, are escaping to nearer places, like Sagar Island, Bakkhali, which are also vulnerable to climate change. For the record, Ghoramara Island is shrinking day by day due to the rapid rise of sea levels (Goyal and Ojha 2010, 2012). The western adjacent side of Indian Sunderban mangroves is more vulnerable than the core area (Das et al. 2020; Ghosh et al. 2015). The people are usually impoverished, and the loss of mangrove cover due to climate change has multiplied the calamity for local communities in Bakkhali, which lies along the coast of the Bay of Bengal. Whereas the mangroves act as carbon sinks, a further decrease in the mangrove cover over the region will decrease the natural absorption strength to the atmosphere, thus contributing to climate change as well (Khairuddin et al. 2016). The Indian Meteorological Department (IMD), 2013, marked all adjacent areas of Sunderban on the base of the climatic vulnerable zone. The areas (western, northern, and northwestern), which are very close to Sunderban mangroves, are recognized as high-risk zones due to the high density of the population, degradation of mangroves, frequent hits from cyclones, floods, and storms.

17.2 Background

The mangrove ecosystem is growing in the intertidal region that is in-between the average sea-level and the high-tide coastlines of tropical and subtropical regions. The mangrove community apprehends a large variety of species of flora and fauna that play a significant role. Mangrove is subjected to the tidal influence to grow in intertidal regions, and also, the mangrove soil supports a dense and tall mangrove plant (Selvam 2003). Mangrove has a unique quality of surviving under extreme saline conditions, and it requires muddy soil, good rainfall, an average maximum temperature of 34 °C, and a minimum temperature of 13.7 °C. The mangrove community serves ecological and non-ecological values, such as giving protection from floods and storm surges, producing timber and non-timber products, filtering pollutants from coastal water, supporting aquaculture to enhance tourism, and creating aesthetic beauty. FAO estimated the mangrove declined by 14,653 hectares from around the globe. It is estimated that mangrove losses are around 1–2% worldwide per year. Rapid developments near coastal areas, expansion of human settlements, and industrialization are the major causes of the deterioration of mangrove forests, and the deterioration has an impact on the production of fisheries, crustaceans, and mollusks. Mangrove is the only natural buffer system, and it can hold huge amounts of sediments and chemicals to prevent water pollution. The deforestation of mangroves has an impact on fisheries despite the fact that it has a negative impact on adjacent seagrasses and coral reefs because they are interlinked (Arancibia et al.). The mangrove ecosystem plays a crucial role in supporting the livelihoods of humans and creating more economic opportunities that support the economy of the country. Mangrove ecosystem depicts as a fragile ecosystem; it is majorly affected by many extreme weather events and anthropogenic events. The mangrove ecosystem is threatened by sea-level rise, a rise in temperature, and changes in rainfall patterns and intensity, all of which are due to climate change. The Hooghly estuary carries many raw materials from ports and industries and municipal waste from the city of Kolkata, and it directly impacts the Sunderban mangrove area, mainly on the western side, because of southwest and northeast winds that led to southwesterly and northeasterly winds (Sarkar and Naskar 2003). Climate change has an impact globally that leads to the loss of mangrove forests; the width of mangrove forests is becoming narrow, leading to the loss of the island (Nanlohy et al. 2015), and this will change the social and economic life of the local communities. In the current situation, climate change is a highlighted word and a threat to the overall world. The population is rising day by day, which is the cause of the exploitation of nature and natural resources. The growing population is one of the major causes of land cover change. The land is converted from forest land into agricultural land, cultivated land, and a settlement area (Mondal et al. 2010). Due to the setup of various industries by clearing forest land, the air quality gets deteriorated, which increases the level of carbon dioxide, carbon monoxide, CFC, methane, and nitrous oxide in the environment (Mondal et al. 2010). The increasing heat in the atmosphere traps the heat in the earth's atmosphere, which results in global warming and the trapping gas known as greenhouse gas (Poonia et al. 2021; Goyal et al. 2018). The changes in average temperature trends due to global warming. Global warming is caused by anthropogenic activities as well as physical activities, such as ocean circulation, the proportion of water vapor, orbital eccentricity, and differences in cloud cover (Mandal et al. 2009). The man-made activities are not controllable, which results in the high intensity of warming in the atmosphere, but the physical cause is equally responsible, directly or indirectly (Hazra et al. 2002). Since the mid-twentieth century, the high concentration of greenhouse gases has resulted in rising temperature (IPCC 2007). In the last 100 years, the temperature has increased by five times (Mondal et al. 2010). The temperature of India in the coming 50 years will increase by 1 °C. In the next century, greenhouse gas emissions will rise by 3.3° centigrade, as estimated from the scenario of business as usual (IPCC 2007). The conservation of the mangrove ecosystem is another way of an adaptation strategy for climate change, as stated by the Intergovernmental Panel on Climate Change (IPCC 2007), whereas mangrove plantations are disappearing at a high rate. Climate change shows a drastic change due to increasing temperatures over the last three decades (Mondal et al. 2010). Global warming increases the risk of sea-level rise, which affects the coastal zone like the Indian Sunderban. The Indian Sunderban is very vulnerable to climate and weather-related events. The cyclonic events suffered Sunderban and Sunderban's people from many disastrous cyclone events (Mandal et al. 2009). The rising sea surface temperature increases the intensity of cyclone patterns over the time period (Hazra et al. 2002). The sea surface temperature is co-relatable to sea-level rise; the increase in sea surface temperature causes the rise of sea level, or vice versa. It was estimated that due to sea-level rise, storm surges, coastal erosion, natural disasters, and flooding, 1.35 million people are critically venerable from decades ago, and there is also a huge loss of economy and property damage for people in Sunderban. In the last 30 years, 7000 people have been displaced from Sunderban Island by the threat of sea-level rise (Mondal et al. 2010). Sea-level rise forced them to become environment-induced refuges (Mondal et al. 2010). The people of Sunderban have been suffering from constant threats that compromise their basic needs like protection from natural calamities, food security, and socioeconomic security (Hazra et al. 2010). However, the mangrove plays a vital role in preventing climate change through the sequestration of carbon dioxide. The loss of mangrove forest means the loss of carbon dioxide sequestration capacity of mangroves that leads to climate change (Nanlohy et al. 2015), which indicates a reduction of resilience from sea-level rise. In terms of carbon sequestration, the mangrove has a significant and effective role as a carbon sink, either for short- or long-term storage, that performs function globally. The destruction of the mangrove ecosystem affects the blue carbon ecosystem globally. The main loss from mangrove ecosystem loss is climate change mitigation, where emissions are increasing day by day. The mangrove ecosystem serves different types of values, like direct use values and indirect use values. The direct use values considered the values that are directly used by humans, like timber and non-timber products (NTFPs), and indirect use values considered the values that serve the ecosystem but can't be used by humans (Reid et al. 2005). The mangroves are the only natural and efficient way for climate change mitigation worldwide. The coastal population is highly dependent and uses the mangrove ecosystem as a primary source (Saw and Kanzaki 2015) like a collection of timber, honey, fodder, traditional medicine, and food like fish, crabs, and shrimps. The main source of livelihood for communities in the Sunderban is aquaculture (shrimp, prawn, stinging catfish, and mollusk flesh) and agriculture like paddy cultivation, but the impacts of climate change like rising sea level, changes in rainfall patterns, changes in the salinity of soil and water, rising temperature, and increasing frequency of extreme weather events have paused their sources of income, and that affects their contribution to the Indian economy. The mangroves are witnessed many natural disasters, mainly in 1980, 1991, and 2009 in the Sunderban area. Some western coastal islands in the Indian Sunderban, like Bakkhali, Namkhana, Sagar, and Kakdwip, have a very low level of adaptation (Sahana et al. 2019), which means these regions are highly vulnerable to climate change. The extreme events are evidenced by the people, and they were badly affected by the loss of lands, assets, earning options, and livelihood. The coastal people are more vulnerable to climate change, so there is a need to address the problems and prepare them with high resilience to combat climate change (Goyal et al. 2022). The first step to increasing awareness and knowledge in local communities is to involve them in planning and managing their natural resources. The National Action Plan on Climate Change (NAPCC) was enforced in 2008 with the aim of eight missions. The National Action Plan on Climate Change (NAPCC) provides opportunities and identifies risks in key sectors with the help of state action plans. The Government of West Bengal identified climate change as a major threat and assessed key sectors that regulate it through the state action plan (SAP). The National Adaptation Program of Action (NAPA) helps to mark vulnerable areas against climate change and addresses the challenges. In the government's policy, the main focus should be on climate change and managing the Sunderban sustainably. The governance structure is required as an adaptation based on Sunderban, with engaging local people's participation and discussion with different actors.

17.2.1 Objectives

- To identify the drivers and uncertainties for mangrove cover loss over Indian Sunderban
- To understand the inundation patterns the Sunderban wetland has seen over the past 30 years (1991–2020) using Google Earth Engine (GEE)
- To identify the slackness in government policy and governance structure for Indian Sunderban

17.3 Material and Methods

The study takes an exploratory approach to determine the requirements for the completion of the study. Both the qualitative and quantitative data have been taken into consideration in carrying out the study. The secondary data from many sources are the key to meeting all the findings. Secondary sources of data, like research papers (research gate, science direct) and journals (UGC approved), reports (soured from government and nongovernment agencies), data, are used to understand the prime drivers and uncertainties of climate change in the Indian Sunderban as well as identify the gap in government policy and the need to restructure for the protection of mangroves. The quantitative external data and systematic literature review are taken into account to envisage the first objectives. For the next goal, the theme paper of government structure and action from the colonial period will be used to understand the problem and shortcoming and increase resilience toward climate change in the Indian Sunderban. These sources formed the database for the secondary analysis of the Indian Sunderban.

17.3.1 Study Area

The Indian Sunderban lies along the coast of Bay of Bengal in the state of West Bengal, which extends over the districts of South 24 Parganas and the southern part of North 24 Parganas. The Sundarbans delta is formed by the confluence of the Ganga, Brahmaputra, and Meghna Rivers. The geographical location of Indian Sunderban is 21°30′N and 22°40′48″ latitudes and 88°1′48″ E and 89°04′48″ E longitudes. The largest Indian mangrove forest covers 2122.421 sq. km out of 6312.76 sq. km area (Samanta and Hazra 2017). The Sunderban area has constructed miscellaneous types of land like preserved and protected areas, i.e., national park, biosphere reserve, and wildlife sanctuary; inhabited lands; cultivated land; forest land; and agricultural land. The inhabited island is further distributed in-between the northwestern part and the southern part; as already discussed above, the northwestern part is positioned at an upper level from the coast, so it less vulnerable to natural calamities ,whereas the southern part is highly exposed and vulnerable (Fig. 17.1).

The Sunderban area has a unique and special character that provides the highest aesthetic value in the world. There are some features of the Indian Sunderban, like geology and geomorphology, geohydrology, hydroclimate, and soil that have held the precious mangrove vegetation for decades.

The Indian Sundarban falls between the latitudes 21° 31′ and 22° 31′ North and 88° 10′ and 89° 51′ East longitudes (Fig. 17.2). Table 17.1 describes the climatic and other general information of the Sundarban wetland. As a part of the Indian Sundarban, the West Sundarban Wildlife Sanctuary has its significance in the international context for providing shelter and protection to various species of wildlife included in the Red Data Book (Ramsar 2022).

17.3.2 Geology and Geomorphology

The study area has major geomorphic areas are creeks, mudflat, alluvial plain, channels of the river, sand flat, estuary, and mangrove vegetation (Ganguly et al. 2006). The area formed by the confluence of the three major Ganga, Brahmaputra, and Meghna Rivers, the Ganga delta, is highly dynamic, and the formation of the delta is very rapid (Morgan and McIntire 1959). The tectonic activity has brought

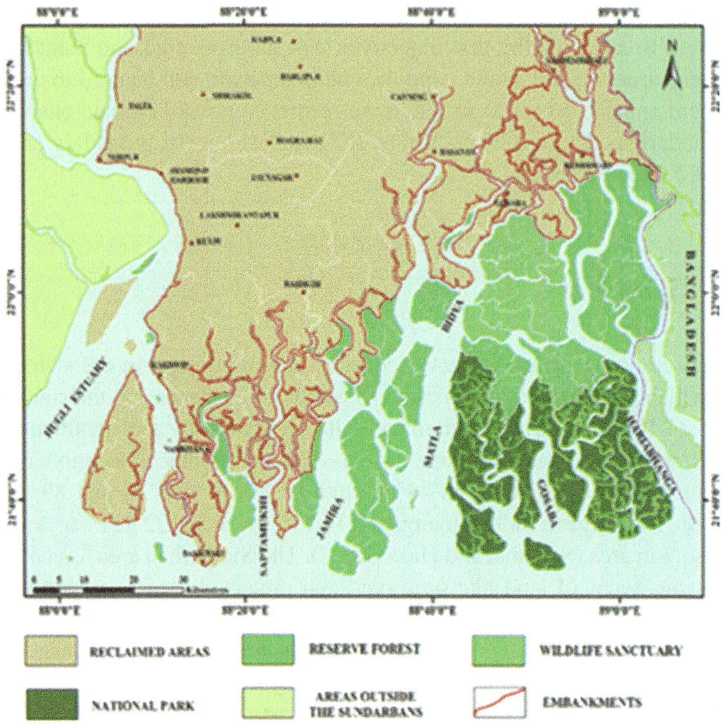

Fig. 17.1 The map of Indian Sunderban (Source: (Rudra 2018))

changes in the geomorphology of the Sunderban, and several river channels carry tons of sediments from its source that help in eroding and accreting.

17.3.3 Geohydrology

The groundwater in the area is neutral to mildly alkaline with a pH range of 7.2 to 8.1. The natural aquifers that contain freshwater are at a depth of 76–360 m, with a thickness of sand and clay beds. The concentrations of magnesium bicarbonate, iron, calcium, and chloride vary from 11–44, 281–640, 0.01–3.4, 14–76, and 14–188 ppm respectively. The groundwater of the area is very suitable for consumption.

17.3.4 Hydroclimate

The Sundarban area has a topical monsoonal climatic condition, whereas the average maximum temperature is 34 °C and the average minimum temperature is 20 °C. The

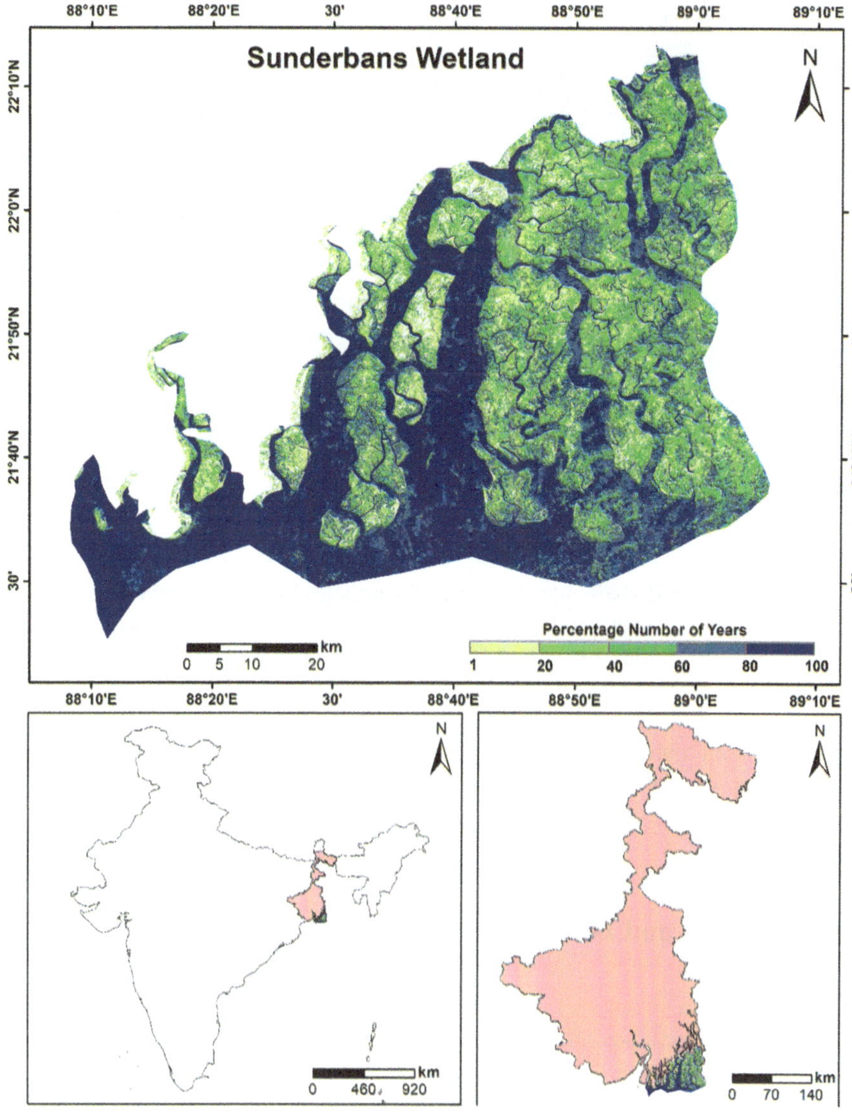

Fig. 17.2 Inundation map of Sundarban wetland. Inundation maps represent the number of years each pixel inundated during the period 1991–2020 (June to September). The legend represents the value of 80–100% (dark blue) of the pixels inundated in all available timesteps and 1–20% (yellow) as not inundated in any available timescale

total rainfall is approximately 1500–2000 mm (Gopal and Chauhan 2006), and the area receives extreme rainfall. The seasonal climate is divided into three phases are pre-monsoon period is between February-May, the monsoon period is between June-September and the post-monsoon period is between October-January (Subramoniam

Table 17.1 Climatic and other general information of the Sundarban wetland

Country	India
Area	423,000 ha
Designation date	30-01-2019
Coordinates	21°46′ N and 88°42′ E
Average precipitation	1904 mm
Maximum temperature	29.8 °C
Mean temperature	26.43 °C

et al. 2022). The relative humidity is about 75–80% over the year due to humid air from the Bay of Bengal. There are mainly two dominant winds that serve the monsoon blowing from the southwest known as the southwesterly and the northeast known as the northeasterly.

17.3.5 Soil

The soil of the Sundarbans has unique characteristics and is different from that of other islands, which are habitats for thousands of species. The soil structures are sandy loam, silt, or clay loam, and pH ranges from 5 to 8. The soil is also called "mangrove soil" (Mandal and Ghosh 1989; Banerjee 2013). The content of sodium and calcium is low in eastern, organic matter ranges 4–10% and the salinity of the soil varies from high to moderate; the land is inundated with saline water. This soil character supports the unique mangrove community, which plays an initial role in the atmosphere.

17.4 Results and Discussion

The precious mangrove cover is slowly depleting from nature. There are some main genesis factors, including climate change in the context of sea-level rise, pollution, and population growth, behind the evanescence of mangrove from the Indian Sunderban over decades. The study was found the mangrove forest cover decreased from the year 1986 to the mangrove cover 2246.839 sq.km afterward in 1996 it was 2201.41 sq.km. Then, in 2001, it was 2168.914 sq.km., and in 2012, it was 2122.421 sq.km. (Hazra et al. 2017). Some islands, like Gosaba, Dulibhasani, Dalhousie, Bhangaduni, and Jambudwip, have witnessed a great loss of mangrove forest cover (Fig. 17.3) (Table 17.2).

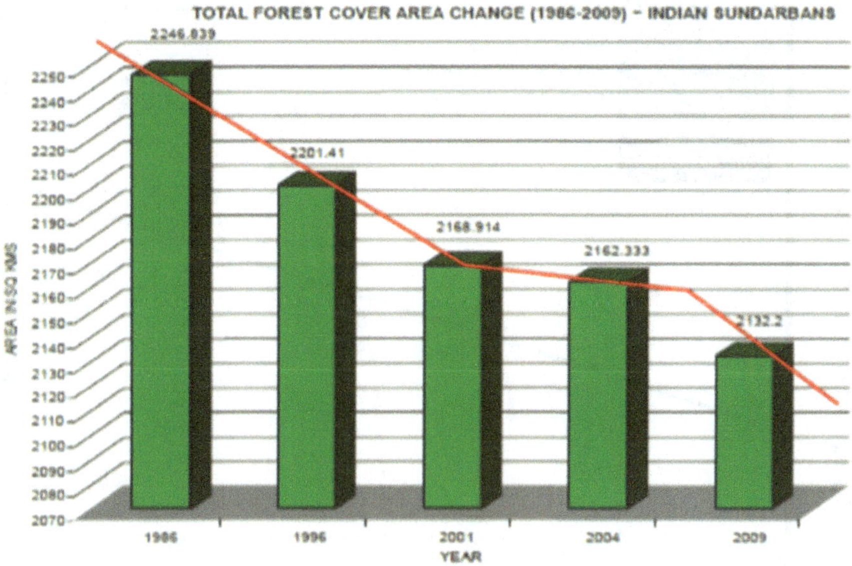

Fig. 17.3 The decrease in forest cover in the Indian Sunderban (Source: Danda 2010)

Table 17.2 The reduction of mangrove covers in 1986–2012

Years	Reduction in mangrove forest cover in sq.km.
1986–1996	45.429
1996–2001	32.5
2001–2012	46.493

17.5 The Main Drivers and Uncertainties of Climate Change

17.5.1 Climate Change and Sea-Level Rise

Climate change plays the main role in the drastic sea-level rise and changes in ocean circulation in the Indian Sunderban. The sea-level rise has an expansive relationship with surface air temperature. With the rising sea level over the past few decades, the air temperature has increased by 0.019 °C per year. The melting of glaciers globally has an adverse effect on local areas like the Indian Sunderban. Several areas in Sunderban, like Namkhana, Jambudwip, Ghoramara, Sagar, Mousuni, Bulchery, Dalhousie, and Dhanchi, were predicted to be vulnerable and cause 68% land loss in the future (Hazra et al. 2010). Since the mid-twentieth century, the Sagar Island has shrunk by 20 square miles, and three islands – Lohachara, Suparibhanga, and Bedford – were covered by mangrove species (Schwartzstein 2019), but now these islands have disappeared due to sea-level rise. The vanishing of the mangrove ecosystem and rising sea level will make the surrounded areas of Sunderban, like

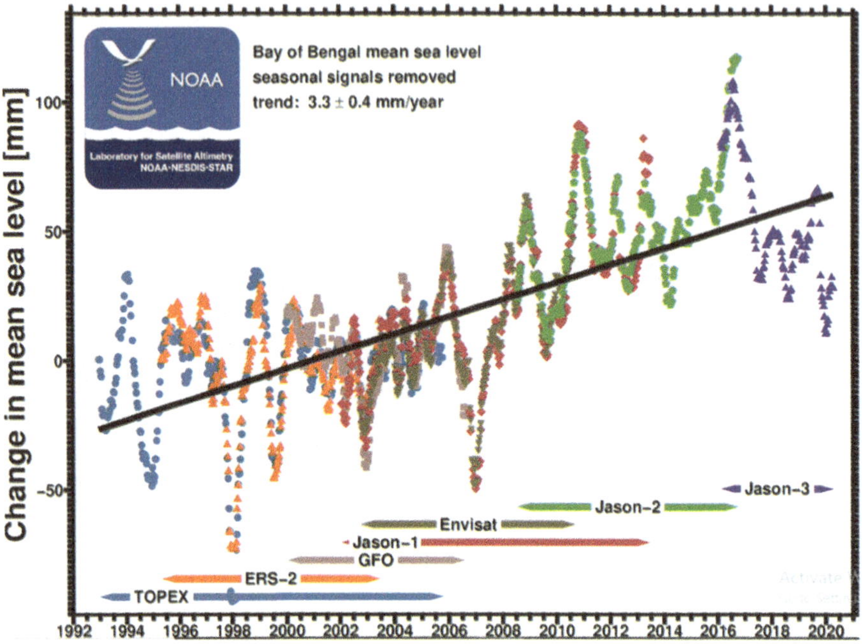

Fig. 17.4 The rising of mean sea-level over the Bay of Bengal (Source: NOAA)

Kolkata, more vulnerable to cyclones, floods, and storm surges (Ghosh and Lalitha 2017) (Fig. 17.4).

The gradual increase in annual sea levels in the Bay of Bengal has effects globally. The highest peak in those years (from the image), which indicates the excess level of water in the system, was due to El Niño phenomena and contribution of the highest amount of rainfall during monsoon periods. Hazra et al. estimated that without seasonal disturbance, the mean sea level would rise by 1.62 cm in 5 years. In the current scenario, the relative sea-level rise will 3.14 mm per year, and by the year 2050, the sea-level elevation will be approximately 1 m. The study found that the mean sea-level rise of 2.6–4 mm per year from 1985 to 2010 influenced tidal creeks along the shoreline. Already, some islands, such as Lohachara and Ghoramara, have fully drowned, and now the Sagar Island is at risk position from constant sea-level rise. The people living on those islands get displaced and travel to other islands of Sunderban, which is badly affects the resources of the island and increases the pressure and conflict between people. The constant, gradual sea-level rise will cause the disappearance one of the most precious and unique mangrove communities in the future. Climate change and sea-level rise result in increasing temperatures, changes in cyclone occurrence, and changes in precipitation. The sea-level rise is also responsible for some associated changes like erosion and accretion patterns on Sunderban Island, which cause land use and land cover change and also changes in the occurrence of storms, cyclones, and floods that result in seasonal variation.

17.5.2 Cyclone and Storm

According to records, the number cyclonic storm has increased by 26% over the Bay of Bengal in the last 120 years (Singh 2009), and coastal erosion is dominant over the southwestern islands (Hazra et al. 2010). From 1891 to 1961, the cyclones occurred 56 times in Sunderban (Fig. 17.5).

The rising intensity of cyclones is possible due to sea-level rise (Hazra et al. 2002). When it was analyzed over the last 120 years, the frequency and intensity of cyclones appeared to be very high (Singh 2007). Over the last 120 years, 26% of intense cyclonic events occurred over the Bay of Bengal, mainly in post-monsoon seasons. With the rising sea level over the past two decades, the intensity of storms and cyclones increased in the past, and the frequency of their occurrence decreased, which is due to global warming. The rising sea level also causes decadal change. Before, the intensity and frequency of cyclones were very high, but now, according to the present study, the intensity of cyclones has increased and the frequency of cyclones has decreased. The increase in the intensity of cyclones and storms influences the rate of precipitation over the Bay of Bengal. From the table, after many years, the disastrous cyclone Aila came on May 25, 2009, in West Bengal. It

Type of Hazard	*Year of Occurrence*	*Blocks affected*	*Impact on Life*	*Impact on Livestock*
Flood	1978	All blocks	Severe	Affected to a great extent
	1986	1. Kultali	Severe	In some areas affected to a great extent
		2. Joynagar-II	Moderate	
		3. Canning-I	Moderate	
		4. Gosaba	Severe	
		5. Basanti	Severe	
		6. Mathurapur-II	Moderate	
		7. Kakdwip	Severe	
		8. Namkhana	Severe	
		9. Sagar	Severe	
		10. Patharpratima	Severe	
Drought	1998-99	1. Kakdwip	Moderate (5 mouza)	Affected to a great extent
		2. Basanti	Do (6 mouza)	
		3. Caning-II	Do (5 mouza)	
		4. Gosaba	Do (23 mouza)	
		5. Mathurapur-I	Do (6 mouza)	
		6. Patharpratima	Do (2 mouza)	
Cyclone	1999	1. Kultali	Severe	Affected to a great extent
		2. Joynagar-II	Moderate	
		3. Canning-I	Moderate	
		4. Gosaba	Severe	
		5. Basanti	Moderate	
		6. Mathurapur-II	Moderate	
		7. Kakdwip	Severe	
		8. Namkhana	Do	
		9. Sagar	Do	
		10. Patharpratima	Do	
Cyclone (Aila)	2009	Almost all the blocks of Sundarban	Severe	Affected to a great extent in most of the blocks

Fig. 17.5 List of natural disaster occurrences over the decades (Source: UNDP 2009)

was one of the climatic disasters that hit the Indian Sundarbans. In a few minutes, it made 4000,000 people homeless, thousands of villages disappeared, and killed thousands of people. With that, it destroyed livelihood patterns and the economy (Mondal et al. 2010).

17.5.3 Erosion and Accretion Patterns

It was estimated that the geomorphic characteristics of Sunderban over 32 years (1980–2012) varied with the interaction of the external variable, sea-level fluctuation (Ghosh et al. 2015), and the outcomes showed a reduction in the land, an increase in salinity, changes in the pattern, and degradation in the mangrove (Fig. 17.6).

The above image shows the erosional and accretion areas over three decades. The erosional area is dominant over the southern and western parts of Sunderban, and the estimated erosional area is 162.879 km^2 (Ghosh et al. 2015), which also indicates those lands are more vulnerable and situated along shorelines. The sea-level rise directly or indirectly changes the size, shape, and structure of islands. Some parts of land, like Mousuni, Namkhana, Gosaba, Chuksardwip Island, and Jambudwip Island (from the image), are strongly or moderately related to the erosion in the context of sea-level rise. It was noticed that the accretion is partially related to sea-level rise, and in the accretion area, adding layers of mud and excess siltation influence sea-level rise gradually (Hazra et al. 2012). Erosional plays a highly powerful role

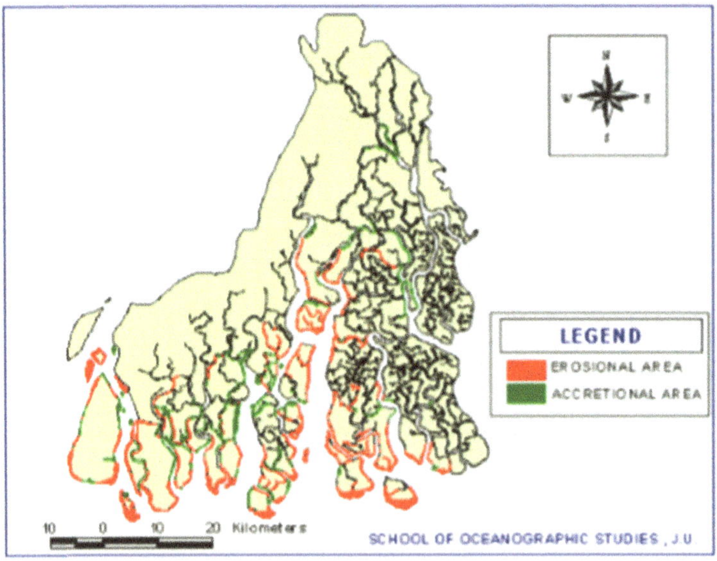

Fig. 17.6 The changes in area through erosion and accretion over Sunderban (Source: Hazra et al. 2002)

that results in the disappearance of Lohachara Island, which was 6.212 km^2 (Ghosh et al. 2015). The accretion area was estimated at 82.505 km^2 over three decades (Ghosh et al. 2015). The study also found there is a direct correlation. The analysis showed that not only erosion but also accretion have a relationship with the fluctuation of sea level (Ghosh 2012). From the analysis, in the Sunderban, the sea-level rise is strongly related to the erosion and accretion rate, or accretion contribute to sea-level rise, which ultimately makes the area vulnerable and increases the pressure for people on the island.

The rate of erosion and accretion has been the primary cause of land-use change for decades. Climate change and sea-level rise are indirectly involved in land-use change pattern, and the change in land use has impacted the whole system. The land loss affects the mangrove cover in the coastal zone, where the settlement largely extends near the shoreline, and their activity for their livelihood is changing the land-use system. The settlement along the shoreline is to be affected more by erosion and accretion from sea-level rise. The reserved and protected forest or area is badly affected by sea-level rise. The major natural parameters like erosion, accretion, and geomorphic changes and anthropogenic parameters like population rise, growing urbanization, and resource utilization (agricultural land, aquaculture, and construction) affected the land-use change over two to three decades by 66% and some inexplicable anthropogenic changes by 2% in the Sunderban Island.

17.5.4 Rise in Population

The Indian Sunderban consists of an area of 9630 sq. km out of that 4264 sq. km of wetland and mangrove reserve forest and 5366 sq. km for human settlement that is divided into 19 blocks: 13 blocks are Jaynagar-1, Jaynagar-2, Kultali, Canning-1, Canning-2, Basanti, Gosaba, Kakdwip, Namkhana, Mathurapur-1, Mathurapur-2, and Patharpratima in South 24 Parganas, and 6 blocks are Haroa, Hasnabad, Hingalganj, Sandeshkhali-1, Sandeshkhali-1, Sandeshkhai-2, and Minakhan in the North 24 Parganas (Hazra et al. 2002). In 1901, the population was 487,377, whereas in 1931, the population increases by 55%, a rise of 755,434. In the partition time, the population was 11,70,922 in 1951. According to the 1991 census, the average density of all the block is 690 persons per sq. km, and the population differs for each block from 120,000 to 250, 000. The population of the Sunderban rose by 18% over one decade, according to the 2011 census. According to the 2011 census, the sex ratio was comparatively high (955 females per 1000 males) in the state West Bengal and India, which shows the population of male children is high than the population of female children. The mangrove forest cover of the Indian Sunderban is more than 5000 km^2 and has been designated as a reclaimed area due to the growing population over the last two centuries (Naskar et al. 2004). The study shows the population marginally decreased by 19.1% over a decade, and it was found that the population growth rate decreased for the previous five decades (Ghosh 2018). From

Table 17.3 The population rise over decades in the Indian Sunderban

Year	Population in numbers	Rise in percent
1901	4,87,377	55
1931	7,55,434	55
1951	1,170,922	160
1991	3,044,397	18
2011	3,738,417	–

the previous decade, in the last decade (2001–2010), the working population rose by 28%, according to the 2011 census (Table 17.3).

The people's livelihood options are very limited, mainly agriculture, aquaculture, timber, and non-timber products, such as honey and medicinal plants. Their inelastic way of running their existence and abject poverty are very deficient due to a lack of alternatives and opportunities. The miserable and woeful condition may not be uprooted because, according to the census of 2011, one of the reasons for population structure is that half of the population belongs from the Scheduled Castes and Scheduled Tribes, and another half of the population belongs to the landless laborers, farmers, fishermen, and honey collectors. And another reason for the harsh condition of physical infrastructure and road construction is lack of communication, infrequent electric supply (mainly in the southern area of Sunderban), and transport (Ghosh et al. 2018). The constant rise of the population is one of the major factors for mangrove loss due to cutting trees, an increase of settlement areas, and over-exploitation of forest resources (Banerjee 2013). In 1991, the table shows 3,044,397, which reflects a havoc rise of 160% due to the reason for the partition of India after independence. In 1991, the population rise for the major reason is industries. The surrounding areas of the Indian Sunderban, such as Haldia, have a large scale of industries, like metal, chemical, fertilizer, leather, etc. The people from the surrounding area and neighboring countries like Bangladesh were in search of food, shelter, and other living opportunities during that period. In Sunderban, migrant people can suffice at a minimum cost rather than living in a big city like Kolkata, and also, the border area is very close to the Indian Sunderban (Sen and Bhadury 2017). In the different years, many migrants and climate-induced migrants came to the Indian Sunderban after partition between Bangladesh and India. The geomorphic feature and land-use change have also been regulated by the population in Sunderban for decades. Indeed, because of exigency and barriers, the people of Sunderban are constantly suffering and struggling from natural calamities, and contrarily at some extent, the people of Sunderban are responsible for bringing them into jeopardy. The population of Sunderban is highly dependent on natural resources, and at the same time, people have been under constant threat from natural calamities, climate change, poverty, disempowerment, and socioeconomic security trade for decades. The vital role of the mangrove ecosystem is loss in daily life with the deterioration of mangrove vegetation.

17.5.5 Pollution

The mangroves of Sunderban Island act as a wall between freshwater and ocean water; without the supply of freshwater, the mangrove vegetation could not grow in Sunderban. The Ganga water from the Himalayas flows through the Hooghly River in terms of freshwater, which gives service to the Sunderban mangrove through the Hooghly-Matla estuarine system. This river has a length of 260 km and also passes from the city of Kolkata, which also covers the Haldia Port and Kolkata Port (Estuaries of India 2002). The last estuary channel is connected through Diamond Harbour, which extends and spreads over more than 20 kms and mixes in the Bay of Bengal. The soil types, geology, and climate of Sunderban and Hooghly are similar (Sadhuram et al. 2005). The Hooghly River is surrounded by many types of industries, like jute mills, pulp, and papers, plastic, rubber, and thermal power plant (Biswas et al. 2007). All the toxic material and wastewater are directly delivered into the Hooghly River for decades, and the municipal waste from the surrounding cities is delivered into the river. The Haldia Port and the Kolkata Port activities are equally responsible for discharging waste materials into the Hooghly River (Bhattacharya et al. 2008) (Fig. 17.7).

The Hooghly estuary act as a huge waste disposal sink (Hazra et al. 2015). Over the past decades, the water quality has highly deteriorated and been impacted by anthropogenic activities, which also increase the threat of environmental degradation. The freshwater from Hooghly River enters through the lower part of Sunderban, facing toward the Bay of Bengal, where the mangrove vegetation lies. According to the study, the western part of Sunderban is more affected and vulnerable to the discharge of toxic chemicals. These toxic chemicals reach the mangrove through wind and ocean circulation, which is the main cause of destruction of the mangrove's unique characteristics and increases the threat to the mangrove community. These toxic chemicals suppress the natural function of mangrove; the toxic and heavy metals penetrate through the roots system of mangrove, blocking the entry of oxygen (Singh 2009). The waste material and garbage in the river reduce availability of freshwater and the supply of sediments (Hazra et al. 2015) to the mangrove community. Whereas the mangrove vegetation needs the minimum inflow of freshwater, if not, the mangrove succession is replaced from freshwater-loving to salt water-loving, but the inflow of freshwater is required for the growth of wealthy and rich mangrove. Being a forested area, the mangrove vegetation not only serves ecosystem service but is also a habitat for numerous species. The eternal changes and threats, both naturally and anthropogenic, put the mangrove plantation in such an extremely vulnerable condition that, one day in near future it will obliterate one of the oldest and most unique World Heritage sites since 1985.

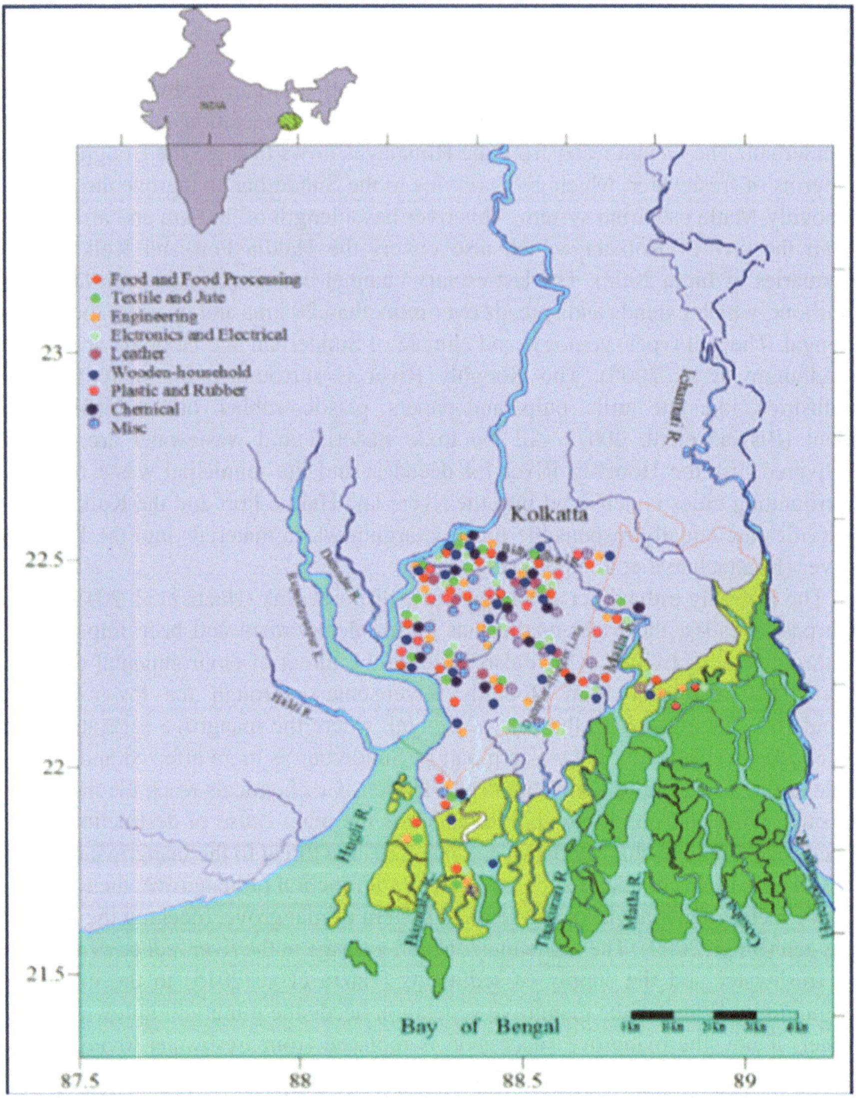

Fig. 17.7 The industrial area surrounded by Sunderban (Source: (Singh 2009))

17.5.6 History of Sunderban

The name of Sunderban Day was mentioned in the Puranas (Sarkar 2011). At very fast speed, the Sunderban land, when it came to marking the land, was originated mainly by Portuguese and Burmese (Sarkar 2011). In the era of the British ruling period, the land of Bengal, Bihar, and Orissa was given on lease in 1765. After that, in 1785, the small portions of land in Sunderban were given a lease to zamindars for many years. At the time of partition between India and Bangladesh, a large Hindu population and Muslims entered from Bangladesh districts. Many came as laborers in search of work from the neighboring district of West Bengal, Midnapur. The Sunderban formed with the main four major communities (Jalais 2010). First are the tribal communities in colonization who were clearing the forest; the second population of people were designated as OBC; the third are Muslim immigrants from Bangladesh; and the fourth are Hindu immigrants who came from Bangladesh as refugees (Jalais 2010). At a slow pace, the land of Sunderban mixed with a different socio-culture atmosphere, which helped to develop a threat to nature.

17.6 Methodology for Inundation Mapping Using Google Earth Engine (GEE)

The schematic flowchart of the methodology is shown in Fig. 17.8.

- **Cloud-masking:** Clouds and cloud shadows are present in the Landsat sceneries, and they need be masked to generate proper composites to enhance accuracy of the inundation maps. The pixels classified as cloud or cloud shadow on the Landsat cloud mask band were masked for each scene using a gap-filling method (Yin et al. 2016). The pixels were then filled with the median value for the pixel from a year before or after the scene's date.
- **Landsat composites:** A gap-filling method was afterward used to cloud masked images. The SWIR band (B7) was chosen for each scenario. Each year's composites are made using all the scenes available from June to September. A median of the corresponding pixel values from all the scenes of that year is evaluated and

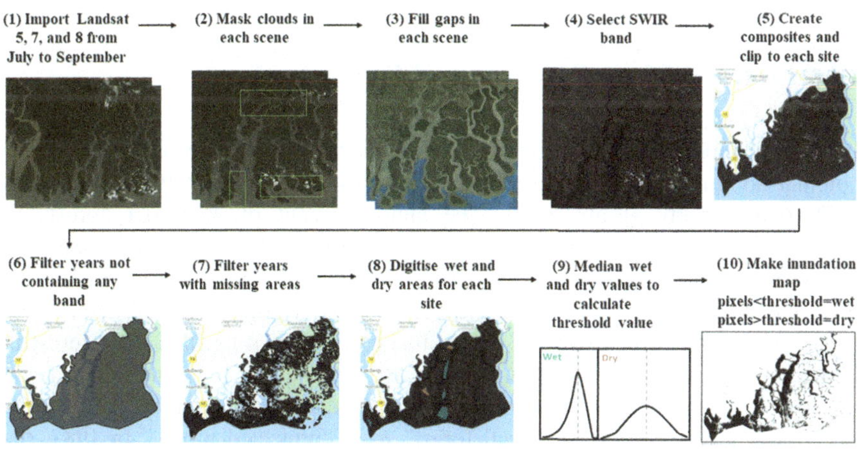

Fig. 17.8 Schematic flowchart of the methodology

designated as the value of the corresponding pixel in the composites to be generated for each pixel in the study area.

- **Filtering bad composites:** The SWIR band was missing for some of the composites created. Those composites were manually filtered by removing them from the image collection. Most of the cloud masking is done via the cloud masking algorithm. However, there were some regions in almost all the coastal sites, where the pixels were classified as a cloud most of the time. The masking algorithm made the pixels transparent in these circumstances. A filtering algorithm was used to filter these pixels. This resulted in a set of composites without even a single masked pixel.
- **Creating inundation maps from the composites:** The inundation maps are made from composites by thresholding the SWIR band pixel values. We manually assessed and digitized permanent wet (e.g., the lake's central region and permanent channels) and dry areas (like barren land or hill region near the wetland) for each site. The median SWIR values for wet ($SWIR_{wet}$) and dry ($SWIR_{dry}$) inundated areas for each composite were calculated using these digitized areas, and a composite specific $SWIR_{threshold}$ (to account for the dynamic seasonal and annual nature of the inundation patterns in wetlands) value was calculated using eq. (17.1) (Wolski et al. 2017).

$$\mathrm{SWIR}_{\mathrm{threshold}} = \mathrm{SWIR}_{\mathrm{wet}} + 0.3\left(\mathrm{SWIR}_{\mathrm{dry}} - \mathrm{SWIR}_{\mathrm{wet}}\right) \quad (17.1)$$

The classifier compares each pixel's SWIR value to its $SWIR_{threshold}$ for each composite. The pixels having SWIR values less than or equal to $SWIR_{threshold}$ are classified as inundated, while pixels having SWIR values larger than or equal to $SWIR_{threshold}$ are classified as dry. As a result, each pixel with a particular SWIR value is classified and transformed into one of two values: 0 for dry pixels and 1 for inundated pixels, and an inundation map is generated using these changed pixels.

17.6.1 Governance Structure and Policy

In colonial times, when they get to know about the revenues from the forest, the colonial administration announced the Sunderban forest as "protected forest" in 1878 (Chakraborty et al. 2009). The announcement of mangrove forests as "protected forest" sparked the struggle of people who were highly dependent on mangrove forests because of the regulation for using the forest. The Forest Department earned an advantage by being designated as "protected forest" rather than "reserved forest." They have the option for lease, converting land from forest to agriculture and timber production (Ghosh and Lalitha 2017). At that period, the mangrove forest has suffered from threats due to man-made activities. In 1943, the tidal forest reclassified as "reserved forest" from "protected forest." In 1987, the Sunderban National Park was designated a World Heritage Site, giving it a high level of protection. The Forest Department suppressed their regular activity in forest protection that further inhibited the opportunities for research. Due to the remaining slackness in the protection of mangrove forests, the National Forest Policy (NFP), in 1988, was redrafted to enhance and improve the condition of the forest, socioeconomic values, and sustainable issues (Bandopadhyay and Nandy 2011). The new changes in National Forest Policy emerged with two specific guidelines for the management of the Sunderban tidal forest, which was earlier led by Joint Forest Management in 1990 through the Government of West Bengal. The specific guidelines are for, first, mangrove forest areas of Sunderban and, second, national park and sanctuaries of the state (Ghosh et al. 2011). Under the 11th 5 years, the State Planning Board of 2007 introduced a specific policy for community participation and involvement. Under this notification, 51 forest protection committees (FPC) and 14 eco-development committees (EDC) were developed to work more specifically in national park and sanctuaries, which are adjoining to villages, and also to increase the involvement of the local community in forest management and distribute the equal benefits (Ghosh et al. 2011) according to the Supreme Court order. But this could not run long enough to further decrease the management of mangrove forest and make the communities receive unequal benefits, rise difficulties, and detach from forest management, which also increases the corruption (Ghosh et al. 2011). In 1973, the Sunderban Development Board (SDB) was controlled under the Planning Department of Government of West Bengal. The Sunderban Development Board mainly focused on the socioeconomic development and development of mangrove plantations in Sunderban area, and the functions were the preparation of an integrated program for utilization, coordination with developmental plans for execution, supervision of the execution of the developmental project, and the evaluation of the progress with reviewing the policies and measures (Bramhachari et al. 2018). The Sunderban Development Board formed the Sunderban Affairs Department (SAD) in 1994, which is under the control of the State Minister to continue the development of the Sunderban. The Sunderban Affairs Department focused on the social, economic, and cultural aspects of people residing in Sunderban (Bramhachari et al. 2018). The Sunderban Affairs Department measures the development of irrigation and agriculture system, provides balance in ecosystem, and preserves them. Some of its main functions are forestry, disaster risk management, water supply, and sanitation, with the help of following government (Bose et al. 2018). After that, the Sunderban Affairs Department was created to manage the governance structure and support people living in climate-vulnerable areas. The Green Bench of Kolkata enforced strict norms for the protection of wetland as Sunderban is one of the Ramsar site, but the Green Bench of the Kolkata was also unable to perform at a time decree in terms of changes in policy for adaptation and translation in development to simplify their use of local communities (Bose et al. 2018). The West Bengal, according to Disaster Management, 2005, proposed disaster management policy and framework in three forms: pre-disaster measures, disaster phase, and post-disaster response (Department of Disaster Management 2005). In the past 10–15 years, transformational changes have been observed in policy due to the introduction of the new structure of governance in West Bengal. The political vibrancy could not protect the implemented plans and program because the plans were made without taking into account the voices of locals in respect to needs in Sunderban (Bose et al. 2018). Efforts were taken from the government to give protection of mangrove forests and adapt to climate change over the pasts, but the results are vague due to lack of communication. The problem was detected by the local people who were living in the land of Sunderban, and they had better knowledge regarding the problems of Sunderban than outsiders. With implementation of plans and programs by a government institution, the locals don't know about the purpose, functions, and use of the plans. The findings also show the scientist have significant roles in policymaking to set purposes according to research and findings through the government (Ghosh 2018). At the local level of an institution like Gram Panchayat and self-help group, structure and initiatives were lacking and fuzzy due to a lack of uniformity (Ghosh 2018). It appears there is a scarcity of scientific knowledge from experts to protect mangrove vegetation, which raises uncertainty in various departments and mainly at the grassroots level and also suffers from a lack of compatibility and motivation. The governance structure is not well defined to meet the goals of plans and programs. The governance requires an authority to design it, and it also needs the body of governance either run by the public and private institution or the cooperation of two formal and informal institutions. The rising disparity between the views of policymakers and local level of institution increases the risk and threat of climate change and introduces new challenges to the people living in the Sunderban area and to others (Ghosh et al. 2018). The policy having a lack of clarity and awareness for the local people to combat for coping strategies of climate change. The lack of institutional linkages has made it impossible to inquire about their basic needs and demands to run their livelihood, and this has also deprived them of the traditional knowledge and experiences (Bose et al. 2018). The traditional knowledge is an auspicious tool for the locals to sustainably manage the mangrove plantation. The people of Sunderban are highly dependent on the natural resources of the forest, and the forest act as a natural wealth for them. Their traditional knowledge and participation are not incorporated into the government policy, which has shown many lack of communication between government and locals for the protection of mangrove forests (Karmakar 2018). The ethnic identity of local communities is needed to be preserved, so their involvement will give a sustainable result in mangrove conservation, but the inconsistency of policy acts as a barrier to community participation. It is essential to raise the awareness and understanding among locals about mangrove destruction from decades ago because many people don't have the idea about the climate change that can uproot the mangrove vegetation in the near future. There is a need for traditional knowledge as well as scientific knowledge with the help of sustainable policy to conserve the precious tidal forest. There is a need for a distinct change in policy and governance structure that has the capacity to cope in harsh conditions (Gleditsch et al. 2007). If the present policies continue, the Indian Sunderban will face inviolable challenges from climate change and population rise. There is a strong need for the intervention of all different actors (national, state, district, and local) for the management. WWF gave a vision as a long-term perspective by 2050 for Indian Sunderban to protect the land from natural and anthropogenic loss; it also aimed to reduce the risk from weather events, restore mangrove plantations, and raise socioeconomic status in the future. There is a need of an interdisciplinary and collaborative approach, which can help Sunderban restore the mangrove plantation in a sustainable way that would propagate a large amount of economic activity from the forest in the future, so that not only national institution but also international institution will take part in raising the finance.

17.7 The Governance Structure Needs to Restructure the Sunderban Land in Future

In the structure, the level of governance comprises a three-tier mechanism, and all the actors are equally important for participation. The Sunderban area is divided into parts for different functions. The area which lies near the shoreline would not permit a settlement area. Understanding the capacity building of the population is necessary for the shoreline area. The increasing population has already expanded in the danger zone area, which enhances the vulnerability. The area, like Gosaba, Basanti, and Patharpratima (Hazra et al. 2011), would be covered by mangrove plantations, which would act as a good adaptation strategy. The initiative the institution would take for the living population of these areas would give them some alternative, or they could access other areas. In this way, the land can restore mangrove vegetation in the future. These areas have active tidal creeks, which can submerge in the ocean. If these islands are not taken into consideration, then it increases the risk in the surrounding areas, like Namkhana and Sagar Island having, which have a large population. The physical infrastructure development would take into account the department of natural disasters because of the intensity of cyclones, floods, and all-weather events in Sunderban. All areas should not be considered for physical development, which results in the clearance of mangrove forests. If the restoration program of mangrove plantations continues in this way, then it would be possible to return to sustainable development in Sunderban. The mangrove cover benefits not only the surrounding areas of Sunderban also the neighboring city of Sunderban. It was estimated that healthy mangroves earn as an ecosystem service in the range of US$ 2000 to 9000 per hectare yearly (Spalding et al. 2010).

17.8 Conclusion

The study shows that the mangrove ecosystem is highly vulnerable to climate change issues. Climate change and climate-induced changes like sea-level rise have pose a threat to the Indian Sunderban for decades. The major divers of climate change are sea-level rise, population rise, and the increase in pollution levels in Sunderban. Over the Bay of Bengal, the sea surface temperature is increasing by 0.019 °C annually, which enhances the effect of sea-level rise of 3.14 mm annually (Hazra et al. 2002). The sea-level rise introduces many changes, like changes in cyclonic activity, erosional, and accretion patterns. It was noticed from the past the intensity of cyclone increase than the frequency of cyclone appears on Indian Sunderban. The changes in erosional and accretion patterns result in changes in land cover. The population rising is one factor that influences sea-level rise by clearing mangrove vegetation and also climate change and sea-level rise impacts the population of Sunderban. The pollution is directly impacted and can destroy mangrove plantations. The climate change and its associated changes are suffering the mangrove ecosystem and slowly destroying its natural precious ecosystem service. The living people of Sunderban are and will always be highly dependent on mangrove service regardless of any alternatives. In the current situation, the changes cannot alter the situation, but changes in the government policy and governance could control the loss of mangrove cover from a long-run perspective. All types of actors are needed, from international to district levels, for the restoration of mangrove in the Indian Sunderban. As we know, the Sunderban has the largest number of mangroves that meet the human needs until death. There is a need for a bottom-up approach that focuses on the grassroots level. Indigenous knowledge is a special tool in terms of mangrove protection. The lack of communication from ground level is unable to bring change in the restoration of mangroves. More community participation, local involvement, and discussion with scientific knowledge are needed in policymaking, which can function successfully and sustainably in the future. Education and empowerment skills are necessary for locals to combat climate change, and they need to understand the uncertainties of climate change in Sunderban, which would ruin their future. The small-scale adaptation program would increase the resilience of mangroves againts natural calamity. The interdisciplinary collaboration needs to develop through policies that address the conservation of mangroves in the Indian Sunderban.

References

Bandopadhyay S, Nandy S (2011) Trend of sea level change in the Hugli estuary, India. Indian J Geo-Mar Sci 40:802–812

Banerjee K (2013) Decadal change in the surface water salinity profile of Indian Sundarbans: a potential indicator of climate change. J Mar Sci Res Dev 01(S11):3

Bhattacharya S, Nageswara Rao K, Subraelu P, Rao TV, Hema MB, Ratheesh R, Rajawat AS (2008) Sea-level rise and coastal vulnerability: an assessment of Andhra Pradesh coast, India through remote sensing and GIS. J Coast Conserv 12:195–207

Biswas SR, Choudhury JK, Nishat A, Rahman MM (2007) Do invasive plants threaten the Sundarbans mangrove forest of Bangladesh? For Ecol Manag 245(1):1–9

Bose S (2004) The Sundarbans biosphere: a study on uncertainties and impacts in active delta region

Bose S, Bhatt MR, Mehta L, Adam HN, Srivastava S, Ghosh U, Movik S, Narayanan NC, Naess LO, Parthasarathy D, Wilson C (2018) Bridging the gaps in understandings of uncertainty and climate change: round table reports, Experience learning series 74. AIDMI, Ahmedabad

Bramhachari R, Ghosh U, Bose S (2018) Living on the edge: climate change and uncertainty in the Indian Sundarbans, STEPS Working Paper 101. STEPS Centre, Brighton

Chakraborty SK, Giri S, Chakravarty G, Bhattacharya N (2009) Impact of eco-restoration on the biodiversity of Sundarbans mangrove ecosystem, India. Water Air Soil Pollut: Focus 9:303–320

Danda A (2010) Sundarbans: future imperfect climate adaptation report. WWF-India, New Delhi. http://awsassets.wwfindia.org/downloads/sundarbans_future_imperfect__climate_adaptation_report_1.pdf

Das J, Jha S, Goyal MK (2020) On the relationship of climatic and monsoon teleconnections with monthly precipitation over meteorologically homogenous regions in India: Wavelet & global coherence approaches. Atmos Res 238:104889. https://doi.org/10.1016/j.atmosres.2020.104889

Ganguly D, Mukhopadhyay A, Pandey RK, Mitra D (2006) Geomorphological study of Sundarban deltaic estuary. J Indian Soc Remote Sens 34(4):431–435

Ghosh A (2012) Living with changing climate—impact, vulnerability and adaptation challenges in Indian Sundarbans. Centre for Science and Environment, New Dehli, p 108

Ghosh N (2018) Climate change and agrarian systems: adaptation in climatically vulnerable regions. Indian J Agric Econ 73(1):38–53. (Keynote Paper from the 73rd Conference of Indian Society of Agricultural Economics)

Ghosh U, Lalitha V (2017) Child health amidst climatic adversities in the Indian Sundarbans. Future Health System, Brighton

Ghosh A, Danda AA, Sriskanthan G, Bandyopadhyay J, Hazra S (2011) Indian Sundarbans delta: a vision. World Wide Fund for Nature-India, New Delhi

Ghosh A, Schmidt S, Fickert T, Nüsser M (2015) The Indian Sundarban mangrove forests: history, utilization, conservation strategies, and local perception. Diversity 7:149. https://doi.org/10.3390/d7020149

Ghosh U, Bose S, Brahmachari R (2018) Sundarbans living on the edge: climate change and uncertainty in the Indian Sundarbans, 101

Gleditsch NP et al (2007) Climate change and conflict: the migration link. International Peace Academy

Gopal B, Chauhan M (2006) Biodiversity and its conservation in the Sundarban mangrove ecosystem. Aquat Sci 68:338–354

Goyal MK, Ojha CSP (2010) Evaluation of various linear regression methods for downscaling of mean monthly precipitation in arid Pichola watershed. Nat Res 01(01):11–18. https://doi.org/10.4236/nr.2010.11002

Goyal MK, Ojha CSP (2012) Downscaling of precipitation on a lake basin: evaluation of rule and decision tree induction algorithms. Hydrol Res 43(3):215–230. https://doi.org/10.2166/nh.2012.040

Goyal MK, Panchariya VK, Sharma A, Singh V (2018) Comparative assessment of SWAT model performance in two distinct catchments under various DEM scenarios of varying resolution, sources and resampling methods. Water Resour Manag 32(2):805–825. https://doi.org/10.1007/s11269-017-1840-1

Goyal MK, Gupta AK, Jha S, Rakkasagi S, Jain V (2022) Climate change impact on precipitation extremes over Indian cities: non-stationary analysis. Technol Forecast Soc Change 180:121685. https://doi.org/10.1016/j.techfore.2022.121685

Hazra S, Ghosh T, Dasgupta R, Sen G (2002) Sea level and associated changes in the Sundarbans. Sci Cult 68(9–12):309–321

Hazra S, Samanta K, Mukhopadhyay A, Akhand A (2010) Temporal Change Detection (2001–2008) of the Sundarban; Unpublished Report. WWF-India, New Delhi

Hazra S, Danda AA, Sriskanthan G, Ghosh A, Bandyopadhyay J (2011) Indian Sundarbans delta: a vision. World Wide Fund for Nature-India, New Delhi, p 27

Hazra S, Akhand A, Chanda A, Dutta S, Hazra S, Sanyal P (2012) Comparative study of heavy metals in selected mangroves of Sundarban ecosystem, India. J Environ Biol 33(6):1045–1049

Hazra S, Bhadra T, Sinha Roy SP (2015) Proceedings of the international seminar on challenges to ground water management: vision 2050. Centre for Ground Water Studies, Kolkata

Hazra R, Szabo S, Tessler Z, Ghosh T, Matthews Z, Foufoula GE (2017) Unravelling the association between the impact of natural hazards and household poverty: evidence from the Indian Sundarban delta. Sustain Sci 12:453–464

IPCC (2007) Climate change: the physical science basis, Report of Working Group I

Jalais A (2010) Forest of tigers: people, politics, and environment in the Sundarbans. Routledge, Abingdon

Karmakar S (2018) Ethnic identity and forest preservation: a sociological enquiry on Sundarbans, West Bengal. J Res Humanit Soc Sci 6(1):20–25

Khairuddin B, Yulianda F, Kusmana C, Yonvitner (2016) Degradation mangrove by using Landsat 5 TM and Landsat 8 OLI image in Mempawah regency, West Kalimantan Province year 1989-2014. Procedia Environ Sci 33:460–464

Mandal AK, Ghosh RK (1989) Sundarbans: a socio bio-ecological study. Bookland Private Limited, Calcutta

Mandal S, Roy S, Ghosh PB (2009) Modelling of the contribution of dissolved inorganic nitrogen (DIN) from litter fall of adjacent mangrove forest to Hoogly-Matla estuary, India. Ecol Model 220:2988–3000

Mondal S, Kanjilal B, Guha Mazumdar P, Barman D, Mukherjee M, Singh S, Mandal A (2010) HealthCare in the Sundarbans (India) challenges and plan for a better future

Morgan JP, McIntire WG (1959) Quaternary geology of the Bengal basin, East Pakistan and Burma. Bull Geol Soc Am 70:319–342

Nanlohy H, Bambang A, Ambariyanto, Hutabarat S (2015) Coastal communities knowledge level on climate change as a consideration in mangrove ecosystems management in the Kotania Bay, West Seram regency. Procedia Environ Sci 23:157–163

Naskar K, Sen Sarkar N, Ghosh A, Dasgupta M, Sengupta B (2004) Status of the mangroves and mangrove ecosystem of Sundarbans in West Bengal: its impact on Estuarine Wet lands fisheries. Bulletin No. 134 CIFRI, Barrackpore, Kolkata, pp 1–40

Panda A (2010) Climate refugees: implications for India. Econ Polit Wkly 45(20):76–79

Poonia V, Goyal MK, Gupta BB, Gupta AK, Jha S, Das J (2021) Drought occurrence in different river basins of India and blockchain technology based framework for disaster management. J Clean Prod 312:127737. https://doi.org/10.1016/j.jclepro.2021.127737

Ramsar (2022) Ramsar sites information service-Sundarban wetland [WWW document]. https://rsis.ramsar.org/ris/2370

Reid W, Mooney H, Cropper A, Capistrano D, Carpenter S, Chopra K (2005) Millennium ecosystem assessment. Ecosystems and human well-being: synthesis. Curr Sci 84(6):25

Rudra K (2018) Rivers of the Ganga-Brahmaputra-Meghna delta: a fluvial account of Bengal. https://link.springer.com/book/10.1007/978-3-319-76544-0

Sadhuram Y, Sarma VV, Murthy TV, Rao BP (2005) Seasonal variability of physico-chemical characteristics of the Haldia channel of Hooghly estuary, India. J Earth Syst Sci 114:37–49

Sahana M, Hong H, Ahmed R et al (2019) Assessing coastal island vulnerability in the Sundarban Biosphere Reserve, India, using geospatial technology. Environ Earth Sci 78:304

Samanta K, Hazra S (2017) Mangrove Forest cover changes in Indian Sundarban (1986–2012) using remote sensing and GIS. In: Hazra S, Mukhopadhyay A, Ghosh A, Mitra D, Dadhwal V (eds) Environment and earth observation. Springer, Cham

Sarkar A (2011) Sundarbaner Itihas. Rupkatha Prakasan, Kolkata

Saw A, Kanzaki M (2015) Local livelihoods and encroachment into a mangrove forest reserve: a case study of the Wunbaik reserved mangrove forest, Myanmar. Procedia Environ Sci 28:483–492

Sarkar NS, Naskar K (2003) Algal Flora of Sundarbans Mangals. Sanctum Books. Daya Publishing House. ISBN: 9788170352860

Schwartzstein P (2019) This vanishing forest protects the coasts and lives of two countries. Retrieved from https://www.nationalgeographic.com/magazine/article/sundarbans-mangrove-forest-in-bangladesh-india-threatened-by-rising-waters-illegal-logging

Selvam V (2003) Environmental classification of mangrove wetlands of India. Curr Sci 84(6):757–765

Sen A, Bhadury P (2017) Intertidal foraminifera and stable isotope geochemistry from world's largest mangrove, Sundarbans: assessing a multiproxy approach for studying changes in sea-level. Estuar Coast Shelf Sci 192:128–136

Singh OP (2007) Long-term trends in the frequency of severe cyclones of Bay of Bengal: observations and simulations. Mausam 58(1):59–66

Singh OP (2009) Simulations of frequency, intensity and tracks of cyclonic disturbances in the Bay of Bengal and the Arabian Sea. MAUSAM 60(2)

Spalding M, Kainuma M, Collins L (eds) (2010) World atlas of mangroves. Earthscan, London, pp xvi–+319

Subramoniam SR, Ravindranath S, Rakkasagi S, Ram H (2022) Water resource management studies at micro level using geospatial technologies, pp 49–74. https://doi.org/10.1007/978-3-030-98981-1_2

UNDP (2009) Human development report: overcoming barriers: human mobility and development. https://hdr.undp.org/content/human-development-report-2009

Wolski P, Murray-Hudson M, Thito K, Cassidy L (2017) Keeping it simple: monitoring flood extent in large data-poor wetlands using MODIS SWIR data. Int J Appl Earth Obs Geoinf 57: 224–234. https://doi.org/10.1016/j.jag.2017.01.005

Yin G, Mariethoz G, McCabe M (2016) Gap-filling of landsat 7 imagery using the direct sampling method. Remote Sens 9:12. https://doi.org/10.3390/rs9010012

Chapter 18
Eco-Restoration for Climate Resilience and Disaster Risk Reduction

Anjali Barwal and Atisha Sood

Abstract Eco-restoration is an approach for coping with and adapting to climate change, but its results are also vulnerable to its effects. When carried out successfully and responsibly, eco-restoration helps to safeguard biodiversity, enhance human health and happiness, increase food and water security, deliver services and products, promote economic growth, and assist in climate resilience and adaptation. Moreover, it is a problem-solving strategy that involves local groups, academics, decision-makers, and land managers in order to restore ecological harm and foster better coexistence between humans and the rest of nature. Eco-restoration is the connection that is required to shift regional, national, and worldwide environmental circumstances from a state of continuing degradation to one of net improvement when paired with conservation and sustainable usage. There has never been a time when restoring degraded ecosystems was more urgently needed. In order to account for complex ecosystem functioning, this chapter discusses problems, potential strategies, and execution, particularly in relation to climate resilience and disaster risk reduction.

Keywords Eco-restoration · Climate change · Resilience · Disaster risk reduction · Disaster management

18.1 Introduction

Ecosystems are living groups of microbes, plants, and animals that interact with their surroundings in a dynamic way. Human activity has the potential to harm, deteriorate, or even destroy these communities. The loss of ecological systems, including

---

A. Barwal (✉)
ECDRM Division, National Institute of Disaster Management (Ministry of Home Affairs), New Delhi, India

A. Sood
ECDRM Division, Ministry of Home Affairs, Government of India, New Delhi, India

A. K. Gupta et al. (eds.), *Ecosystem Restoration: Towards Sustainability and Resilient Development*, Disaster Resilience and Green Growth,
https://doi.org/10.1007/978-981-99-3687-8_18

forests, wetlands, marine and coastal systems, and dry lands, is a significant contributor to disaster risk and vulnerability. Due to the loss of biodiversity and ecological system services, degraded land and seascapes cost one-tenth of the world's gross product, jeopardize the well-being of 3.2 billion people—that is, 40% of the world's population—and cover over 2 billion hectares (IPBES 2018; Simonson et al. 2021).

The issues and challenges facing the planet are enormous. The COVID-19 pandemic, food and water insecurity, and the climate emergency are all having negative effects on billions of people worldwide. Ecosystems are a significant resource as we tackle these problems. It is crucial to keep them safe and sustainably manage their resources. But it won't be enough to simply increase the preservation and wise usage of our surviving beautiful habitats and water resources; we also need to restore the earth's devastated ecological systems and the enormous advantages they give (Goyal and Ojha 2010, 2012; UNEP 2021).

Eco-restoration is becoming more widely acknowledged as a vital tool for reducing the effects of environmental disasters and climate change as well as preparing for them. Globally, restoration can assist in the transition from decades of incremental environment degradation to neutral land use and eventually to net ecological improvement (Gann et al. 2019; Das et al. 2020). The entire planet must accomplish its goal to repair at least 1 billion deteriorated acres of land over the next 10 years, an area approximately the size of China, in order to combat the triple threat of climate change, loss of biodiversity, and pollution (WMO 2021).

By 2050, it is anticipated that between 50 and 700 million people will migrate due to land deterioration, climate change, and increasing trends of natural as well as man-made disasters. Rehabilitation can address some of the primary causes of environmentally prompted population movements by lowering resource scarcity, improving revenue creation, and supporting climate resilience and disaster risk reduction (Goyal et al. 2018; IPBES 2018; Poonia et al. 2021).

For tropical and mega-diverse nations, ecological restoration still poses significant obstacles, including the requirement to develop strategies that are both economically and technically realistic as well as government regulations and management systems that can gauge effective implementation. The setting and the diagnosis of the area in relation to reference ecosystems have a significant impact on the designing, implementation, and analyzing of restoration activities (e.g., forests, savannas, grasslands, wetlands) (Bustamante et al. 2019).

18.2 Eco-Restoration for Climate Resilience

Along with other related changes like rising ocean acidification and atmospheric carbon dioxide levels, climate change has an impact on ecosystems via altering mean conditions and climate variability. At the same time, ecosystems can also help with climate change adaptation and mitigation. It is necessary to investigate and quantify the mechanisms, potential, and limitations of such natural climate change remedies.

By strengthening preparedness and reducing susceptibility to adverse weather conditions, ecosystem restoration can significantly contribute to people's ability to adapt to climate change. As part of a comprehensive adaptation plan, restoration, conservation, and sustainable management of ecosystem and biodiversity services can be combined with more conventional techniques and methods (Kapos et al. 2019). Communities can frequently benefit from the restoration of coastal ecosystems by learning how to adapt to climate risks, including sea level rise, storm surges, and related flooding. Recent data demonstrates that healthy mangroves can operate as a powerful barrier against the devastation caused by tsunamis and can lower wave heights by 5% to 30% (Spalding ct al. 2014). Inland eco-restoration can also lessen climate-related risks including flooding, soil erosion, and landslides brought on by exceptionally heavy rains. Restoration of the forest on slopes slows erosion brought on by heavy rain. In a similar way, the restoration of upland forests aids in controlling the flow of water, regulating supplies through periods of both heavy precipitation and dry spells (Kapos et al. 2019).

18.3 Eco-Restoration for Disaster Risk Reduction

Natural disasters have had a devastating impact on economies, properties, livelihoods, and human lives over the past decade. Ecosystem-based Disaster Risk Reduction (Eco-DRR) is a strategy where disasters are mitigated, prevented, or buffered by systematically utilizing the regulating functions of ecological systems (such as forests, wetlands, and mangroves). Ecosystem-based solutions can offer services for reducing catastrophe risk with economic and cultural value, all of which help communities become more resilient to disasters and the effects of climate change.

Eco-DRR is the sustainable and green method to minimize the disaster and extreme weather event damages. To mitigate the disaster risks, it is important to concentrate on natural solutions (such as maintaining wetlands, forests, and floodplains and restoring mangroves to the seashore to lessen the impact of waves and storm surge and prevent flooding). Additionally, it must guarantee that both new and current infrastructure is climate resilient.

Sadly, mangroves and coastal zones have lost size and health due to deforestation, land use changes, and human activities like aquaculture and tourism. Experts already believe that this degradation plays a big role in the lethal nature of Amphan in 2020 and Yaas in 2021 cyclone consequences. The "right place, right tree" concept offers technical assistance for intelligent greening and public involvement, which can greatly aid in sustainable eco-restoration for climate resilience and DRR. Hence, a systematic and integrated approach to planning is very important for the proper participation of many different organizations, including national, local, and municipal planning bodies, departments, etc.

Simultaneously, traditional coping mechanisms and wisdom must be combined with contemporary methods. The impacts of natural and climate change-related disasters can be lessened by combining traditional and scientific management of coastal ecosystems with mangroves and other plants that follow the triple-tier mechanism and habitat. The entire coastal zone may be made more productive and sustainable under such a management structure (Sen 2021).

Eco-Restoration and Disaster Risk Reduction at Indian Sundarbans
The Sundarban Biosphere Reserve (a world heritage site, according to UNESCO) in India (9630 sq km) is the only home of mangrove tigers and is located in the largest delta in the world. It is covered in the longest continuous mangrove forest on the planet. This region is forced to take the brunt of natural disasters, which are occurring more frequently as a result of the whims of climate change and the effects of sea level rise along the Ganges-Brahmaputra-Meghna delta. There have been steps undertaken to lessen the effects of cyclones on the socioeconomically disadvantaged population of the delta and their way of life. Eco-DRR is a common strategy used globally to reduce resource damage. Mangroves have been shown to provide significant ecosystem benefits in this situation. One of the most important of these is to act as a natural bio-shield to protect the coastline from the impacts of climate change, furies of cyclones, storm surges, and tsunami. During the COVID-19 pandemic period, between May 16–21, 2020, this zone was struck by a severe super storm, AMPHAN reaching the peak wind speed of 260 km/h. This caused enormous devastation, and according to the official report, 1200 square kilometers (i.e., or 28% of the Sundarban region) of the 4263 square kilometres of the mangrove reserve forest were badly affected. In addition to preventing 60% of earthen embankment breaches, the mangrove bio-shield acted as a wind breaker and has significantly reduced flooding (Chowdhury et al. 2021).

18.4 Solutions and Practical Applications

In order to improve climate resilience, we primarily focus on the potential and difficulties related with the practical management, restoration, and ecosystem preservation. Under the broad framework of nature-based solutions (NbS), where the backdrop is mitigation of climate change, there has been an increase in interest in the ability to save, restore, and utilize ecosystems as tools to combat climate change.

By 2050, approximately two-thirds of the planet's population will reside in cities, making them a crucial location for addressing the impacts of climate change. Cities have many characteristics that make them vulnerable to the effects of climate change, such as low plant cover, increased impermeable cover, pollution production, urban heat, growing requirements for freshwater resources, and concentration of infrastructure and population in vulnerable areas, like coastal zones, river floodplains, and deforested hillsides. Urban areas' amplification effects on climate change hazards can be reduced through eco-restoration and NbS. These strategies include increasing the amount of vegetation and green space, building rainwater harvesting ponds and other buildings that restore natural hydrologic function, and reestablishing natural protective habitats along the coastlines.

18.5 Decade on Ecosystem Restoration

The United Nations (UN) declared the current decade (2021–2030) to be the "Decade on Ecosystem Restoration" due to the urgent need to combat the effects of climate change and human-induced ecosystem degradation. This international movement aims to stop, prevent, and reverse ecosystem damage on a global scale. The main commitment of this goal is to:

(a) Focus on solutions that allow species to thrive or reestablish where their numbers have been depleted.
(b) Model the impact of climate and land use change on genetic diversity and provide early warning signs of ecosystems in danger of collapse.
(c) Create accurate habitat maps, land use projections, and decision-support tools to inform landscape-scale restoration for biodiversity net gain, water and soil security, and poverty alleviation.

The conclusion of the UN Decade on Ecosystem Restoration and the accomplishment of emission reduction objectives in line with the Paris Climate Agreement goal to keep global warming below 2 °C should both occur in 2030 in order to prevent catastrophic climate change. Sustainable land governance, which includes restoration, is a successful method of reducing global warming (Bastin et al. 2019; Strassburg et al. 2020). However, the only ultimate approach is eco-restoration. The benefits of healthy and restored ecosystems are shown in Fig. 18.1.

Also, since 2019, the United Nations Environment Program (UNEP) and Partners for Resilience (PfR) have worked together to develop and implement scalable Eco-DRR models. They have done this by partnering with various governments and their respective communities to build their capacity and shape Eco-DRR policy interventions. Through the three main pillars of Eco-DRR—ecosystem restoration/protection, disaster risk reduction, and climate smart livelihoods—a model for scaling up community resilience has been devised (shown in Fig. 18.2). With a focus on water-related, risk-sensitive wetlands restoration and capacity building efforts, India places a higher emphasis on ecosystem restoration and protection (UNEP 2022).

Fig. 18.1 Benefits of healthy and restored ecosystems

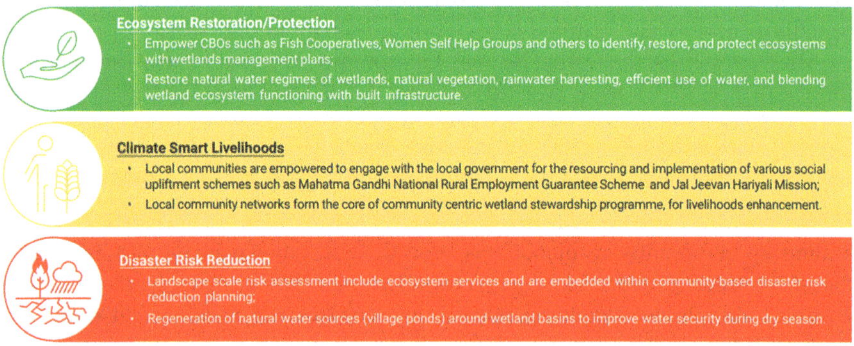

Fig. 18.2 A model devised for Eco-DRR upscaling through capacity-strengthening and the participation of communities and (local) governments (Source: UNEP 2022)

18.6 Conclusion

As a result of ongoing climate change, people and ecosystems will either have to adapt to a world that is significantly warmer than it is now or significant steps will have been taken to minimize warming within the next few decades. Eco-restoration can be a significant factor in societal adaptation and climate resilience, but it will only be beneficial if used in combination with a decline in fossil fuel emissions. In order to provide ecosystem services and mitigate climate and disaster-related risks, the integration of Eco-DRR initiatives into management plans should place sufficient emphasis on institutional structures, monitoring, and community involvement. The development of a training manual and the dissemination of Eco-DRR information to regional disaster management agencies should make it easier for them to incorporate this strategy into their planning and operations. It is also possible to lessen the effects of natural and climate change-related disasters by integrating traditional and modern scientific ecosystem management with mangroves and other vegetation that use a triple-tier mechanism and habitat.

References

Bastin JF, Finegold Y, Garcia C, Mollicone D, Rezende M, Routh D et al (2019) The global tree restoration potential. Science 365(6448):76–79. https://doi.org/10.1126/science.aax0848

Bustamante M, Silva JS, Scariot A, Sampaio AB, Mascia DL, Garcia E, Nobre C (2019) Ecological restoration as a strategy for mitigating and adapting to climate change: lessons and challenges from Brazil. Mitig Adapt Strateg Glob Chang 24(7):1249–1270

Chowdhury A, Naz A, Iyer AS, Bhattacharyya S (2021) Ecosystem based disaster risk reduction at Indian Sundarbans: a lesson learned from AMPHAN supercyclone. In IOP conference series: earth and environmental science 796, 1, 012042. Bristol: IOP Publishing

Das J, Jha S, Goyal MK (2020) On the relationship of climatic and monsoon teleconnections with monthly precipitation over meteorologically homogenous regions in India: Wavelet & global coherence approaches. Atmos Res 238:104889. https://doi.org/10.1016/j.atmosres.2020.104889

Gann GD, McDonald T, Walder B, Aronson J, Nelson CR, Jonson J, Dixon KW (2019) International principles and standards for the practice of ecological restoration. Restor Ecol 27(S1):S1–S46

Goyal MK, Ojha CSP (2010) Evaluation of various linear regression methods for downscaling of mean monthly precipitation in arid Pichola watershed. Nat Res 01(01):11–18. https://doi.org/10.4236/nr.2010.11002

Goyal MK, Ojha CSP (2012) Downscaling of precipitation on a lake basin: evaluation of rule and decision tree induction algorithms. Hydrol Res 43(3):215–230. https://doi.org/10.2166/nh.2012.040

Goyal MK, Panchariya VK, Sharma A, Singh V (2018) Comparative assessment of SWAT model performance in two distinct catchments under various DEM scenarios of varying resolution, sources and resampling methods. Water Resour Manag 32(2):805–825. https://doi.org/10.1007/s11269-017-1840-1

IPBES 2018, summary for policymakers of the assessment report on land degradation and restoration, intergovernmental science-policy plattorm on biodiversity and ecosystem services (2018)

Kapos V, Wicander S, Salvaterra T, Dawkins K, Hicks C (2019) The role of the natural environment in adaptation: background paper for the global commission on adaptation. Global Commission on Adaptation, Rotterdam and Washington, DC

Poonia V, Goyal MK, Gupta BB, Gupta AK, Jha S, Das J (2021) Drought occurrence in different river basins of India and blockchain technology based framework for disaster management. J Clean Prod 312:127737. https://doi.org/10.1016/j.jclepro.2021.127737

Sen S (2021) Combating tropical cyclones Amphan, Yaas and after: eco-restoration of coastal zones. Harvest 6(1):33–38

Simonson WD, Miller E, Jones A, García-Rangel S, Thornton H, McOwen C (2021) Enhancing climate change resilience of ecological restoration– a framework for action. Perspect Ecol Conserv 19(3):300–310

Spalding M, McIvor A, Tonneijck FH, Tol S, van Eijk P (2014) Mangroves for coastal defence: guidelines for coastal managers & policy makers. Wetlands International and the Nature Conservancy, Wageningen

Strassburg BBN, Iribarrem A, Beyer HL, Cordeiro CL, Crouzeilles R, Jakovac CC et al (2020) Global priority areas for ecosystem restoration. Nature 586(7831):724–729. https://doi.org/10.1038/s41586-020-2784-9

UNEP (2021) Becoming #GenerationRestoration: ecosystem restoration for people, nature and climate. United Nations Environment Programme, Nairobi

UNEP (2022) Up-scaling community resilience through ecosystem-based disaster risk reduction. United Nations Environment Programme, Nairobi

WMO (2021) Ecosystem restoration is vital for climate, biodiversity. World Meteorological Organization, Geneva; https://public.wmo.int/en/media/news/ecosystem-restoration-vital-climate-biodiversity

Chapter 19
Ecosystems and Nature-Based Solutions (NbS) for Health Protection and Epidemic Resilience

Atisha Sood and Anjali Barwal

Abstract There has been significant concern that a loss of access to natural environments, particularly those with varied plants and soil microbes, magnifies personal and public health issues. "Nature-based solutions" (NbS) seek to make use of nature to tackle issues like climate change, health, water resources, food security, and disaster risk reduction. As a result, experts interested in both public and environmental health have advocated a variety of nature-based solutions to promote public health. Despite significant advances in both extent and scale regarding environmental protection, along with biological restoration, natural resource debasement continues, threatening both ecosystem functions and human well-being. The NbS idea is widely understood to include the advantages of ecosystem restoration and rehabilitation, carbon neutrality, enhanced environmental quality, public health and wellness, and proof for such benefits. In regard to climate change action and sustainable resolutions to improve ecosystem resilience and adaptable capability for health systems, NbS as a comprehensive concept might be advantageous.

Keywords Ecosystem · Nature-based solutions · Health protection · Epidemic · Resilience

19.1 Introduction

Environmental scientists, restoration naturalists, public health professionals, and health service researchers and doctors are some of those worried about climate change wreaking havoc on habitat, ecological systems and balance, and public health even now (Liddicoat et al. 2018; Goyal et al. 2018). Urbanization and climate

A. Sood (✉)
ECDRM Division, Ministry of Home Affairs, Government of India, New Delhi, India

A. Barwal
ECDRM Division, National Institute of Disaster Management (Ministry of Home Affairs), New Delhi, India

A. K. Gupta et al. (eds.), *Ecosystem Restoration: Towards Sustainability and Resilient Development*, Disaster Resilience and Green Growth,
https://doi.org/10.1007/978-981-99-3687-8_19

change have brought about a variety of social concerns for urban regions, including setbacks and deterioration of natural habitats, soil capping, flooding, and drought, all of which present additional challenges to ecosystem functioning, biodiversity, ecosystem services (ES) delivery (like clean air, water, and soil), and, as a result, human and public health and well-being. Since it has been commonly established that regular exposure to diversified, healthier surroundings may have modest to extraordinary advantages on physiological health and mental well-being, researchers and practitioners from a variety of fields have collaborated on these concerns to address real-world challenges (Martinez-Juarez et al. 2015; Wall et al. 2015).

Furthermore, scholarly and widespread research has emerged with an ever-growing following and coalitions composing academics, policymakers, and laypeople alike curious as to how biodiversity loss and access to natural ecosystems may harm people's health, especially for children living in urban areas (Poonia et al. 2021; Das et al. 2020; Williams 2017). According to the hygiene hypothesis, some of these health consequences can be traced to the misuse of antibiotics and fungicides, exposure to polluted air, or insufficient physical activity (Velasquez-Manoff 2012). Mental and physical health issues may be minimized or improved by increasing the accessibility of public to agricultural, natural, and restored environments near their homes, workplaces, and schools, according to research.

19.2 Nature-Based Solutions

Nature-based solutions necessitate humans working with nature, as an element of nature, to address social difficulties, with advantages for both human well-being and biodiversity. Policymakers, governing authorities, scientists, and nongovernmental groups (NGOs) have all endorsed the name NbS. Various definitions of NbS have arisen over time, including those of two major governmental and nongovernmental groups that have significantly affected the NbS debate latterly, namely, the International Union for Conservation of Nature (IUCN) and the European Commission (EC) (Bauduceau et al. 2015). However, there is no consensus on the differences and similarities between these two NbS classifications. Furthermore, ecosystem-based adaptation/mitigation (EbA/EbM), blue infrastructure (BI), green infrastructure (GI), blue/green infrastructure/green/blue infrastructure (BGI/GBI), and ecological engineering (EE) are all clearly linked to the NbS idea (Table 19.1). Even though the extent of these notions' parallels and differences are explored in popular research, it is unclear if NbS is significantly distinct from all the other terms. Regardless of the differences in definitions, NbS are being tested, created, and implemented all over the world and have therefore been scientifically evaluated. Many studies have demonstrated that NbS help to conserve biodiversity, produce extra environmental-economic-social benefits, and serve as a foundation for climate change adaptation and mitigation (Nesshöver et al. 2017; Cohen-Shacham et al. 2016; Goyal and Ojha 2010).

Table 19.1 Terminologies related to nature-based solutions

Terminology	Definition
Nature-based Solutions (NbS)	Climate change, biodiversity degradation, water security, water pollution, food security, human health, and disaster risk management are all examples of concerns that need sustainable management and use of nature to address socio-environmental difficulties
Ecological Engineering (EE)	To integrate society with its natural surroundings, a sustainable ecosystem is developed for the mutual interests of human society and the natural environment
Green Infrastructure (GI)	A network of natural and semi-natural regions, as well as other environmental characteristics, that has been intentionally planned and managed to provide a diverse variety of ES
Blue Infrastructure (BI)	Refers to water-related urban infrastructure, described as a network that provides the "ingredients" for overcoming urban and climatic difficulties by constructing with nature
Blue/Green infrastructure/Green/ Blue Infrastructure (BGI/GBI)	A network of interconnected natural and built landscape components, such as water bodies and green and open areas, that serve many roles
Ecosystem-based Adaptation/Mitigation (EbA/EbM)	The use of biodiversity and ecosystem services as part of a larger adaptation/mitigation plan to assist people in adapting to/mitigating the negative consequences of climate change Policies and actions that take into account the role of ES in lowering society's susceptibility to climate change in a multi-sectoral and multi-scale manner

19.3 Role of NbS for Addressing Climate Change

In order to limit global warming to well below 2° and preferably to 1.5° C to meet the Paris Agreement goal, the world economy would have to lower the greenhouse gas (GHG) emissions substantially and rapidly, achieving net-zero carbon dioxide by roughly 2050, and this would need continuous net carbon dioxide removal beyond this. Nature-based solutions play a significant role in lowering GHG emissions from agriculture, forests, and land use, which now accounts for about 22% of yearly GHG emission, as well as conserving and increasing the carbon sink on land and in water. Improved management of land and sea utilized for working (i.e., cropland, timberlands, pastures, and aquatic systems) reduces CO_2, methane, and NO_2 emissions and sequesters carbon; preserving indigenous vegetation throughout rural landscapes and tree plantations in urban areas increases CO_2 expulsion from the atmosphere; and efficaciously defending undamaged forest areas, vegetation, and wetlands retains vital deposits of carbon trapped in the biosphere. Estimates of how much nature-based remedies may chill the earth differ greatly (Cohen-Shacham et al. 2019). Recognizing that major climate change has already happened and that more warming is unavoidable, participants to the Paris Agreement have a commitment to "increasing adaptation capacity, improving resiliency, and lowering vulnerability towards climate change."

Even though there isn't one gauge that can quantify adaptation progress, which is both an undertaking and a conclusion, there exist several instances of effective adaptations at the local level. Therefore, there is a significantly growing disparity between climate change adaptation demands and practical action (Melanidis and Hagerman 2022; Goyal and Ojha 2012). This one is particularly the case in poor countries, where international climate money is scarce and mitigation costs are 5 to 10 times higher than present public funding flows. Nature-based solutions can assist in closing this gap. Since time immemorial, humans have collaborated with Mother Nature in order to help them adapt to environmental changes. Nature-based solutions make use of this process and contribute to many elements of adaptation (IPCC 2022; Anderegg et al. 2020).

There is strong evidence that they can limit direct exposure to the effects of climate change. Reviving natural ecosystems, for example, could support healthy ground and vegetations, which reduces dangers of droughts, floods, and landslides by enhancing water penetration and depository, stabilizing slope and beaches, and diminishing energy of the waves. Blue and green infrastructure, including bioswales, green-roofs, and built wetlands, could assist to temper heat waves and prevent floods in metropolitan settings (Berrang-Ford et al. 2021). Also, there are indications that nature-based solutions may minimize sensitivity, or how much climate change affects individuals. Nature-based agriculture approaches, like agroforestry, could, for example, sustain and improve production in drier and more variable climate.

19.4 Co-Benefits of NbS for Ecosystems Restoration and Public Health

Biodiversity is recognized as one of the basic necessities for ecosystems to function well for human health and well-being, and it is entrenched like in policy-centric settings. Climate change is understood to be altering the natural world and how it interacts with human health. Climate, for example, has an impact on ecological processes of the human. Climate change impacts ecosystem functioning in terms of the number and quality of functions that benefit human physical health.

Climate and biodiversity serve as critical "boundary conditions" for people's health and well-being. These boundary conditions have an impact on various other variables that affect persons' health and well-being via their natural surroundings and associated ecosystem activities. Health is controlled not just by exterior ecosystem-related processes and causes but also by internal ones, such as acknowledging that the human body itself harbors complex and biodiverse ecosystems with varying effects on physical health. The abundance, kind, and quality of the natural environment, which is supported by 'external' biodiversity, are examples of external variables.

Other external determinants include social relationships (e.g., family and neighborhood), access to health care, and income (e.g., through diet). All are inextricably linked to larger socioeconomic and political situations.

A number of studies have shown that nature-based interventions might indeed help enhance people's overall health in restored ecosystems in terms of the following indicators:

- Higher resistance to contagious (respiratory) illnesses (Liddicoat et al. 2018)
- Increased cardiorespiratory endurance (Williams 2017)
- Increased anti-inflammatory capacity (Deans 2017)
- Increased resistance to endoparasites (Wall et al. 2015)
- Lower prevalence of neurological and musculoskeletal problems (Robinson and Breed 2019)
- Decreased allergen sensitivity
- Lowered susceptibility to obesity and adult-onset diabetes
- Illness prevention (Mills et al. 2017)

Also, there is evidence that nature-based initiatives can improve the mentioned factors of psychological health or mental well-being:

- Decreased anxiousness (Chun et al. 2017)
- Lowered stress-related cortisol levels (Ibrahim et al. 2017; Ward Thompson et al. 2012)
- Lesser vulnerability to autistic disorders (Clarke et al. 2013)
- Lowered neurodevelopmental disorders
- Increased perceived sense of restorativeness (Breitenbecher and Fuegen 2019)
- Enhanced cognitive ability (Bratman et al. 2012)
- Higher social cohesion (Chou et al. 2017)
- Lower intensity of depression

19.5 Conclusion

Nature-based solutions aid in the transition to a circular economy by promoting greater ecological sustainability of the physical environment. NBS has a broad variety of applications in the physical environment, and they may also be used to supplement existing grey infrastructures. Nature-based solutions can contribute significantly to attaining net-zero carbon emissions by the end of 2100, but only when supplemented with other climate solutions, such as drastic reductions in GHG emissions across all sectors of the economy. Attaining zero-net carbon emissions and making the transition to a nature-positive economy will also necessitate systemic change in how humans function as communities and run our economic systems, requiring a shift to a dominant worldview based on valuing the quality of life and human well-being over materialistic things and relationship with nature over conquests. There are many signs that this transformation is occurring, such as the increase of grassroots climate and environmental advocacy.

Nature-based solutions, when carefully implemented to guarantee that diverse values of the natural world are maintained, have the potential to expedite this transition while simultaneously lowering warming, creating resilience, and safeguarding biodiversity.

References

Anderegg WR, Trugman AT, Badgley G, Anderson CM, Bartuska A, Ciais P, Randerson JT (2020) Climate-driven risks to the climate mitigation potential of forests. Science 368(6497):eaaz7005

Bauduceau N, Berry P, Cecchi C, Elmqvist T, Fernandez M, Hartig T, Tack J (2015) Towards an EU research and innovation policy agenda for nature-based solutions & re-naturing cities: final report of the horizon 2020 expert group on 'Nature-based solutions and re-naturing cities'

Berrang-Ford L, Siders AR, Lesnikowski A, Fischer AP, Callaghan MW, Haddaway NR, Abu TZ (2021) A systematic global stocktake of evidence on human adaptation to climate change. Nat Clim Chang 11(11):989–1000

Bratman GN, Hamilton JP, Daily GC (2012) The impacts of nature experience on human cognitive function and mental health. Ann N Y Acad Sci 1249:118–136

Breitenbecher KH, Fuegen K (2019) Nature and exercise interact to influence perceived restorativeness. Ecopsychology 11:33–42

Chou SY, Grossman M, Saffer H (2017) An economic analysis of adult obesity: results from the behavioral risk factor surveillance system. In: Determinants of health: an economic perspective. Columbia University Press, pp 666–693

Chun MH, Chang MC, Lee SJ (2017) The effects of forest therapy on depression and anxiety in patients with chronic stroke. Int J Neurosci 127:199–203

Clarke G, Greenham S, Scully P, Fitzgerald P, Moloney RD, Shanahan F (2013) The microbiome—gut-brain axis during early life regulates the hippocampal serotonergic system in a sex-dependent manner. Mol Psychiatry 18:666–673

Cohen-Shacham E, Walters G, Janzen C, Maginnis S (2016) Nature-based solutions to address global societal challenges, vol 97. IUCN, Gland, pp 2016–2036

Cohen-Shacham E, Andrade A, Dalton J, Dudley N, Jones M, Kumar C, Walters G (2019) Core principles for successfully implementing and upscaling nature-based solutions. Environ Sci Pol 98:20–29

Das J, Jha S, Goyal MK (2020) On the relationship of climatic and monsoon teleconnections with monthly precipitation over meteorologically homogenous regions in India: Wavelet & global coherence approaches. Atmos Res 238:104889. https://doi.org/10.1016/j.atmosres.2020.104889

Deans E (2017) Microbiome and mental health in the modern environment. J Physiol Anthropol 36:1. https://doi.org/10.1186/s40101-016-0101-y

Goyal MK, Ojha CSP (2010) Evaluation of various linear regression methods for downscaling of mean monthly precipitation in arid Pichola watershed. Nat Res 01(01):11–18. https://doi.org/10.4236/nr.2010.11002

Goyal MK, Ojha CSP (2012) Downscaling of precipitation on a lake basin: evaluation of rule and decision tree induction algorithms. Hydrol Res 43(3):215–230. https://doi.org/10.2166/nh.2012.040

Goyal MK, Panchariya VK, Sharma A, Singh V (2018) Comparative assessment of SWAT model performance in two distinct catchments under various DEM scenarios of varying resolution, sources and resampling methods. Water Resour Manag 32(2):805–825. https://doi.org/10.1007/s11269-017-1840-1

Ibrahim MH, Somers JA, Luecken LJ, Fabricius WV, Cookston JT (2017) Father–adolescent engagement in shared activities: effects on cortisol stress response in young adulthood. J Fam Psychol 31:485–494

IPCC (2022) Climate change 2022: mitigation of climate change. In: Shukla PR, Skea J, Slade R, Al Khourdajie A, van Diemen R, McCollum D, Pathak M, Some S, Vyas P, Fradera R, Belkacemi M, Hasija A, Lisboa G, Luz S, Malley J (eds) Contribution of working group III to the sixth assessment report of the intergovernmental panel on climate change. Cambridge University Press, Cambridge, New York, NY

Liddicoat C, Bi P, Waycott M, Glover J, Breed M, Weinstein P (2018) Ambient soil cation exchange inversely associates with infectious and parasitic disease risk in regional Australia. Sci Total Environ 626:117–125

Martinez-Juarez P, Chiahbai A, Taylor T, Quiroga-Gomez S (2015) The impacts of ecosystems in human health and well-being: a critical review. J Outdoor Recreat Tour 10:63–69

Melanidis MS, Hagerman S (2022) Competing narratives of nature-based solutions: leveraging the power of nature or dangerous distraction? Environ Sci Pol 132:273–281

Mills JG, Weinstein P, Gellie NJC, Weyrich LS, Lowe AJ, Breed MF (2017) Urban habitat restoration provides a human health benefit through microbiome rewilding: the microbiome rewilding hypothesis. Restor Ecol 25:866–872

Nesshöver C, Assmuth T, Irvine KN, Rusch GM, Waylen KA, Delbaere B, Haase D, Jones-Walters L, Keune H, Kovacs E et al (2017) The science, policy and practice of nature-based solutions: an interdisciplinary perspective. Sci Total Environ 579:1215–1227

Poonia V, Goyal MK, Gupta BB, Gupta AK, Jha S, Das J (2021) Drought occurrence in different river basins of india and blockchain technology based framework for disaster management. J Clean Prod 312:127737. https://doi.org/10.1016/j.jclepro.2021.127737

Robinson JM, Breed MF (2019) Green prescriptions and their co-benefits: integrative strategies for public and environmental health. Challenges 10:9. https://doi.org/10.3390/challe10010009

Velasquez-Manoff M (2012) An epidemic of absence: a new way of understanding allergies and auto-immune diseases. Scribner/Simon & Schuster, New York, NY

Wall DH, Nielsen UN, Six J (2015) Soil biodiversity and human health. Nature 528:69–76

Ward Thompson C, Roe J, Aspinall P, Mitchell R, Clow A, Miller D (2012) More green space is linked to less stress in deprived communities: evidence from salivary cortisol patterns. Landsc Urban Plan 105:221–229

Williams F (2017) The nature fix: why nature makes us happier, healthier, and more creative. W.W. Norton, New York, NY

Chapter 20
Freshwater Ecosystem Conservation for Social Protection, Business, and Local Economy

J. S. Sudarsan, Mayur Shirish Jain, Padma Parija, and S. Nithiyanantham

Abstract With rapid urbanization and industrial development, rivers are being turned into drains. Awareness and cultural orientation toward the importance of rivers are lacking among people. There need to be robust policies to tap into this huge potential of riverfront development (RFD). Therefore, most RFD projects must be planned without giving attention to outcomes related to intensifying natural risks of flooding and relegated water quality. This research study was taken up to develop effective RFD plans and policies. In this paper, the Mula-Mutha River in Pune, a city of Maharashtra, India, was considered. Data about river quality was collected along with drafting the policy framework based on questionnaire input and expert interviews. About 38% of people consider RFD to impact ecological balance positively, and 49% believe it will exponentially affect the city's economic development. It can be inferred that this kind of study in all the water bodies is mandatory based on the outcome of this RFD research study.

Keywords Urbanization · Riverfront development (RFD) · Questionnaire survey · Expert survey

---

J. S. Sudarsan
School of Energy and Environment, NICMAR University, Pune, Maharashtra, India

M. S. Jain (✉)
Department of Civil Engineering, Indian Institute of Technology Indore, Indore, Madhya Pradesh, India
e-mail: mayur.jain@iiti.ac.in

P. Parija
L and T Reality, Powai, Mumbai, Maharastra, India

S. Nithiyanantham
Department of Physics (Ultrasonic/NDT and Bio-Physics Divisions), Govt Arts and Science College (Bharathidasan University, Thiruchirapalli), Thiruvarur, Tamil Nadu, India

A. K. Gupta et al. (eds.), *Ecosystem Restoration: Towards Sustainability and Resilient Development*, Disaster Resilience and Green Growth,
https://doi.org/10.1007/978-981-99-3687-8_20

20.1 Introduction

In recent decades, quick urbanization, serious utilization of water assets, ways of life, and expanded industrialization have ensured climate-related issues, such as water shortages, contamination, strong waste, and biodiversity (Subramoniam et al. 2022). From a metropolitan viewpoint, the riverfront of any city is a significant road for diversion openings and open green spaces inside the city improvement (Dutta et al. 2018). Urban riverfronts, such as riverfront improvement, trenches/lakefront turn of events, or even seafront advancement, have been disregarded for quite some time. To improve the nature of the city's urban life and upgrade social development, lakefronts, seashores, and riverside promenades are being created with the view that such arranged spaces would pull in venture capital, the travel industry, and add to the economic development of the city (Poonia et al. 2021; Das et al. 2020; Burgers 2000). Rivers and riverfronts, which were significant components of the societal fabric, have proficient rapid changes due to alterations in their use, the pressure of urbanization, and the decaying health of river systems (Binti et al. 2011). Riverfront is also included in open spaces and considered in urban fabric for the development aspects of a city. RFD in India confines ecological, economic, and social elements (Goyal et al. 2018; Al-Shams et al. 2013). There has been a suitable thought of the characteristic capability of the land for forming into a biodiversity zone for moderating the normal legacy of the river bowl just as the neighborhood prerequisites of offices at the metropolitan level, similar to enormous level city open spaces of variable structures with sporting offices. There have been trials to spread these recovery projects from humble and straightforward variants to aggressive complex ones (Kushwah and Mehmood 2017).

The main agenda of the research was to understand and analyze the policy downfalls in RFD in India, progressively giving policy ideas to overcome shortcomings in RFD. Rivers in India are present in every part. Researchers have developed policy recommendations from this research to help us improve RFD, particularly in the Mula and Mutha Riverfront in Pune. Essentially, the riverfront is where the urban turn of events and water associate and are characterized as a one-of-a-kind and vital asset where land, water, air, sun, and beneficial plants interface with one another (Goyal et al. 2022; Chen and Zhang 2007). Waterway banks normally repress the centers around which towns develop. They comprise the noteworthy regions that are huge travel industry destinations. They are viewed principally as open spaces that flaunt social collaborations and network correspondences (Lu et al. 2013). It is likewise characterized as a coordinating area of water and people. Any advancement in the fronts of water bodies is termed a riverfront improvement or development. There is always a need to tend to RFD for various reasons. It can be converted into the city's focal point by merging individuals' fascination, vacation destination, production of infrastructural and recreational offices, and changing the city's overview to be more reasonable as far as natural improvement and comprehensive growth are concerned (Shang 2018). Henceforth, riverfront improvement plans and ensuing waterways must advance the production of an equalized social and social biological system to ensure smooth and maintainable use of the same (Patil et al. 2019; Goyal and Ojha 2010, 2012).

For the sustainable development of the urban riverfront, we must consider each regeneration: developing different partnership models that are liable for the entire management and in the design, construction, and implementation stages (Wakefield 2007; Piryani et al. 2017) and developing finance models that gather public and private sector sources about the project area. Neighborhood riverfronts have procured expanding importance to the financial advancement of urban areas and thus managed themselves as critical hubs in globalization procedures and the travel industry improvement, something that the Mula and Mutha Riverfronts need to work on (McCarthy 2004; Light et al. 2020). A significant part of the severe repositioning of Asian urban communities worldwide is the limit of postmodern riverfronts for urban and ecological recoveries. As these urban areas move from modern to support economies, the nature of their urban spaces becomes progressively significant. The perceivability of riverfront territories and their entrance to different city pieces become advantages for redeveloping these spots (Shah and Roy 2017; Panuccio 2019). This paper explores the sustainable principles of urban riverfronts, focusing on the Mula and Mutha Riverfronts based on their design, functionality, and management components (Hussein 2014).

It was indicated that the present promenade is more sustainable than it was a few decades ago (Follmann 2015). Understanding the physical sustainability through basic promenade designs and social sustainability in terms of functionality and management is essential (Damanik and Pratiwi 2017). The riverfront has been changed. A public-driven process, be it planning, design, or management, has always been followed. This has helped with enhanced space functionality and made the planning process fast and convenient (Trivedi and Awasthi 2021). The interpretation of literature spanned across the Internet over the years inspired the methodology and research technique used in this work, such as (1) parameters for riverfront development, (2) statistical analysis, (3) comparative study for the policy of riverfront development, etc. Our findings triggered the understanding of the parameters, like flood mitigation, tourism generation, water quality, creating recreational spaces, public amenities, financing patterns, land use, economic growth, etc., which lead to the successful riverfront.

20.2 Research Methodology

The research has three major segments, showcased in Fig. 20.1, using which the study's objective was fulfilled. An in-depth analysis of RFD projects was studied, along with policies governing the same. This research is peddled by primary and secondary data, which depend on literature reviews and comparative analysis of surveys.

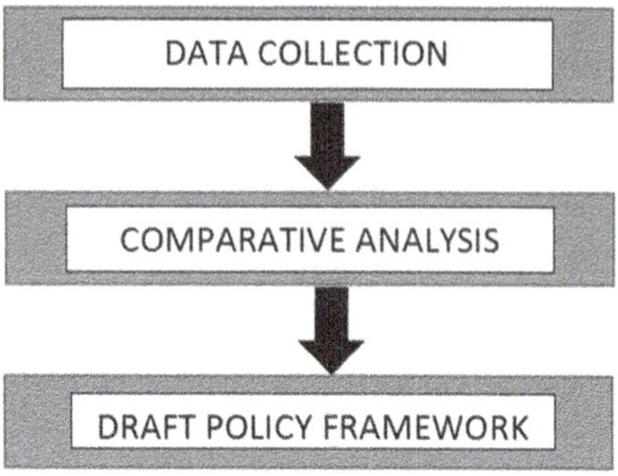

Fig. 20.1 Research methodology flowchart

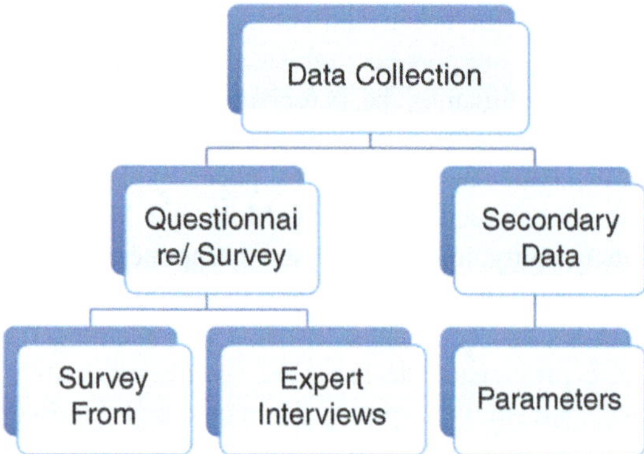

Fig. 20.2 Data collection flowchart

20.2.1 Data Collection

This research study collects data in two ways: first, secondary data is collected utilizing riverfront development parameters from national case studies, and second, expert interviews and survey forms are used, with responses recorded for statistical analysis. Figure 20.2 shows a flowchart we followed to achieve the study's goal.

20.2.1.1 Secondary Data: Parameters

The importance of studying the parameters is to understand how the RFD policies work and to see where we have gaps in our policies and how we can fill them. The following parameters, illustrated in Fig. 20.3, were found after an in-depth investigation of the literature spanned across the years and the secondary data, which were based on national case studies.

Fig. 20.3 RFD parameters

National Studies

The riverside development projects in India have various limits that have been extrapolated from the case studies. Various rivers flow through India, passing through many cities and towns, but the criteria for developing the riverfront here necessitate a separate set of guidelines. Because the development guidelines are the same worldwide, the parameters deduced are the same as in international case studies. Economic growth spurred by riverfront development, water quality and treatment, property use around the river in the public interest, and public amenities boosts revenue and expands the scope of neighborhood development. Finance is essential for progress, as it entails various political and economic constraints. These factors contribute to a better and more sustainable neighborhood, which must be investigated, regulated, and applied in a specific way to benefit everyone in and around the city. The results are parameters affecting riverfront development, such as land use, accessibility and connectivity, economic growth, flood mitigation, water quality, financing patterns, and public amenities after examination of secondary data from national case studies (Patil et al. 2019).

20.2.1.2 Questionnaire Survey

A total of 75 responses were acquired through a Google form survey described here. This survey was done among a group of ordinary individuals who reacted positively. This study aimed to learn about the public's perceptions regarding riverfront development.

20.3 Results and Discussions

The main aim of this study was to understand, analyze, and compare the RFD policies achieved by doing a thorough literature study, otherwise called the primary source of data collected. It aids in framing the methodology incorporated and understanding the research technique adopted in this paper. Then, analyzing the secondary data substantiating the questionnaire survey and conducting expert interviews aid in achieving the main objective of recommendation policies governing RFD in India.

20.3.1 Outcome of Questionnaire Survey

To substantiate the outcome of the survey form responses for further statistical analysis, they are summarized below according to the questions asked in the survey form shown in Fig. 20.4.

20.3.1.1 Perception of People Toward the Development Model

This question was formulated to understand whether people know the different financing models available to finance a project. The three models, public, private, and public-private, are the most commonly used. It is clear that people are aware of different models and have a better understanding of how these models work and that it is viable to go for a public-private model. Adapting this model will share the risk among public and private parties. This model will also help massive projects like

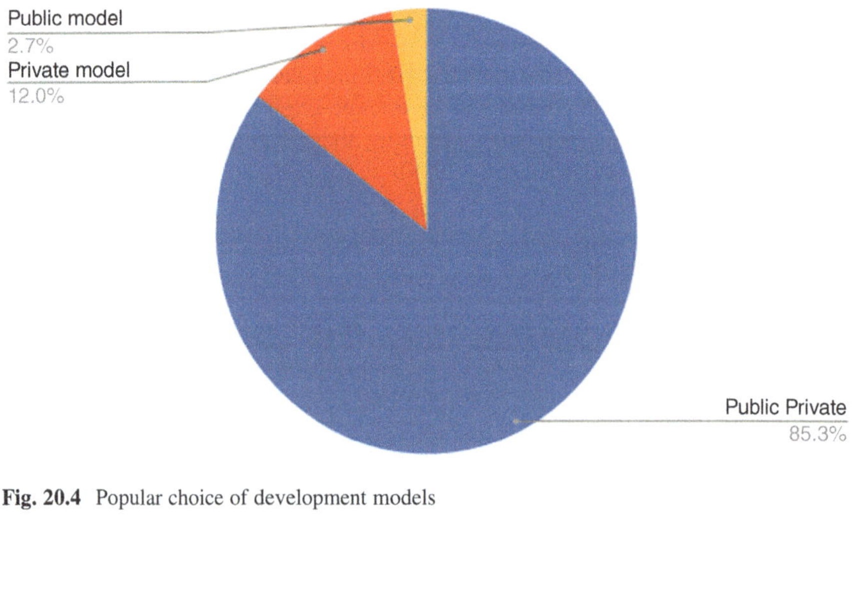

Fig. 20.4 Popular choice of development models

riverfront development to be viable in financing. A concession agreement can be formed to help both parties profit and provide better service quality to the people (Vaništa Lazarevic and Antonic 2016; Pagés Sánchez and Daamen 2020). The private parties can look over the operation and maintenance. Also, the quality of service provided by the private parties will be good, which will also help generate revenue. Therefore, the financing parameter can be taken care of by using a public-private model, which will help in risk sharing but will also help in generating revenue.

20.3.1.2 Public Impression of the Effect on Ecological Balance

A river is now one of the most critical common assets, especially for human existence and its current situation. Since the dawn of civilization, rivers have demonstrated their importance in shaping and influencing the development of human settlements, resulting in the growth of massive urban communities worldwide. It may very well be seen that none of the riverfront redevelopment plans incorporate ecological plan standards into the riverfront plan. The inclusion of human interference in natural activities through rapid development has contributed to the massive transformation of riverbanks, as most of them no longer resemble their original state. Riverbank development in urban areas has made the change processes even faster and more drastic, as depicted in Fig. 20.5. It very well may be concurred that distinguishing the most appropriate site is essential before directing any riverfront redevelopment. This is where all the significant and delicate territories are distinguished and saved.

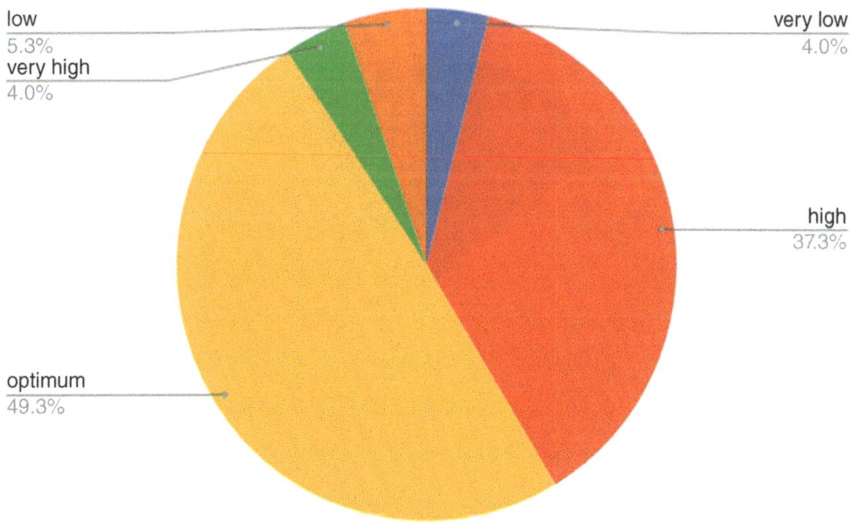

Fig. 20.5 Effect of RFD on ecology

20.3.1.3 Public Opinion on the Value of Real Estate Development

The location of a real estate property is a critical factor that determines the value of a property (Fig. 20.6). Rivers and water bodies are appreciated natural assets for human life, the environment, and national development. Appreciation of water resources as a national heritage will add to more long-term sustainable property development. RFD will increase demand for residential properties near that area, leading to increased construction of real estate properties in the same.

20.3.1.4 Perception of People on the Footfall in the Locality

Footfall is the new currency for shopping centers in this continuously digital world. Amenities, such as sports facilities, performing arts halls, and recreational parks, on riverfronts aim to enhance residents' quality of life near the river area, leading to an increase in the footfall of the riverfront area. A tourist attraction is one of the activities that will increase footfall in the RFD area, as shown in Fig. 20.7.

20.3.1.5 Public Opinion on the Economic Development of the City/Town

Figure 20.8 suggests that proper management of the riverfront is necessary for the day. Positively, there will be preparatory work, pay, and yield impacts from the development of the dams and the advancement of parks and sporting offices. Be that as it may, such effects are not durable. What is enduring is expanded financial movement that outcomes from upgraded financial matters. The external cash from guests into the nearby economy through affirmation tickets, transportation,

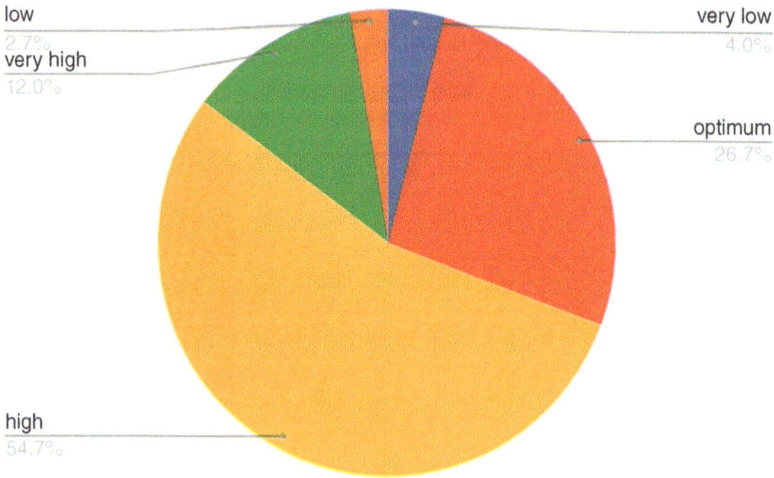

Fig. 20.6 Influence on real estate due to RFD

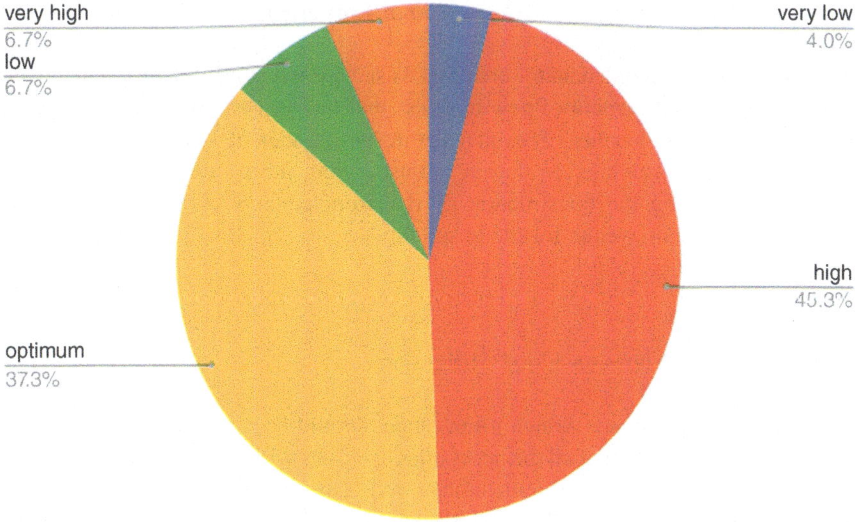

Fig. 20.7 RFD shall help make the city a tourist spot

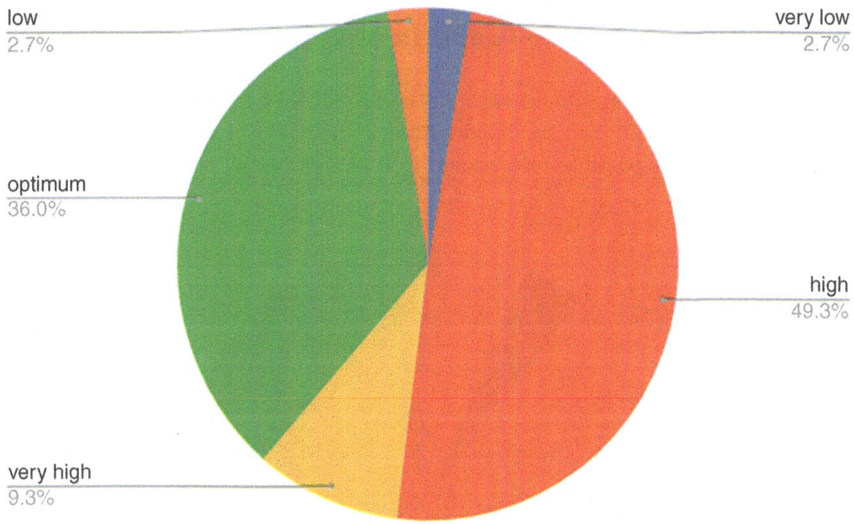

Fig. 20.8 Votes suggest that RFD is required for the economy of the city

enrollment charges, lodging stays, and cafe and retail spending during the riverfront visit will prompt the neighborhood's financial improvement.

20.3.1.6 Public View on the Impression of Employment Generation

RFD has much maintenance, creating vast opportunities at different levels (Fig. 20.9). Encouragement of new business will have an enormous impact on the employment of the locality. The increase in the number of local shops will give employment opportunities to locals. Construction labor will have short-term employment during the construction of riverfront embankments, spillovers, etc. RFD is an excellent avenue for the hospitality sector, which will increase employment in the locality.

20.3.1.7 Public Views on Flood Mitigation

Floods are hard to foresee, which means states should be set up to react unequally. Figure 20.10 illustrates that all the necessities and rules will successfully alleviate the effect of atmosphere calamities, including floods. Springs are fundamentally used to alleviate the effect of dry spells, so the accompanying portrayals center around the other three proposals.

20.3.1.8 Public Views on the Tourism Development

India's tourism contributes to international and national economic growth, and RFD offers recreational and tourism opportunities. Figure 20.11 demonstrates that in order to implement possible recreational activities properly, an assessment of the

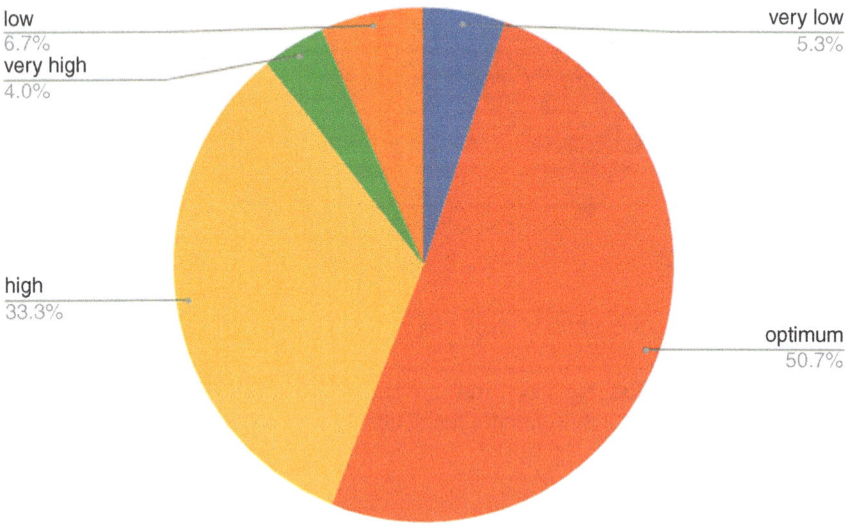

Fig. 20.9 Chances of employment

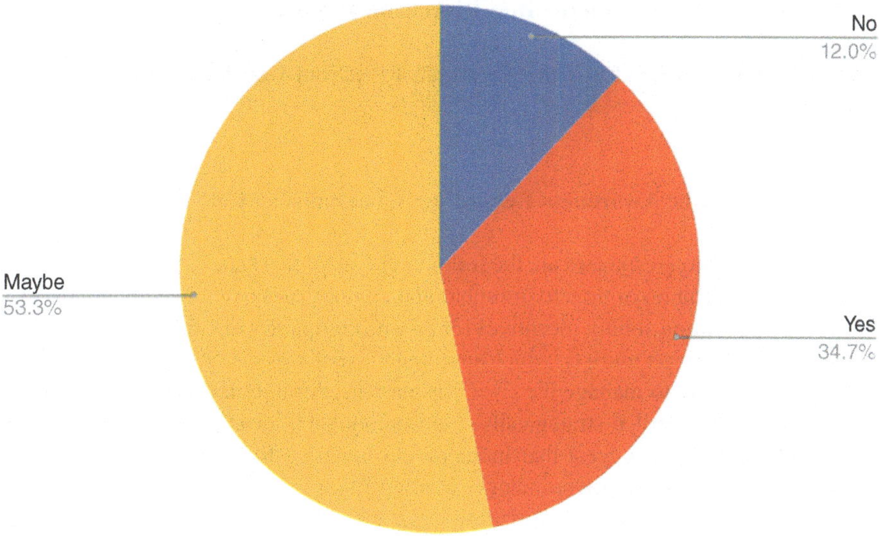

Fig. 20.10 RFD might control floods as per the popular public opinion

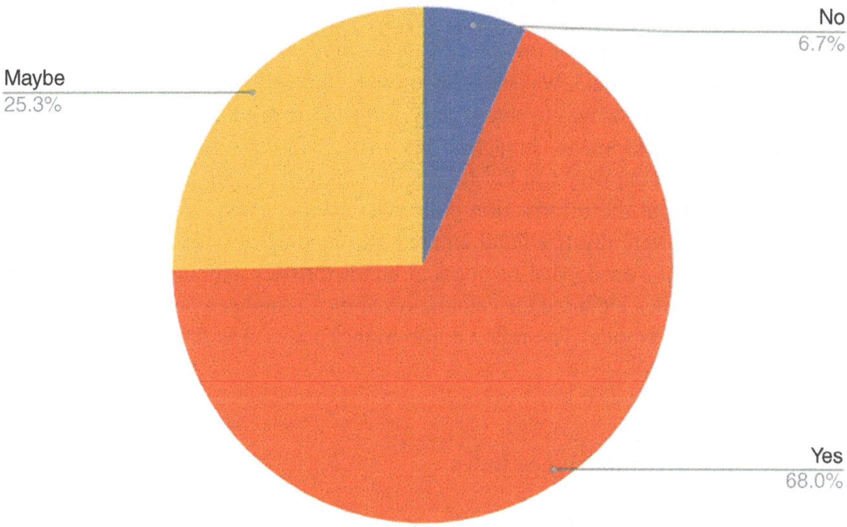

Fig. 20.11 Positive response to the presence of recreation resource elements

recreation resource element is required. Tourism is no longer limited to natural sites but is increasingly viewed as a possible destination with well-developed infrastructure and increased accessibility, as well as its tourist market.

20.3.1.9 Public View on the Best Riverfront Project

Most people responded that the Sabarmati Riverfront Development Project is the best in India.

20.3.1.10 Public Awareness Regarding Sabarmati Project

The most common responses are the following: (1) An RFD could be successful with careful consideration of the environment and ecology and could become financially viable by providing zoning incentives for private companies in public spaces. (2) It is a great way to attract tourists, which simultaneously helps the city in case of a flood. An excellent way to manage the river was passing through the city. (3) It is unique but needs to be handled strategically. Fine-tuning such ideas is critical and crucial. (4) And the authors suggest that India should invest in RFD projects, and citizens should not feel that the government is wasting funds.

20.3.2 Comparative Analysis

SPSS (Statistical Package for the Social Sciences), a statistical program for analyzing complex statistical data, was used to compare survey responses. The purpose of the comparative analysis was to statistically confirm the respondents' assessments of the riverside's development characteristics so that India's riverside development standards might be improved in the future. The analysis in this section was confirmatory in that it put the information collected in the survey form to the test. The survey responses were digitized and entered into a computer database (Hersh et al. 2012). To offer an overview of respondent characteristics and answer patterns, the analysis of the replies included the calculation of means and percentages by response category for all questions. The relative importance index was the tool used for this analysis.

20.3.2.1 Relative Importance Index

Relative importance index (RII) analysis allows for identifying most of the essential criteria based on responses. It is also an appropriate tool to prioritize indicators rated on Likert-type scales.

From the ranking given in Table 20.1, real estate value, town or city economic development, and locality footfall are the most critical factors shown by the relative importance index. So, to substantiate the output of RII, the following ranking was given: (1) real estate value, (2) economic development, and (3) footfall in the locality

Table 20.1 Output of relative importance index test

Parameters	Relative importance index	Ranking
1. What will be the effect of RFD on the overall ecological balance?	66.40	4
2. What will be the effect of RFD on the value of the real estate in the locality?	73.60	1
3. What will be the impact of RFD on footfall in the locality?	68.80	3
4. What will be the influence of RFD on the economic development of the town/city?	72.00	2
5. What will be the impact of RFD on employment in a locality?	64.80	5
6. So you think RFD is one of the best ways of flood mitigation?	43.73	6
7. Do you think RFD will attract more tourists to the city?	31.47	7

are considered the most influencing parameters among the rest considered and analyzed.

20.3.3 Expert Interview

A total of 10 expert interviews were conducted online among researchers, scientists, builders, academicians, NGOs, etc. The experts are identified based on their expertise in this RFD study. The objective for the interviewees' feedback was two-way. Firstly, to obtain reactions from the respondents about riverfront development in India, including their experience and thoughts on future implementations of riverfront development and policies related to riverfront development, as well as any obstacles along the development process. Secondly, it provided a platform for interviewees to offer insights and alternative perspectives or views on how they visualized future riverfront development in India, especially concerning policies and practices.

In this research study, the Mula-Mutha River in Pune, Maharashtra, was studied, and data on river quality was acquired via questionnaire survey and feedback, as well as exemplary expert interviews. A thorough cross-checking of responses was performed to provide analytics of the current situation and how it may be improved using statistical analysis. There is a need to monetize water bodies and improve the condition they are currently in. The data collected from background study and on-ground research helped us draft some recommendations and a skeleton of policies that could be implemented.

20.3.4 Draft Policy Framework

Recommendations for policies for best practices for riverfront development in India are as follows:

Policy 1: Unity of organizations, effective monitoring of illegal constructions, proper mapping of rivers, cultural orientation toward rivers, and long-term planning are the needs of the hour for effective RFD in India.

The main reasons that constrain effective administration and management and administration for riverfront resources and development are (1) multiplicity of organizations, (2) ineffective monitoring of illegal construction, (3) encroachments, (4) pollution, (5) lack of proper mapping of rivers, and (6) lack of cultural orientation toward rivers.

Policy 2: The central government, state government, MOEFCC, urban local government, and special purpose vehicles constituted for RFD should work in perfect coordination for efficient policy implementation in India.

Different parties involved in riverfront development in India are (1) the central government, (2) the state government, (3) MOEFCC, (4) the urban local government, and (5) special purpose vehicles constituted for riverfront developments. All are simultaneously working.

The biggest hurdles for riverfront development in India are (1) Lack of public consultations, NGOs and citizen groups also oppose the projects, and (2) All Rules and regulations are in place. Strict enforcement is lacking, which leads to several environmental issues, (3) Encroachment along the river body and (4) Blue line and red line, and these lines are not adequately publicized.

Policy 3: Numerous laws exeunt under municipalities and development authorities; the need is for enforcement of each law, such as public forums, since it is the duty of every resident living nearby the river and also of the city.

RFD does not include real estate. Supporting infrastructure should only include STPs or revenue generation, like small food courts. Small commercial development may lead to economic benefits. However, the administration does not enforce the law. These infrastructures will help increase river pollution. Industries can be relocated to other areas; they should not be near rivers. Also, housing projects do not do much good for RFD; perhaps if we permit housing near rivers, then sewage should be treated as much as possible.

Policy 4: Riverfronts can help local communities on a social level. Social dimensions include improving people's quality of life through actions to improve health, sanitation, public amenities, and capacity building.

Policy 5: Channelization as per local, state, and central laws and regulations is needed—stringent policies with heavy penalties for the bodies involving polluting the rivers.

Policy 6: Revamp sewage treatment plants on the riverfront to eliminate potential threats to the river's health and aquatic flora and fauna.

Dumping sewage into rivers or seas without treatment poses a threat to aquatic vegetation and wildlife. As a result, sewage must be treated before being discharged into a body of water, as it can affect humans and aquatic life.

20.4 Conclusion

The goal of this research was to look at riverfront development in India, identify the characteristics of riverfront development, and propose strategies for it. The main findings of this study were based on (1) expert interviews, (2) survey form responses, and (3) secondary data; for example, after conducting a comparative analysis, the most influencing parameters for successful RFD were the effects of (1) real estate development, (2) economic growth, and (3) footfall in the area. The features discovered were then proposed as policies for RFD best practices in India. A series of proposals to utilize as policies for a successful riverside in India that would minimize difficulties that arose during previous development, such as advising respective governing bodies, should be included in the RFD controlling policy. More studies may be done to see if the recommendations were implemented, how successful they were in riverfront development projects, and how the people involved were influenced or affected by them.

Conflict of Interest

The authors have no conflict of interest.
Data Availability Statement The data can be available properly from the editor.

References

Al-Shams AR, Ngah K, Zakaria Z, Noordin N, Hilmie MZ, Sawal M (2013) Waterfront development within the Urban design and public space framework in Malaysia. Asian Soc Sci 9(10): 77–87

Binti MD, Yassin A, Bond S, Mcdonagh J (2011) Developing guidelines for riverfront developments for Malaysia. Pac Rim Prop Res J 17(4):511–530

Burgers J (2000) Urban landscapes on public space in the post-industrial city. J Housing Built Environ 15:145–164

Chen C, Zhang H (2007) Idea space-riverside region planning creation and practice. China Architecture and Industry Press, Beijing, pp 36–40

Damanik FK, Pratiwi WD (2017) Consideration of tourism riverfront development elements for Pekanbaru City transformation. J Reg City Plan 28(2):140–150

Das J, Jha S, Goyal MK (2020) On the relationship of climatic and monsoon teleconnections with monthly precipitation over meteorologically homogenous regions in India: Wavelet & global coherence approaches. Atmos Res 238:104889. https://doi.org/10.1016/j.atmosres.2020.104889

Dutta V, Sharma U, Iqbal K, Kumar R, Pathak AK (2018) Impact of river channelization and riverfront development on fluvial habitat: evidence from Gomti River, a tributary of Ganges, India. Environ Sustain 1:167–184

Follmann A (2015) Urban mega-projects for a 'world-class' riverfront e the interplay of informality, flexibility and exceptionality along the Yamuna in Delhi, India. Habitat Int 45(3):213–222

Goyal MK, Ojha CSP (2010) Evaluation of various linear regression methods for downscaling of mean monthly precipitation in arid Pichola watershed. Nat Resour 01(01):11–18. https://doi.org/10.4236/nr.2010.11002

Goyal MK, Ojha CSP (2012) Downscaling of precipitation on a lake basin: evaluation of rule and decision tree induction algorithms. Hydrol Res 43(3):215–230. https://doi.org/10.2166/nh.2012.040

Goyal MK, Panchariya VK, Sharma A, Singh V (2018) Comparative assessment of SWAT model performance in two distinct catchments under various DEM scenarios of varying resolution, sources and resampling methods. Water Resour Manag 32(2):805–825. https://doi.org/10.1007/s11269-017-1840-1

Goyal MK, Gupta AK, Jha S et al (2022) Climate change impact on precipitation extremes over Indian cities: non-stationary analysis. Technol Forecast Soc Change 180:121685. https://doi.org/10.1016/j.techfore.2022.121685

Hersh B, Pechorzewski D, Yu SX (2012) Redeveloping waterfront brownfields; ideas, plans and experiences for regeneration of shipyards on three continents. CCIM Foundation, Chicago, IL

Hussein RMR (2014) Sustainable urban waterfronts using sustainability assessment rating system. Int J Archit Environ Eng 8(4):488–498

Kushwah N, Mehmood A (2017) Analysis Prospect of sustainable riverfront development in region around Khan River, Indore. Int J Curr Res 9(05):51599–51603

Light D, Cretan R, Voiculescu S, Jucu IS (2020) Changing tourism in the cities of post-communist central and Eastern Europe: introduction. J Balk Near East Stud 22:465–477

Lu WN, Ma HP, Yin C (2013) Water ecological restoration and riverfront landscape design in Xinye County, Henan Province. Henan Agric 24:48–49

McCarthy J (2004) Tourism-related waterfront development in historic cities: Malta's Cottonera project. Int Plan Stud 9(1):43–64

Pagés Sánchez JM, Daamen TA (2020) Governance and planning issues in European waterfront redevelopment 1999–2019. In: Carpenter A, Lozano R (eds) European port cities in transition, 1st edn. Springer, Cham, pp 127–148

Panuccio P (2019) Smart planning: from city to territorial system. Sustainability 11:7184

Patil US, Rathod ST, Bhandari S, Gaikwad S, Jadhav V (2019) Review study on Mula Mutha River rejuvenation project. J Emerg Technol Innov Res 6(6):86–89

Piryani R, Madhavi D, Singh VK (2017) Analytical mapping of opinion mining and sentiment analysis research during 2000–2015. Inf Manag 53:122–150

Poonia V, Goyal MK, Gupta BB, Gupta AK, Jha S, Das J (2021) Drought occurrence in different river basins of India and blockchain technology based framework for disaster management. J Clean Prod 312:127737. https://doi.org/10.1016/j.jclepro.2021.127737

Shah S, Roy AK (2017) Social sustainability of urban waterfront- the case of Carter road waterfront in Mumbai, India. Procedia Environ Sci 37:195–204

Shang J (2018) Riverfront landscape based on the idea of river restoration: a case study of the riverfront landscape of Duoba New District in Xining City. J Build Constr Plann Res 6:151–161

Subramoniam SR, Ravindranath S, Rakkasagi S, Ram H (2022) Water resource management studies at micro level using geospatial technologies, pp 49–74

Trivedi A, Awasthi MK (2021) A review on river revival. Int J Environ Clim Change 10(12): 202–210

Vaništa Lazarevic E, Antonic B (2016) Socio-urban regeneration strategy in the context of migration crisis in Europe. Serbian Sci Today 1:453–484

Wakefield S (2007) Great expectations: waterfront redevelopment and the hamilton harbour waterfront trail. Cities 24(4):298–310

SPRINGER NATURE

GPSR Compliance

The European Union's (EU) General Product Safety Regulation (GPSR) is a set of rules that requires consumer products to be safe and our obligations to ensure this.

If you have any concerns about our products, you can contact us on ProductSafety@springernature.com

In case Publisher is established outside the EU, the EU authorized representative is:

Springer Nature Customer Service Center GmbH
Europaplatz 3
69115 Heidelberg, Germany

The manufacturer's authorised representative in the EU is Springer Nature Customer Service Centre GmbH, Europaplatz 3, 69115 Heidelberg, Germany. If you have any concerns regarding our products, please contact ProductSafety@springernature.com

Printed and bound by CPI Group (UK) Ltd, Croydon, CR0 4YY
30/09/2026
02225319-0014